Tailings Dams: Design, Characterization, Monitoring, and Risk Assessment

Tailings Dams: Design, Characterization, Monitoring, and Risk Assessment

Guest Editors

Masoud Zare-Naghadehi
Javad Sattarvand

MDPI

Basel • Beijing • Wuhan • Barcelona • Belgrade • Novi Sad • Cluj • Manchester

Guest Editors

Masoud Zare-Naghadehi
Universal Engineering Sciences (UES)
Atlanta, GA
USA

Javad Sattarvand
Department of Mining and Metallurgical Engineering
University of Nevada, Reno
Reno, NV
USA

Editorial Office
MDPI AG
Grosspeteranlage 5
4052 Basel, Switzerland

This is a reprint of the Special Issue, published open access by the journal *Minerals* (ISSN 2075-163X), freely accessible at: https://www.mdpi.com/journal/minerals/special_issues/TailingsDams_minerals.

For citation purposes, cite each article independently as indicated on the article page online and as indicated below:

Lastname, A.A.; Lastname, B.B. Article Title. *Journal Name* **Year**, *Volume Number*, Page Range.

ISBN 978-3-7258-6602-1 (Hbk)
ISBN 978-3-7258-6603-8 (PDF)
https://doi.org/10.3390/books978-3-7258-6603-8

Contents

About the Editors

Masoud Zare-Naghadehi

Masoud Zare-Naghadehi is a Registered Professional Engineer (P.E.) in the states of Georgia, North Carolina, and Tennessee. His research and professional practice centers on geotechnical risk assessment, engineering data analytics, geotechnical design and monitoring, systems analysis, and probabilistic methods. He has published a number of well-received technical articles that have attracted notable attention across academia and industry. With a track record of having led technically demanding consulting projects, he emphasizes the essential link between theoretical principles and practical engineering solutions.

Javad Sattarvand

Javad Sattarvand is an executive mining leader and Professor Emeritus of Mining and Metallurgical Engineering at the University of Nevada, Reno. He received his Ph.D. in Mining Engineering from RWTH Aachen University, Germany, in 2009. His research focuses on mine automation and robotics, artificial intelligence and digital-twin-based mining systems, optimization of open-pit operations, numerical modeling, and mining safety. Throughout his academic career, Dr. Sattarvand has led numerous federally and industry-funded research projects, supervised over 50 graduate students and postdoctoral researchers, and published extensively in leading mining and engineering journals. He has contributed actively to the profession through peer review for international journals and service on technical and scientific committees of mining societies.

Preface

Tailings dams rank among the largest man-made structures, impounding slurries from mineral processing with inherent geotechnical challenges. Recent incidents highlight the imperative for robust design, monitoring, and risk management to safeguard communities and environments.

This volume reprints ten contributions from the *Minerals* Special Issue "Tailings Dams: Design, Characterization, Monitoring, and Risk Assessment." The articles explore dewatering methods, geophysical assessments, failure analyses, and closure strategies, delivering actionable knowledge for hazard reduction and sustainable practices.

As guest editors, we extend gratitude to the authors for their contributions to advancing tailings dam engineering.

Masoud Zare-Naghadehi and Javad Sattarvand
Guest Editors

minerals MDPI

Article

Characterization of Tailings Dams by Electrical Geophysical Methods (ERT, IP): Federico Mine (La Carolina, Southeastern Spain)

Julián Martínez [1,*], Rosendo Mendoza [1], Javier Rey [2], Senén Sandoval [3] and M. Carmen Hidalgo [2]

1 Departamento de Ingeniería Mecánica y Minera, EPS de Linares y CEACTEMA, Universidad de Jaén, Campus Científico Tecnológico, Linares, 23700 Jaén, Spain; rmendoza@ujaen.es
2 Departamento de Geología, EPS de Linares y CEACTEMA, Universidad de Jaén, Campus Científico Tecnológico, Linares, 23700 Jaén, Spain; jrey@ujaen.es (J.R.); chidalgo@ujaen.es (M.C.H.)
3 Everest Geophysics SL, Colmenarejo, 28270 Madrid, Spain; senen@everestgeophysics.com
* Correspondence: jmartine@ujaen.es

Abstract: This work analyzed the effectiveness of two electrical geophysical methods in characterizing tailings dams. A large flotation cell used for sludge thickening in the concentration plant of the Federico Mine (closed in 1985) within the old mining district of La Carolina (southeastern Spain) was selected for this research. In addition to the direct information provided by the geology of the study area and the surface exposure of the waste deposits, information regarding the construction of this mining structure was available, which helped in the interpretations of the geophysical survey data. In this study, two geophysical surveying methods were used simultaneously: Electrical resistivity tomography (ERT) and induced polarization (IP). Six profiles were acquired, processed, and interpreted. The length of the profiles allowed the obtaining of data reaching maximum investigation depths ranging between 7 and 65 m. These profiles provided information for a detailed analysis of the internal characteristics of the deposited materials. The lateral and vertical observed variations are linked to different degrees of moisture content. The study also defines the geometry of the top of the bedrock and the tectonics that affect the pouring/dumping hole. Old flotation sludge has resistivity values that range between 1 and 100 Ωm (i.e., wet waste 1–30 Ωm, dry waste 30–100 Ωm), while phyllites in the rocky substrate have resistivities larger than 200 Ωm and can even reach va-lues greater than 1000 Ωm. Between the waste and unaltered phyllites, there is a supergene alteration zone (weathering) with resistivity values between 100 and 200 Ωm. The IP method was used to detect the presence of metals in the accumulated waste in the pond by analyzing the presence of large chargeability anomalies. Anomalies were detected in four of the profiles, which ranged from low (i.e., between 0 and 8 mV/V) to medium (i.e., between 8 and 18 mV/V) and high values (i.e., 18 and >30 mV/V).

Keywords: tailings dam; electrical resistivity tomography; induced polarization; La Carolina Mining district; Spain

1. Introduction

Mineralized veins in the mining district of Linares and La Carolina (SE Spain) are enriched in metallic sulfides, mainly galena (PbS), and have been exploited by underground mining since pre-Roman times until the third quarter of the 20th century, when the last mine closed. The materials extracted from the mine were subjected to a mineral concentration process in treatment plants located nearby. Gravimetric methods were used until the 1950s. From then onward, flotation techniques became the preferred method causing accumulations of generated wastes in tailings ponds and dams [1,2].

The metal crisis of the 1980s, along with the depletion of some deposits, caused mi-ning company closures and the dismantling of their facilities, leaving waste on the ground without the necessary corrective measures being taken to avoid mobilization of the metals contained in the waste. These conditions pose an environmental hazard, due to the lack of preventive measures in the design (conditioning and waterproofing of the site). Failure to seal and revegetate these abandoned sites have exposed the structures to the action of atmospheric agents [3–7]. Mobilization of the metal(oid)s present in such wastes affects the surrounding soils and channels, as has been shown, for example, in Goseong (Korea), where high levels of Cu, Pb, As, and Zn have been measured and detected in crops [8].

In the Linares-La Carolina mining district, there are 33 tailings dams that accumulate 10,500,000 m^3 of waste. This amount is extremely high, especially considering that in the Region of Andalusia, 105 tailings dams have been inventoried out of a total of 344 metal ore processing dams in Spain [9].

Currently, any remediation effort to be carried out requires detailed knowledge of the characteristics of the accumulated waste. Good knowledge is needed about its limits and variations both in depth and laterally and its relationship with the underlying substrate to recognize possible leachate leakage areas, both at the foot of the dam and by direct infiltration into the ground affecting the groundwater system.

Geophysical methods have been successfully used in mining research: Gravimetry [10,11], magnetometry [10,11], electrical [12], electromagnetic [13], and seismic [14,15]. In recent years, these techniques have been applied to evaluate and resolve environmental problems associated with mining activity and the generated waste. In Spain, the electrical resistivity tomography (ERT) technique has been used in a variety of mining districts. For example, in Cartagena (SE Spain), multiple studies have recently evaluated the mobility of metals in tailings dams [16,17], using a combination of ERT, mechanical drilling, and geochemical techniques [18]. In the mining districts of Linares and La Carolina (southern Spain), different tailings ponds and slag heaps have been characterized by applying all of these techniques [3–7]. Similarly, in the Iberian Pyrite Belt (SW Spain), ERT has been used to evaluate tailings ponds created during the concentration of massive sulfides and the generation of acidic waters [19–21].

These methodologies have also been used in other mining districts worldwide. Such is the case in Kettara (Morocco), where electrical geophysical methods have been applied to study a sludge pond created from treating pyrrhotite [22]. In the Zaruma-Portovelo region (Ecuador), ERT has been applied to study the proposed site of a gold mining sludge pond, which was rejected because of possible landslide problems in the granites where this pond would have been located [23]. Additionally, in the Haveri area of Finland, a combination of geophysical and geochemical techniques has been used to evaluate the generation of acidic waters from an Au-Cu mine tailings pond [24].

2. Description of the Study Area

2.1. Geological Setting

The study area is located in the municipality of La Carolina, Jaén (Figure 1a), in the Sierra Monera mountain massif, specifically on the southeastern slope, which is the southeastern limit of the Hesperian massif [25,26]. Two large geological units are distinguished regionally: The Palaeozoic basement (Figure 1c) and post-Hercynian cover.

The Palaeozoic basement consists of a metasedimentary series composed of phyllites and quartzites from the Ordovician to the Carboniferous [27–29], and a granitic intrusion was emplaced in this area at the end of the Hercynian orogeny [26].

Subsequent to this intrusion, an extensional episode occurred that favored hydrothermal fluid circulation though a network of fractures, which gave rise to the Philonian reservoir in La Carolina. The veins exhibit mineral paragenesis that mainly consists of sulfides with galena, chalcopyrite, sphalerite, and pyrite predominating, accompanied by quartz, ankerite, and calcite [29].

On the Palaeozoic basement, the post-Hercynian series is horizontally positioned and consists of Triassic, Miocene (not present in the work zone), and Quaternary materials, which sometimes hide the discordantly arranged mineralized zones.

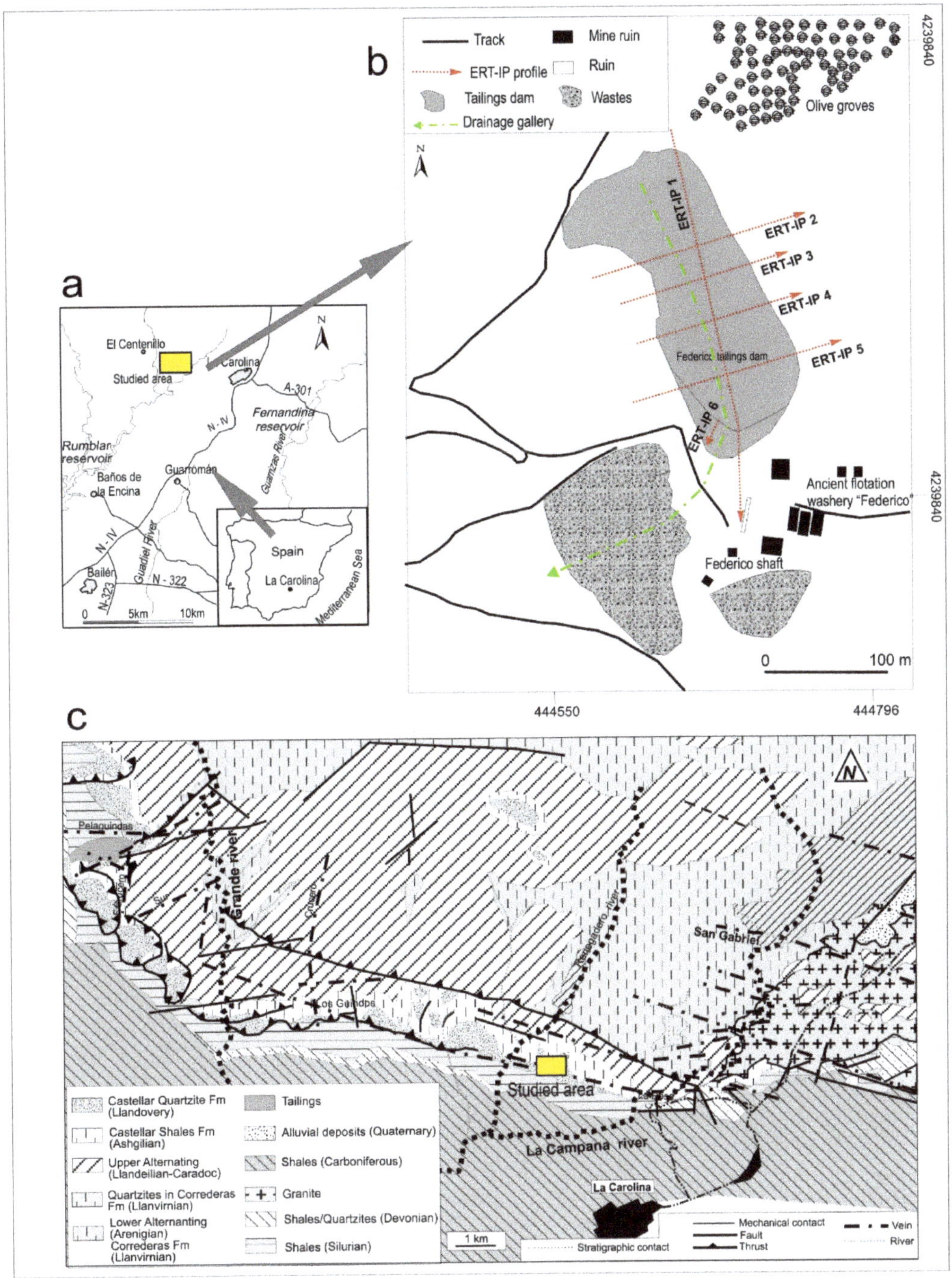

Figure 1. (**a**) Location of the Federico mine dam (white box). (**b**) Federico Dam and location of the investigation profiles. (**c**) Geological cartography (based on the work of Castelló and Orviz in 1976) showing the position of the studied sector.

2.2. Mining Technique

Underground mining methods were used in this mining district [1] and included shrinkage stoping and cut-and-fill stoping. These methods were the most common methods for the Philonian deposits that presented subvertical veins in hard embedded rocks (granites) and were the second-most common used for the mineralization that occurred more in the upper part of the district and were sometimes disseminated (phyllites). The shrinkage stoping technique consists of stripping minerals by drilling and blasting through ascending horizontal strips from the bottom up. The overturned minerals are left in the stope to serve as a working platform and as a provisional support for the gables until it is finally abandoned once the chamber has been excavated [30]. In the cut-and-fill stoping method, the void created by extracting the ore along a horizontal strip over the seam is filled with loose material from waste rock, which can sometimes come from outside the mine and serves as a support for the host rock and thus ascends during chamber excavation.

2.3. Tailings Dam

The Federico mine dam (Figures 1 and 2) is located at a site with the same name, with UTM coordinates 444.900/4.239.700 and a height of 620 m. This tailing dam contains 600,000 m^3 of slate and quartzite sludge.

The mineral concentration that took place in the La Carolina district was structured in two stages. The stripped materials (with high grades of 18% to 20% galena after a previous separation operation in the mine) were subjected to a gravimetric process from which, on the one hand, the first galena concentrate was obtained and, on the other hand, some mixed and high-grade wastes were then treated by a second flotation process. The flotation method required milling the mixture to sizes below 1 mm (approximately 50–70 µm) to release the minerals, and a fine concentrate of galena, as well as a residue that was pumped into ponds and tailings dams.

These structures, as in the case of the Federico mine (Figure 1b), were built using a depression in the ground where a gravel retaining wall was available to form the dam. The flotation waste was pumped from the plant by discharging through a gutter placed at the top of the dam through multiple outlets that were opened or closed to regulate the deposition of the discharged sludge. The accumulated mud formed a beach area next to the wall and a lake of process water. Water level management was carried out either through a central chimney connected to a bottom drainage gallery (in ponds on horizontal terrain) or, as in the case of Federico, through a lower drainage gallery built of masonry on a gravel base, to which a drainage tube was connected through which the process water was freely discharged to take it out of the structure [18] (Figure 2a–d).

In other similar structures of the La Carolina district, the waste metalloid contents have been analyzed. Such is the case for the Aquisgrana sludge dam, where concentrations of 544, 987, 8065, and 5186 mg/kg were obtained for As, Mn, Pb, and Zn, respectively [31].

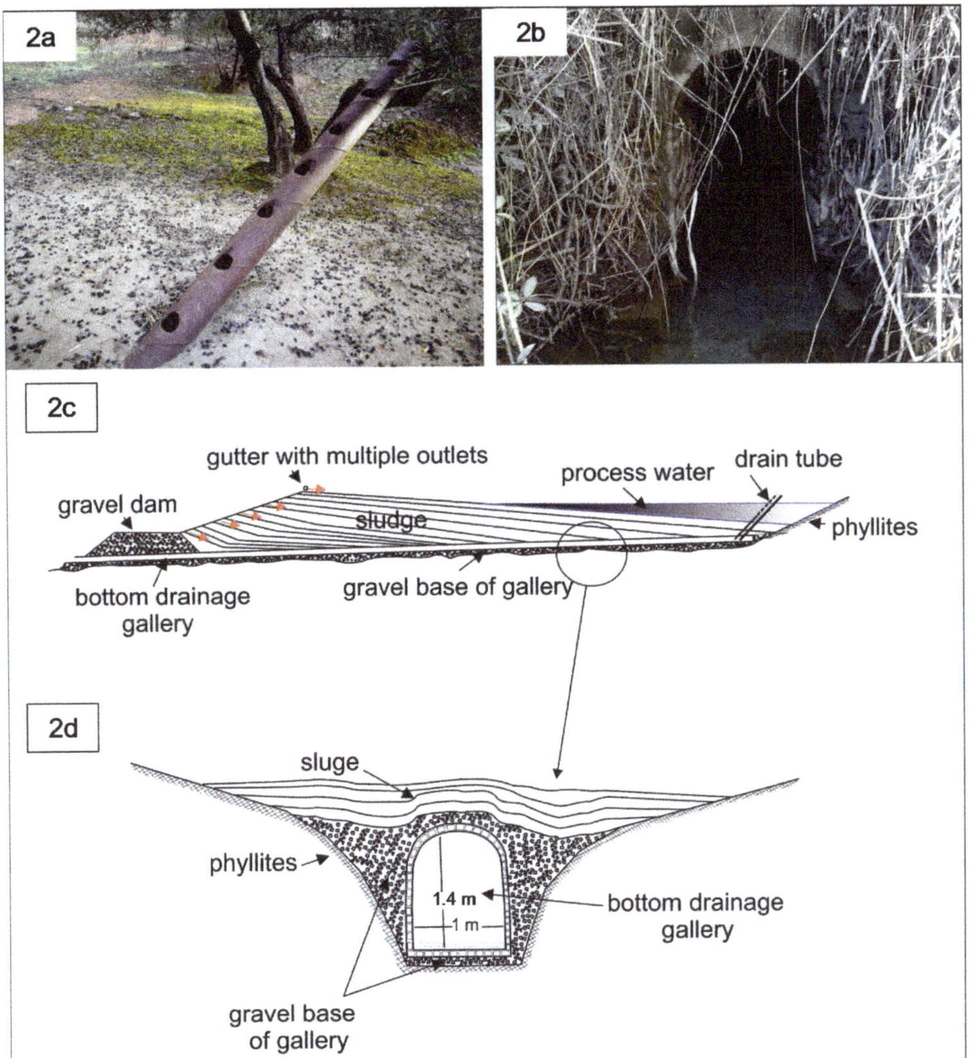

Figure 2. (**a**) Drainpipe connected to the bottom drainage gallery of the Federico dam. (**b**) Bottom discharge-type gallery of a tailings dam. (**c**) Construction diagram of the Federico flotation tailing dam. (**d**) Details of the bottom drainage gallery of the dam, which was constructed with masonry and covered with gravel.

3. Materials and Methods

Electrical Resistivity Tomography (ERT) and Induced Polarization (IP)

Ground resistivity measurements can be made with either an alternating current (AC) or a direct current (DC). Direct current (DC) resistivity methods use a controlled source of electrical current to produce an electrical voltage in the ground. Current is introduced into the ground through a pair of electrodes and the potential field is measured using another pair of electrodes.

The ERT technique can be considered an evolution (or combination) of other well-stablished geoelectrical techniques, such as vertical electrical soundings (VES) and the constant separation traversing (CST). It is a noninvasive technique that produces numerical models of the subsurface representing the 2D or 3D distribution of electrical resistivity in the ground [32–35]. The technique consists of acquiring large datasets of apparent resistivity in the field, filtering them, and finding a resistivity distribution of the ground that produces a similar response to that measured in the field. The models are derived by an iterative inversion process.

In recent years, this technique has expanded rapidly with the development of automated data acquisition systems capable of handling large amounts of data that are quickly processed with appropriate tools and that facilitate the generation of increasingly complex geological models [36].

In the data collection process, numerous electrodes are placed along a profile, with separations determined as a function of the degree of resolution and depths to be investigated. Thus, the smaller the separation between electrodes, the greater the resolution obtained and the lower the depth investigated, while increasing interelectrode separations enable deeper depths of investigation but with lower resolution [33]. We have chosen the Wenner–Schlumberger arrangement, which has a good penetration capacity and a good sensitivity for both vertical and horizontal structures. Other electrode arrangements such as the Wenner or dipole–dipole configurations, are more appropriate for sub-horizontal structures and vertical structures, respectively.

The technique is being used with good results in different earth science fields: Stratigraphy [6,37,38], hydrogeology [39], environment [7,16,17,20,38], and mining [40,41].

The induced polarization (IP) method is used to measure the electrical charging capacity of a given body or material in the subsoil and can detect those materials that have the ability to store a current for a given time [32,42]. To apply this method, a known direct current is applied between two electrodes (current electrodes) for a predetermined period of time (usually longer than one second). When the current is abruptly terminated, the voltage decay between another pair of electrodes (electrode potentials) is measured. A parameter called apparent chargeability can be estimated by measuring the area under the voltage decay curve. This value is normalized by the normal potential, Vo, which is obtained during the period that the known voltage is applied to the current electrodes [43].

The equipment used in this work was a digital resistivity meter RESECS manufactured by Deutsche Montan Technologie (DMT) (Figure 3a), which uses RESECS32 software for field data acquisition. The data sequence is selected to measure the ground under the profile with a dense cloud of measurement points. At each point, two parameters are determined: The apparent electrical resistivity (measured in Ωm) and the apparent chargeability (measured in ms or mV/V). The resistivity meter controls the current application and measures voltages using nonpolarizing electrodes with spiral lead cores immersed in a saturated lead chlorite gel (Figure 3b).

In this work, six ERT resistivity and IP profiles were acquired (Figure 1b) with a Wenner–Schlumberger configuration. For each profile, 64 electrodes were laid out. The lengths of the profiles were 320 m for ERT-IP 1, 192 m for ERT-IP 2, 128 m for ERT-IP 3, 128 m for ERT-IP 4, 192 m for ERT-IP 5, and 36 m for ERT-IP 6. The electrode separation "a" was 5, 3, 2, 2, 3, and 0.5 m, respectively (Figure 3c, profile ERT-IP 6). The parameters used to measure the resistivity and the chargeability were as follows: Applied voltage = 800 V; time of current injection = 1024 ms (with a delay of 10 ms to allow for stabilization and voltage measurement); interval between measurements of 990 ms (where the voltage decay curves were measured); no filters; and one repetition cycle (in profiles 1 to 5). In profile ERT-IP 6, an 880 V current was used; time of application = 2000 ms (with a delay of 50 ms) and interval of 1000 ms (for voltage decay readings); a Bessel filter and two repetition cycles.

Figure 3. (**a**) Deutsche Montan Technologie (DMT) RESECS resistivity meter. (**b**) Nonpolarizable electrodes in the profile ERT-1. (**c**) Profile ERT-6 on the dam slope. (**d**) Crusts observed on the dam slope.

To generate the resistivity and chargeability values, RES2DINV software (Loke and Barker, 1996) was used, which uses the least squares method with forced smoothing and a modified quasi-Newton optimization technique, in which a subsoil model is constructed using rectangular prisms. The resistivity values for each prism are determined and the differences between the observed and calculated apparent resistivity values are minimized [44–46]. The field apparent resistivity and apparent chargeability values were previously processed and filtered with PROSYSII software (IRIS Instruments). At this stage, the topographic information (topographic variations along the profile) was incorporated into the data.

4. Discussion and Results

Electrical Resistivity Tomography (ERT) and Induced Polarization (IP)

Six ERT and IP profiles were acquired (Figure 1b) on a tailings dam in the abandoned Federico Mine. One was longitudinally oriented (ERT-IP 1) along the major axis of the tailings dam. Four were transversely oriented (minor axis of the tailings dam, ERT-IP 2, 3, 4, and 5). The last one was oriented along the maximum slope of the dam (ERT-IP 6). Figures 4–6 show the derived resistivity and chargeability models.

The results of the study allow the geometry of the top of the bedrock basement to be defined and provides information about the tectonic fractures. The flotation sludge has resistivity values that range between 1 and 100 Ωm (wet waste 1–30 Ωm, dry waste 30–100 Ωm), while the phyllites of the rocky substrate have resistivities larger than 200 Ωm and can even reach values over 1000 Ωm. Between the waste and the unaltered phyllites, there is a supergene alteration zone (weathering) with resistivity values between 100 and 200 Ωm.

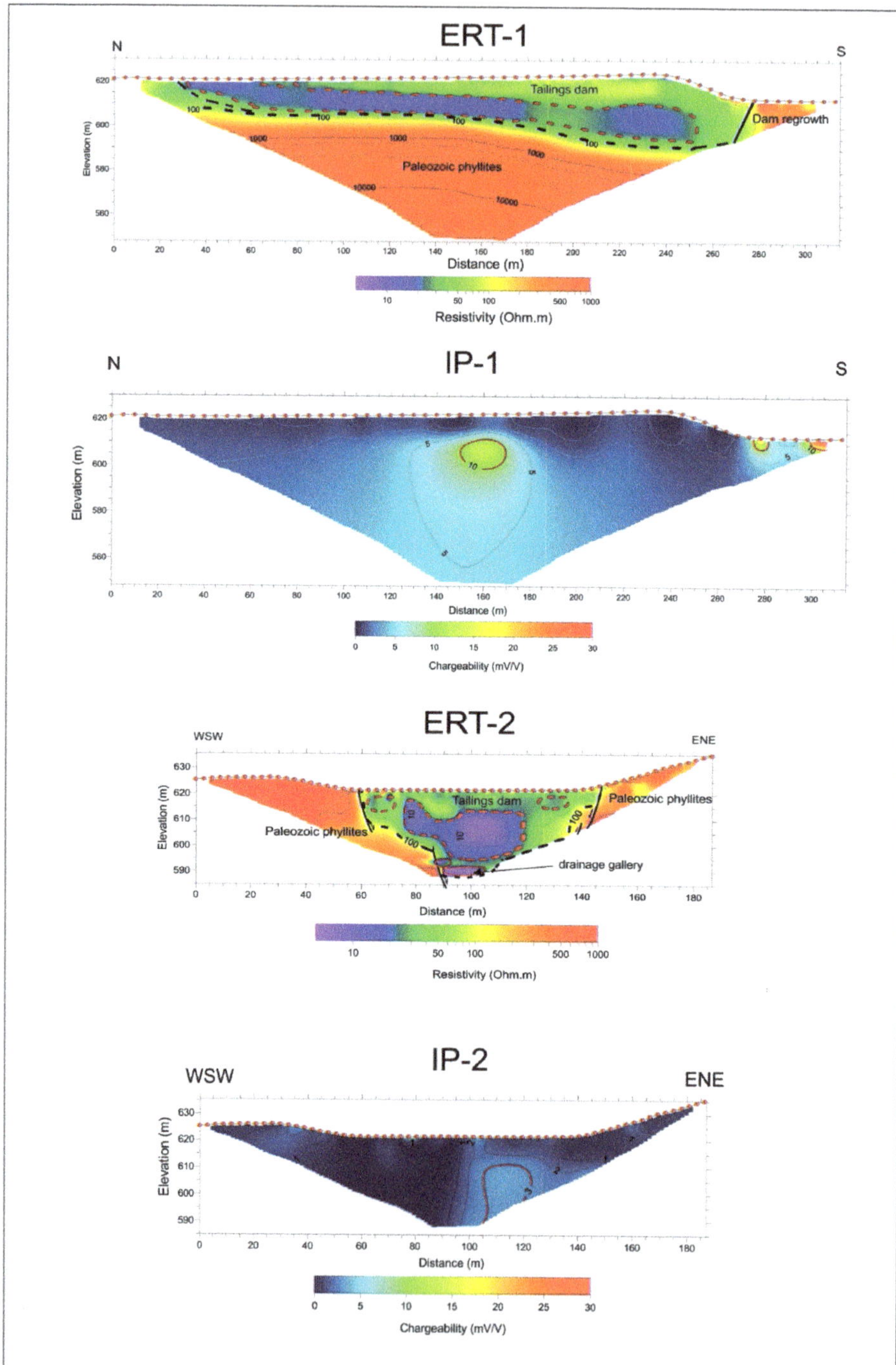

Figure 4. Resistivity models ERT-1 and ERT-2, which show the underlying Palaeozoic (phyllites) with high resistivity values (red colors), and flotation waste with low resistivities (blue and green colors) that indicate wet areas (dashed red line). In induced polarization (IP) models IP-1 and IP-2, low chargeability values are shown as blue and yellow colors and red are associated with the level enriched in metal(oid)s.

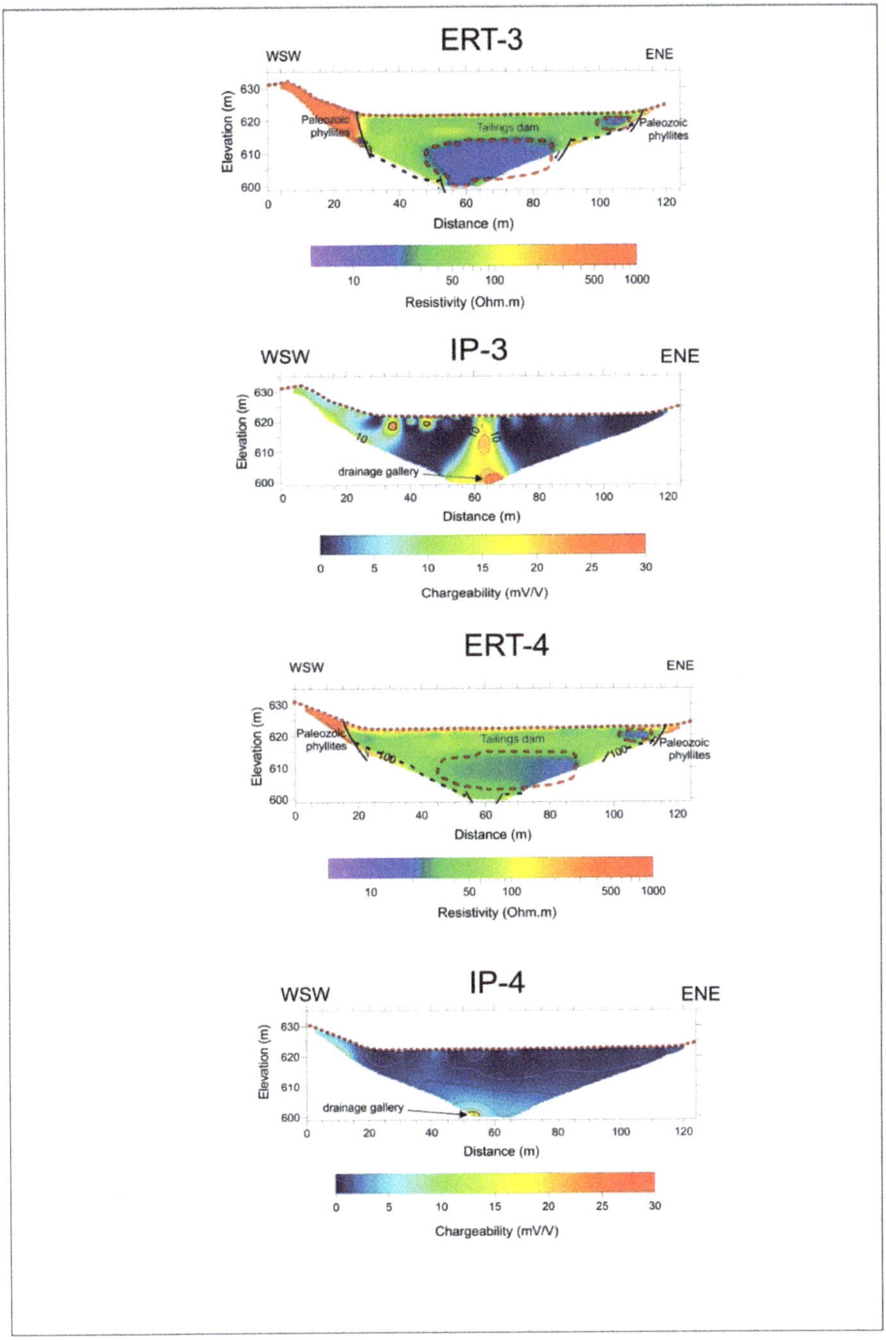

Figure 5. Resistivity models ERT-3 and ERT-4, in which the underying Palaeozoic (phyllites) with high resistivity values (red colors) on the sides of the profiles and flotation waste with low resistivities (blue and green colors), are observed, which indicate wet areas (dashed red line). In IP models IP-3 and IP-4, chargeability values are shown in blue for the sluge. The yellow- and red-color anomalies in IP 3 and IP 4 are associated with the bottom drainage gallery.

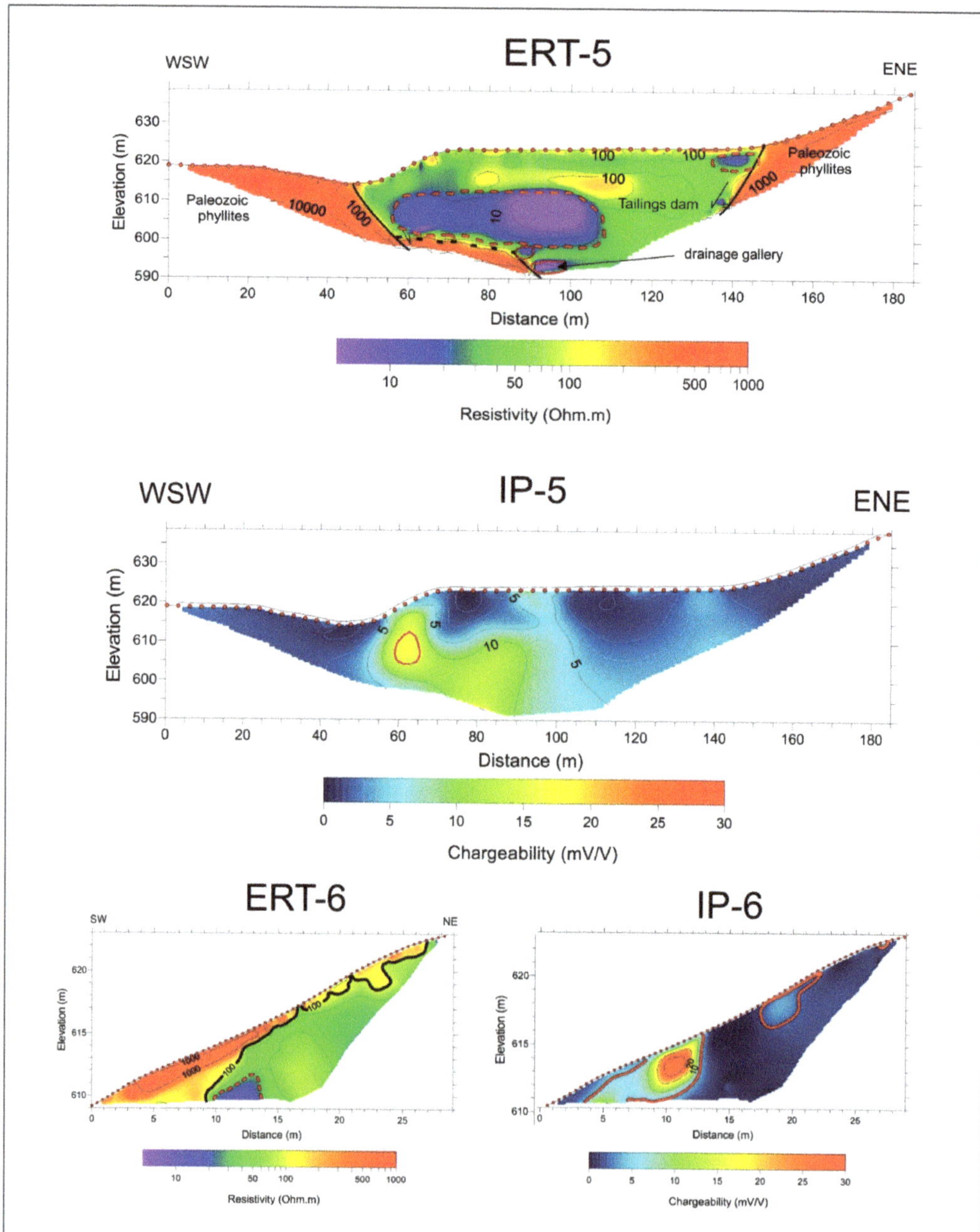

Figure 6. Resistivity models ERT-5 and ERT-6, the underlying Palaeozoic (phyllites) with high resistivity values (red co-lors) on the sides of the profiles, and flotation waste with low resistivities (blue and green colors) are observed, which indicate wet areas (dashed red line). ERT-6 (dam slope) exhibits a different response to the waste, with low values (in blue and green) in areas with specific moisture levels. Yellow colors are associated with crust. In IP model IP-5, low chargeability values are observed, with the medium chargeability zone associated with wet sludge. IP-6 exhibits a strong chargea-bility anomaly probably associated with levels of disseminated sulphides and metal(loid)s-enriched crust.

In the six analyzed profiles, IP measurements were also conducted. The apparent chargeability measurements were then inverted to obtain subsurface chargeability models (Figures 4–6). The inversion results differentiate three ranges of chargeability: Low (0–8 mV/V), medium (8–18 mV/V), and high/very high (18–30 mV/V). Low chargeability va-lues are probably associated with the absence of dispersed metals. They are characterized by a weak polarization phenomenon leading to the low observed chargeability values.

The ERT and IP 1 profile (Figure 4) has a length of 320 m in the N–S direction and reaches a depth of 64 m. The final RMS error (for the simultaneous inversion of resistivity and chargeability data) is 5.52%. The obtained resistivity distribution shows two distinct units: The deepest is characterized by high resistivity values (between 200 and more than 1000 Ωm) that correspond to the Palaeozoic substrate, which is composed mainly by phyllites with quartzite intercalations. These materials are at the base of the profile, are uniform, and exhibit a slight slope toward the South. Overlying these materials, resistivities between 100 and 200 Ωm are found. They are associated with the weathering of the phyllites. At the surface, flotation waste lays directly over these materials with resistivities below 100 Ωm. They are composed by silts and fine sands with thicknesses up to 35 m toward the S, adjacent to the dam wall. Within this waste, two different electrical responses are observed: A zone with resistivities between 1 and 30 Ωm, where we assume that the waste contains a high amount of moisture; and values between 30 and 100 Ωm, which are characteristic of dry waste. The resistive materials observed at the southern end of the profile correspond to the mine gravels used to form the dam wall.

The IP 1 model (Figure 4) shows generally low chargeability values (0 to 8 mV/V). A moderate-amplitude anomaly (8–18 mV/V) appears centered inside the waste body associated with a possible metallic sulfide-enriched zone. At the southern end, two maxima are observed that could be related to objects buried in the gravel dam wall.

The ERT and IP 2 profile (Figure 4) has a length of 180 m and reaches a depth of 36 m. The final RMS error is 1.60%. As in the previous profile, the phyllitic basement waste is clearly differentiated from the bedrock. This waste can also be described according to the resistivity values influenced by the moisture content. A very low-resistivity anomaly surrounded by more resistive (less humid) sludge appears in the center of the model. It should be noted that the low-resistivity values present in the lower part of the deposit (9.5 Ωm) are associated with the bottom drainage gallery on a gravel base (Figure 2). The deeper phyllites with high resistivities were reworked, creating a set of fractures asso-ciated with the trough that serves as the discharge hole.

The IP 2 profile (Figure 4) shows a dominating low-chargeability distribution without any remarkable anomaly.

The ERT and IP 3 model (Figure 5) has a length of 120 m and reaches a depth of 24 m. The final RMS error is 2.53%. A wet core (low resistivities) is present within the waste. There is another conductive zone toward the ENE, at the edge of the model. Given that the wet core is located at the end of the dam, it may be associated with leaching zones. This model also shows the possible presence of the bottom drainage gallery with low resistivity values (Figure 2). It should be noted that in all models, an increase in resistivity (between 80 and 150 Ωm) is observed within the dry zone of the waste, which we associate with the formation of crusts that have resulted from changes in the characteristics of the deposited slurries that facilitated oxidation (Figure 3d). At the sides of the model, Palaeozoic resistive phyllites are observed, probably associated with a fracture-type contact.

The IP-3 model (Figure 5) has an anomaly with high chargeability values located in the center of the waste extending toward the bottom (although it seems to project from the bottom upward). This anomaly is associated with the presence of the bottom drainage gallery and filling gravels that surround the gallery (Figure 3a,b). This can also represent an area that could be significantly enriched in metals at the initial location of the discharge of the flotation fines. The two small peaks observed toward the WSW can also be asso-ciated with metal-enriched areas.

The ERT and IP 4 model (Figure 5) has a length of 120 m and reaches a depth of 24 m. The final RMS error is 8.64%. This profile exhibits the same electrical pattern, with a wet core of low resistivity (<20 Ωm) enveloped by dry wastes (20–100 Ωm). The most conductive zone, which is toward the ENE end of the profile, is observed close to the phyllite contact. It could be a possible water circulation zone within the waste. The higher resistivity values that we associate with crusts are also noteworthy, which in this profile are clearly manifested on the dam surface (Figure 3d). Phyllites (high resistivities) are detected at both ends of the profile.

The IP-4 model (Figure 5) exhibits low chargeability values. A small anomaly in the lower part of the model is associated with the bottom drain.

The ERT and IP 5 model (Figure 6) has a length of 180 m and reaches a depth of 36 m. The final RMS error is 10.93%. It presents a resistivity distribution similar to those of the other profiles and clearly differentiates the phyllites (resistivities greater than 200 Ωm) from the flotation waste (resistivities less than 200 Ωm). The phyllites are interrupted by a series of fractures that constitute the trough. Several high-resistivity zones are observed in the central core and in the lateral zone toward the ENE and in the bottom gallery (supported and surrounded by gravels, Figure 2) within the waste depending on the moisture levels. It should also be noted that the resistivities between 80 and 150 Ωm obtained in the waste are associated with crust formation, which in this profile are observed both inside and on the surface (Figure 3d).

The IP 5 model (Figure 6) exhibits two patterns in the chargeability distribution. Low values characterize both the waste and phyllites. On the other hand, high values between 8 and 18 mV/V are observed both toward the WSW and in the background of the waste. We associate these anomalies with zones enriched in metallic sulfides and wet waste.

In addition to the longitudinal and cross-sectional profiles, a high-density profile (ERT and IP 6; Figure 3c, Figure 6) was aquired with an electrode spacing of 0.5 m on the slope of the dam to study the lateral behavior of the waste. The RMS error is 5.07%. Two very sharp resistivity changes are observed. One is associated with the contact between the unaltered waste (5 to 100 Ωm) and the wet core (5 to 20 Ωm). The second is the contact between high resistivities associated with the crusts observed inside and on the surface of the waste (Figure 3d).

The detailed model, IP 6 (Figure 6), exhibits low background chargeability values and a zone with high and very high values (8–30 mV/V) on the surface of the lower part of the dam slope. The latter can be associated with a local concentration of metal sulfides.

5. Conclusions

In this work, two electrical geophysical methods (ERT and IP) were used in an integrated manner to analyze their capability to model the inner structure of tailings dams. For this study, the abandoned dam at the Federico mine (La Carolina, southeastern Spain) was chosen. It contains flotation waste from an old washing plant, one of the most important in the mining district. The ERT and IP methods were used and their results were compared with direct field observations and other available information regarding the mining structure.

In the electrical tomography models, the mining wastes of the base rock (Palaeozoic phyllites) are clearly defined. The method allows us to deduce the morphology of the phyllitic substrate, which is controlled by the presence of fractures and which forms a trough that serves as a repository for flotation wastes. ERT also allowed us to interpret changes in the interior of the waste. Wet and dry areas were detected, in addition to crusts and levels associated with higher metal sulfide contents. The presence of dam drainage at the bottom of the dam is also detected with this method.

On the other hand, the IP method complements the information obtained with ERT and shows three chargeability levels associated with the bottom of the waste, levels with sulfide enrichment associated with crusts, and a level associated with the bottom drainage gallery.

These indirect electrical methods are of interest for examining tailings ponds and dams, as they allow the definition of both their morphology and deposition on the ground, as well as their internal structures, and constitute a step that would be conducted prior to conducting mechanical soundings.

Author Contributions: Conceptualization and investigation, J.M., R.M. and J.R.; software, J.M. and S.S; writing and review, J.M., S.S. and M.C.H. All authors have read and agreed the published version of the manuscript.

Funding: This research received no external funding.

Institutional Review Board Statement: Not applicable.

Informed Consent Statement: Not applicable.

Data Availability Statement: Please refer to suggested Data Availability Statements in section "MDPI Research Data Policies" at https://www.mdpi.com/ethics.

Acknowledgments: Careful reviews by Mario Zarroca, Roberto Rodríguez and Isaac Corral and anonymous referees by improved the manuscript.

Conflicts of Interest: The authors declare no conflict of interest.

References

1. Gutiérrez-Guzmán, F. *Minería en Sierra Morena;* Ilustre Colegio de Ingenieros Técnicos de Minas de Linares: Granada, Spain, 2007.
2. Contreras, F.; Dueñas, J. *La Minería y la Metalurgia en el Alto Guadalquivir: Desde sus Orígenes Hasta Nuestros Días;* Instituto de Estudios Giennenses, Diputación de Jaén: Jaén, Spain, 2010.
3. Martínez, J.; Rey, J.; Hidalgo, M.C.; Benavente, J. Characterizing Abandoned Mining Dams by Geophysical (ERI) and Geochemical Methods: The Linares-La Carolina District (Southern Spain). *Water Air Soil Pollut.* **2012**, *223*, 2955–2968. [CrossRef]
4. Martínez, J.; Rey, J.; Hidalgo, M.C.; Garrido, J.; Rojas, D. Influence of measurement conditions on the resolution of electrical resistivity imaging: The example of abandoned mining dams in the La Carolina District (Southern Spain). *Int. J. Miner. Process.* **2014**, *133*, 67–72. [CrossRef]
5. Martínez, J.; Hidalgo, M.C.; Rey, J.; Garrido, J.; Kohfahl, C.; Benavente, J.; Rojas, D. Multidisciplinary characterization of a tai-lings pond in the Linares-La Carolina mining district, Spain. *J. Geochem. Explor.* **2016**, *162*, 62–71. [CrossRef]
6. Rey, J.; Martínez, J.; Hidalgo, C.; Rojas, D. Heavy metal pollution in the Quaternary Garza basin: A multidisciplinary study of the environmental risks posed by mining (Linares, southern Spain). *Catena* **2013**, *110*, 234–242. [CrossRef]
7. Cortada, U.; Martínez, J.; Rey, J.; Hidalgo, M.C.; Sandoval, S. Assessment of tailings pond seals using geophysical and hydro-chemical techniques. *Eng. Geol.* **2017**, *223*, 59–70. [CrossRef]
8. Ji, K.; Kim, J.; Lee, M.; Park, S.; Kwon, H.J.; Cheong, H.K.; Jang, J.-Y.; Kim, D.-S.; Yu, S.; Kim, Y.-W.; et al. Assessment of exposure to heavy metals and health risks among residents near abandoned metal mines in Goseong, Korea. *Environ. Pollut.* **2013**, *178*, 322–328. [CrossRef]
9. IGME. Inventario Nacional de Depósitos de Lodos. Available online: http://www.igme.es/actividadesIGME/lineas/RMeIA.htm. (accessed on 31 January 2021).
10. Marangoni, Y.; Mantovani, M. Geophysical signatures of the alkaline intrusions bordering the Paraná Basin. *J. S. Am. Earth Sci.* **2013**, *41*, 83–98. [CrossRef]
11. Abud Louro, V.; Mantovan, M. i3D inversion and modeling of magnetic and gravimetric data characterizing the geophysical anomaly source in Pratinha I in the southeast of Brazil. *J. Appl. Geophys.* **2012**, *80*, 110–120. [CrossRef]
12. Martínez-Pagán, P.; Gómez-Ortiz, D.; Martín-Crespo, T.; Manteca, J.I.; Rosique, M. The electrical resistivity tomography method in the detection of shallow mining cavities. A case study on the Victoria Cave, Cartagena (SE Spain). *Eng. Geol.* **2013**, *156*, 1–10.
13. Smith, R. Electromagnetic Induction Methods in Mining Geophysics from 2008 to 2012. *Surv. Geophys.* **2013**, *35*, 123–156. [CrossRef]
14. Olivier, G.; Brenguier, F.; Wity, T.; Lynch, R. Monitoring the stability of tailings dam walls with ambient seismic noise. *Lead. Edge* **2017**, *36*, 350a1–350a6. [CrossRef]
15. Min, L.; Ming, X.; Shimin, L.; Douglas, R. In Situ Seismic Investigations for Evaluating Geotechnical Properties and Liquefaction Potential of Fine Coal Tailings. *J. Geotech. Geoenviron. Eng.* **2020**, *146*, 04020014.
16. Martínez-Pagán, P.; Faz Cano, A.; Aracil, E.; Arocena, J.M. Electrical resistivity imaging revealed the spatial properties of mine tailing ponds in the Sierra Minera of Southeast Spain. *J. Environ. Eng. Geophys.* **2009**, *14*, 63–76. [CrossRef]

17. Martínez-Pagán, P.; Faz, A.; Acosta, J.A.; Carmona, D.M.; Martínez-Martínez, S. A multidisciplinary study for mining landscape reclamation: A study case on two tailing ponds in the Region of Murcia (SE Spain). *Phys. Chem. Earth* **2011**, *36*, 1331–1344. [CrossRef]
18. Gabarrón, M.; Martínez-Pagán, P.; Martínez-Segura, M.A.; Martínez-Martínez, S.; Faz, A.; Acosta, J.A. Electrical Resistivity Tomography as a Support Tool to Estimate Physicochemical Properties of Mining Tailings Pond. *Preprints* **2020**, 2020010377. [CrossRef]
19. Martín-Crespo, T.; Martín-Velázquez, S.; Gómez-Ortíz, D. A geochemical and geophysical characterization of sul_de mine ponds at the Iberian Pyrite Belt (Spain). *Water Air Soil Pollut.* **2010**, *217*, 387–405. [CrossRef]
20. Gómez-Ortiz, D.; Martín-Velázquez, S.; Martín-Crespo, T.; De Ignacio-San José, C.; Lillo, J. Application of electrical resistivity tomography to the environmental characterization of the abandoned massive sulphide mine ponds (Iberian Pyrite Belt, SW Spain). *Near Surf. Geophys.* **2010**, *8*, 65–74. [CrossRef]
21. Gómez-Ortiz, D.; Fernández-Remolar, D.C.; Granda, A.; Quesada, C.; Granda, T.; Prieto-Ballesteros, O.; Molina, A.; Amils, R. Identification of the subsurface bodies responsible for acidity in Rio Tinto source water; Spain. *Earth Planet. Sci. Lett.* **2014**, *391*, 36–41. [CrossRef]
22. Lghoul, M.; Teixido, T.; Peña, J.A.; Hakkou-Azzouz, R.; Kchikach, A.; Guerin, R.; Jaffal, M.; Zouhri, L. Electrical and Seismic Tomography Used to Image the Structureof a Tailings Pond at the Abandoned Kettara Mine, Morocco. *Mine Water Environ.* **2012**. [CrossRef]
23. Zarroca, M.; Linares, R.; Velásquez-López, P.C.; Roqué, C. Application of electrical resistiviy imaging (ERI) to a tailing dam project for artisanal and small-scale gold mining in Zaruma-Portovelo, Ecuador. *J. Appl. Geophys.* **2015**, *113*, 103–113. [CrossRef]
24. Placencia-Gómez, E.; Parviainen, A.; Hokkanen, T.; Loukola-Ruskeeniemi, K. Integrated geophysical and geochemical study on AMD generation at the Haveri Au–Cu mine tailings, SW Finland. *Environ. Earth Sci.* **2010**, *61*, 1435–1447. [CrossRef]
25. Tamain, G. Recherches Géologiques et Minières en Sierra Morena Orientale (Espagne). Ph.D. Thesis, Université Paris-Sud, Paris, France, 1972. (In French).
26. Castelló, R.; Orviz, F. *Mapa y Memoria Explicativa de la Hoja n° 884 (La Carolina) del mapa Geológico de España, Escala 1:50,000*; Instituto Geológico y Minero de España: España, Spain, 1976.
27. Ríos, S. Estudio del Metalotecto Plumbífero del Ordoviense (La Carolina-Santa Elena, Sierra Morena Oriental, Provincia de Jaén. Ph.D. Thesis, Politécnica Madrid, E.T.S. Ingenieros de Minas, Madrid, Spain, 1977.
28. Fontboté, J.M. *Mapa Geológico y Memoria Explicativa de la Hoja 70 (Linares), Escala 1:200,000*; Ministerio de Industria: Madrid, Spain, 1982; 29p.
29. Lillo, J. Geology and Geochemistry of Linares-La Carolina Pb-Ore Field. Southeastern Border of the Hesperian Massif. Ph.D. Thesis, University of Leeds, Leeds, UK, 1992.
30. Hoek, E.; Kaiser, P.K.; Bawden, W.F. *Support of Underground Excavations in Hard Rock*; Bolkema Publishers: Rotterdam, The Netherlands, 1995.
31. Rojas, D.; Hidalgo, M.C.; Kohfahl, C.; Rey, J.; Martínez, J.; Benavente, J. Oxidation Dynamics and Composition of the Flotation Plant Derived Tailing Impoundment Aquisgrana (Spain). *Water Air Soil Pollut.* **2019**, *230*, 158. [CrossRef]
32. Telford, W.M.; Geldart, L.P.; Sheriff, R.E. *Applied Geophysics*; Cambridge University Press: Cambridge, UK, 1990; 770p.
33. Sasaki, Y. Resolution of resistivity tomography inferred from numerical simulation. *Geophys. Prospect.* **1992**, *40*, 453–464. [CrossRef]
34. Store, H.; Storz, W.; Jacobs, F. Electrical resistivity tomography to investigate geological structures of earth's upper crust. *Geophys. Prospect.* **2000**, *48*, 455–471. [CrossRef]
35. Dahlin, T.; Zhou, B. A numerical comparison of 2D resistivity imaging with 10 electrode arrays. *Geophys. Prospect.* **2004**, *52*, 379–398. [CrossRef]
36. Griffiths, D.H.; Barker, R.D. Two-dimensional resistivity imaging and modelling in areas of complex geology. *J. Appl. Geophys.* **1993**, *29*, 211–226. [CrossRef]
37. Maillet, G.M.; Rizzo, E.; Revil, A.; Vella, C. High resolution electrical resistivity tomography (ERT) in a transition zone environment: Application for detailed internal architecture and infilling processes study of a Rhône River paleo-channel. *Mar. Geophys. Res.* **2005**, *26*, 317–328. [CrossRef]
38. Martínez, J.; Benavente, J.; García-Aróstegui, J.L.; Hidalgo, M.C.; Rey, J. Contribution of electrical resistivity tomography to the study of detritical aquifers affected by seawater intrusion-extrusion effects: The river Vélez delta (Vélez-Málaga, southern Spain). *Eng. Geol.* **2009**, *108*, 161–168. [CrossRef]
39. André, F.; van Leeuwen, C.; Saussez, S.; Van Durmen, R.; Bogaert, P.; Moghadas, D.; de Resseguier, L.; Delvaux, B.; Vereecken, H.; Lambot, S. High-resolution imaging of a vineyard in south of France using ground-penetrating radar, electromagnetic induction and electrical resistivity tomography. *J. Appl. Geophys.* **2012**, *78*, 113–122. [CrossRef]
40. Irawan, D.; Sumintadireja, P.; Saepuloh, A. 2-D subsurface Imaging Tecniques for Deep ore mineral mapping using geoelectrical and induced polarization (IP) methods. *Proc. Earth Planet. Sci.* **2013**, *6*, 139–144. [CrossRef]
41. Martínez, J.; Rey, J.; Sandoval, S.; Hidalgo, C.; Mendoza, R. Geophysical prospection using ERI and IP techniques to locate vein deposits. *Remote Sens.* **2019**, *11*, 2923. [CrossRef]
42. Sumner, J.S. *Principles of Induced Polarization for Geophysical Exploration*; Elsevier Scientific Publishing: Amsterdam, The Netherlands, 1976.

43. Zhadanov, M.S. Direct current and induced polarization methods. In *Foundations of Geophysical Electromagnetic, Theory and Methods*; Elsevier: Amsterdam, The Netherlands, 2018; Volume 2018, pp. 439–493.
44. Loke, M.H.; Barker, R.D. Rapid least-squares inversion of apparent resistivity pseudosections by a quasi-Newton method. *Geophys. Prospect.* **1996**, *44*, 131–152. [CrossRef]
45. Loke, M.H.; Dahlin, T. A comparison of the Gauss-Newton and quasi-Newton methods in resistivity imaging inversion. *J. Appl. Geophys.* **2002**, *49*, 149–162. [CrossRef]
46. Loke, M.H. Tutorial: 2-D and 3-D Electrical Imaging Surveys, Revision Date: 5th November 2014. Available online: www.geotomosoft.com (accessed on 19 February 2015).

minerals

MDPI

Review

Lessons from Tailings Dam Failures—Where to Go from Here?

David John Williams

Geotechnical Engineering Center, School of Civil Engineering, The University of Queensland, Brisbane, QLD 4066, Australia; D.Williams@uq.edu.au

Abstract: Different regions worldwide have adopted various approaches to tailings management, as a result of the site settings and local practices as they have evolved. Tailings dam failures have continued to occur in both developing and developed countries, necessitating a range of tailings management approaches. These failures, while rare, continue to occur at a frequency that exceeds both industry and society expectations, and there is much to be learned from well-documented cases. Tailings management continues to be overly reliant on a net present value approach using a high discount factor, rather than a whole-of-life approach that may result in safer and more stable tailings facilities and may also facilitate the eventual mine closure. There is a need for the further development and implementation of new tailings management technologies and innovations, and for the application of whole-of-life costing of tailings facilities. Changes in tailings management will most readily be achieved at new mining projects, making change across the minerals industry a generational process.

Keywords: foundation failure; improved management; lessons; liquefaction; site settings; slurry tailings; tailings dam failures

1. Introduction

The conventional deposition of slurry tailings behind dams that are raised progressively has led to an unacceptably high rate of catastrophic tailings dam failures, resulting in fatalities, damage to infrastructure and environmental harm. The rate of tailings dam failures is about two orders of magnitude higher than that of water dams, which are generally raised in a single lift to provide the required water storage.

Tailings dam failures that lead to high fatalities can prompt dramatic change, such as occurred in Chile, following the catastrophic El Cobre tailings dam failure during an earthquake. This change was initiated by tailings professionals, who questioned whether stable tailings dams could be constructed and operated in their highly seismic setting, and who made changes that ensured this was possible. The Global Industry Standard on Tailings Management (GISTM) and accompanying guides transpired as a result of the catastrophic tailings dam failure near Brumadinho in Brazil in January 2019. For change to be effective, active recognition and engagement is needed by all stakeholders.

This paper reviews the drivers of conventional tailings management, including the dominance of the net present value accounting approach, and the importance of each site's climatic, topographic and seismic settings that must be accommodated both during tailings operations and post-closure. Selected tailings dam failures are used to illustrate the main modes of failure of conventional tailings dams, and how they may be avoided, and the "as low as reasonably practicable" approach to reducing risk is discussed. The key features and requirements of the GISTM and the accompanying guides are reviewed, as is a comprehensive range of alternatives to the conventional deposition of slurry tailings behind dams. From these discussions, the expectations and implications of the GISTM, and the range of tailings management alternatives available, a path forward in tailings management is offered.

2. Conventional Tailings Management

2.1. Drivers of Conventional Tailings Management

The commonly held perception, supported by net present value (NPV) accounting, is that transporting tailings as a slurry to a surface dam is the most economical solution. Discounted long-term and rehabilitation costs "become" insignificant. Filtering tailings is perceived to be too expensive, despite reducing the storage volume and making it easier to rehabilitate to a high level. This has led to the widespread adoption of surface tailings facilities to store slurried tailings, delivered by robust and inexpensive centrifugal pumps and pipelines. Initially, small storage areas are constructed, leading to soft and wet tailings deposits, storing entrained water.

Operating costs increase over time as capital expenditure is avoided or delayed. This results in unintended increased storage volumes that must accommodate the entrained water, and more difficult rehabilitation of the tailings facility.

2.2. Importance of Site Settings

The key settings of a given mine site are the climate, the topography and the seismicity of the site. The site rainfall/precipitation and evaporation dictate the potential for the tailings to be exposed to desiccation. A dry climate makes slurry tailings disposal easier (e.g., in semi-arid regions of Australia, South Africa, and Southwestern USA). A wet climate has the potential to maintain the wetness of tailings (e.g., in the wet tropics, including Brazil). A near-neutral water balance can be tipped over into net positive by tailings deposition, or net negative post-closure by evaporation from stored water (e.g., for Canadian oil sands tailings).

The topography of the site dictates the volume of "free storage" available in valleys and the tailings dam's height. High seismicity will often govern tailings dam/storage design (e.g., in Chile and Peru). High seismicity may well need to be considered post-closure (in perpetuity) everywhere.

3. Lessons from Tailings Dam Failures

3.1. Selected Tailings Dam Failures

For the purposes of this paper, a number of tailings dam failures have been selected, as shown on the timeline in Figure 1 [1–12].

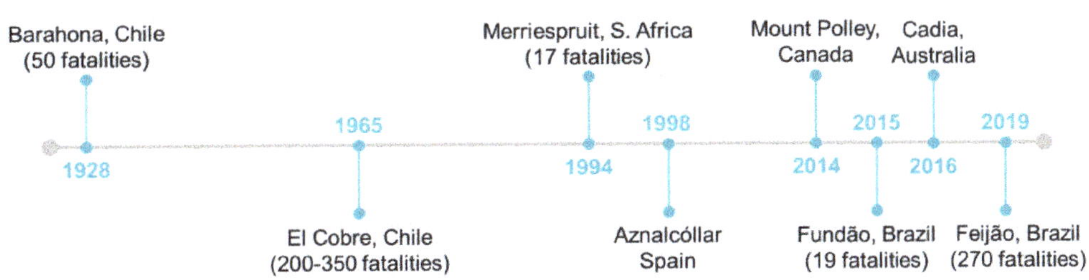

Figure 1. Timeline of selected tailings dam failures.

3.2. Some Root Causes of Tailings Dam Failures

Tailings dams that fail are marginally stable. Most tailings dams are stable and do not fail. Failures typically arise from a combination of causes, with water being a key element. Upstream constructed tailings dams are more prone to failure than downstream dams. Tailings deposited as a slurry are potentially susceptible to seismic or flow liquefaction, unless compacted and/or consolidated and desiccated. Weak foundation layers may cause tailings dam failures. Excessive tailings dam failures threaten the mineral industry's financial and social licenses to operate, and the control of their operations.

3.3. Seismic Liquefaction Failures of Upstream Sand Dams in Chile

"Construction with, or on, liquefiable materials should be avoided", according to Dr Izzat Idriss [13]. *"You would not build a water dam using liquefiable materials or on a liquefiable foundation, so why would we do this for tailings dams?"* This is not to say that the upstream raising of a tailings dam cannot be made to work.

Among the fatal failures of steep upstream sand tailings dams in Chile, due to seismic liquefaction, are the Barahona No. 1 tailings dam failure in 1928 (Figure 2a) [1], and the El Cobre old tailings dam failure in 1965 (Figure 2b) [2]. The Barahona tailings dam underwent liquefaction due to a magnitude 8.2 earthquake, resulting in 50 fatalities, and did not lead to any changes to tailings management practices or regulations.

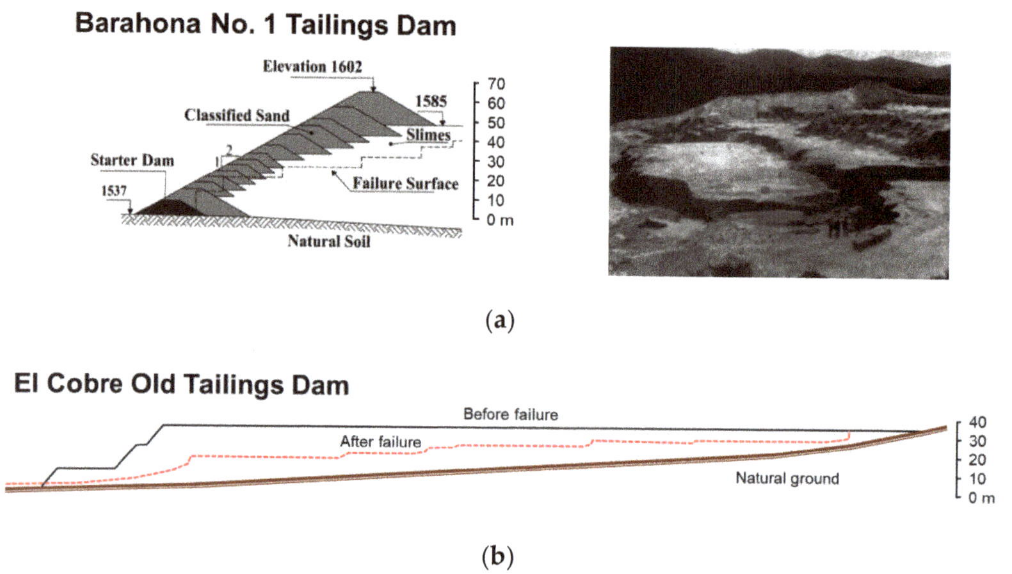

Figure 2. Failures of steep upstream sand tailings dams in Chile due to seismic liquefaction: (**a**) Barahona No. 1 tailings dam [1], and (**b**) El Cobre old tailings dam [2].

The El Cobre tailings dam also underwent liquefaction due to a magnitude 7.4 earthquake, inundating the town of El Cobre, resulting in 200 to 350 fatalities (the number is uncertain, due to the unknown number of occupants of the town at the time of the failure, and an inability to recover all of the bodies).

The El Cobre failure led to tailings practitioners in Chile questioning whether a safe tailings dam could be built to sustain large earthquakes. The industry responded rapidly by flattening and compacting the downstream slopes of sand dams from the natural 2 (horizontal) to 1 (vertical angle) to 4 to 1 (essentially a doubling of the margin of stability), and a change to the downstream construction, using either sand or earth fill. Later, a geomembrane was added to the exposed upstream face of the dam to limit the ingress of water from the contained slimes into the dam, and, in the case of sand dams, central sand cyclone stations replaced the series of cyclones along the crest of the dam. Central cyclone stations allowed greater control of the fines content of the sand (limiting this to less than 20%) and avoided weak spots being formed in the dam if one of the crest cyclones failed. Regulations to formalize these changes were implemented 5 years later and beyond, and these have also been adopted for tailings dams in seismically active Peru.

Typical cross-sections of downstream sand dams in Chile post-1965 are shown for the Las Tórtalas tailings dam (Figure 3) and the Quillayes tailings dam (Figure 4) [3].

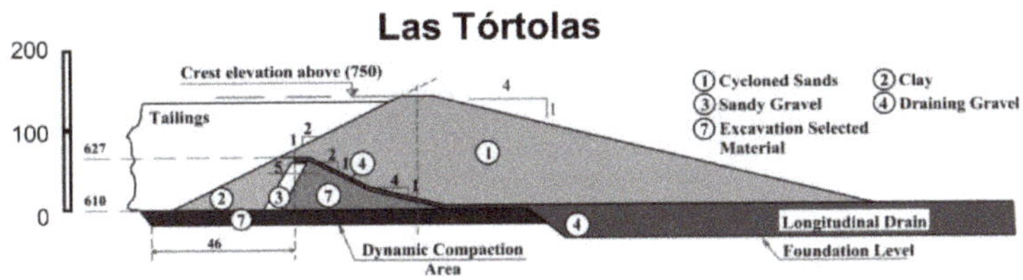

(**a**)

(**b**)

Figure 3. Las Tórtalas downstream sand dam [3]: (**a**) typical cross-section, and (**b**) flattening and compacting the downstream slope.

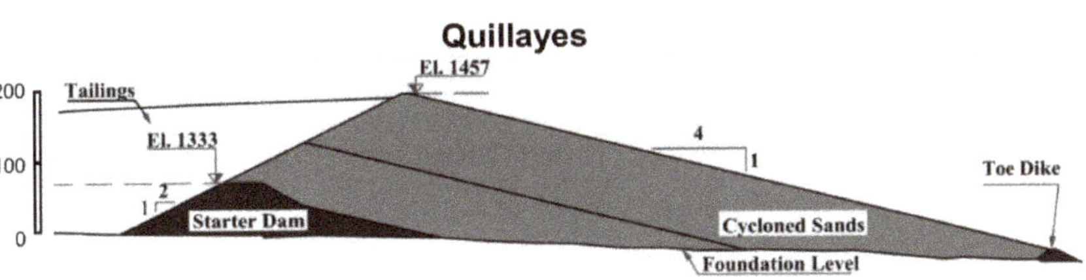

(**a**)

(**b**)

Figure 4. Quillayes downstream sand dam [3]: (**a**) typical cross-section, and (**b**) upstream geomembrane.

Of the approximately 740 tailings dams in Chile, about 100 are currently active, and these are mostly downstream sand dams. Around 470 tailings dams are inactive, and these are mostly upstream sand dams; about 170 are abandoned and are also mostly upstream sand dams. Active downstream sand dams in Chile have performed well since 1965, due to improved construction methods. Only one dam has failed, after the 8.4magnitude earthquake in 2010, resulting in four fatalities.

With the exception of a few inactive upstream sand dams in central Chile, most inactive and abandoned upstream sand dams have performed well, since they have drained down in the dry Chilean climate. This is a "good" story about the capability of the minerals industry in Chile, and later in Peru, to allow for the high seismicity, which governs the stability of tailings dams in their region.

3.4. Overtopping/Flow Liquefaction Failures of a Tailings Dam

The failure of the Merriespruit tailings dam in South Africa in 1994 followed a rainfall event that led to overtopping and erosion of the dam, due to its having inadequate free-board. The resulting flow liquefaction of the tailings engulfed the village of Merriespruit downstream, resulting in 17 fatalities [4], as depicted in Figure 5 [5].

Figure 5. Failure of the Merriespruit tailings dam in South Africa, due to overtopping and flow liquefaction [5].

3.5. Limits to the Upstream Raising of a Tailings Dam

Upstream raising of a tailings dam that is partially on tailings relies on the tailings beach achieving sufficient consolidation and desiccation to form an adequate foundation for the raises. The higher the tailings dam constructed upstream, the further it extends over earlier tailings beaches, and the more likely it is that it will extend over earlier inundated tailings that may not provide an adequate foundation, as shown schematically in Figure 6. Hence, there are limits to how high the upstream raising of a tailings dam can be continued without having a wet and soft tailings layer at depth.

Figure 6. Schematic of the upstream raising of a tailings dam.

3.6. Limits to the Raising of a Tailings Dam on Weak Foundation Layers

As the height of a tailings dam is progressively raised, the pre-consolidation pressure (highest historical stress) of the foundation layers may eventually be exceeded. When this occurs, the foundation layer reverts from over-consolidated to normally consolidated. The stiffness of the now normally consolidated layer is perhaps a tenth of the previously over-consolidated layer; hence, deformations under additional loading are now 10 times higher. Such deformations can threaten the dam freeboard, potentially leading to overtopping and failure of the tailings dam. Examples of foundation failures of tailings dams include Aznalcóllar in Spain in 1998 [6], Mount Polley in Canada in 2014 [7], and Cadia in Australia in 2016 [8], as shown in Figure 7.

(**a**)

(**b**)

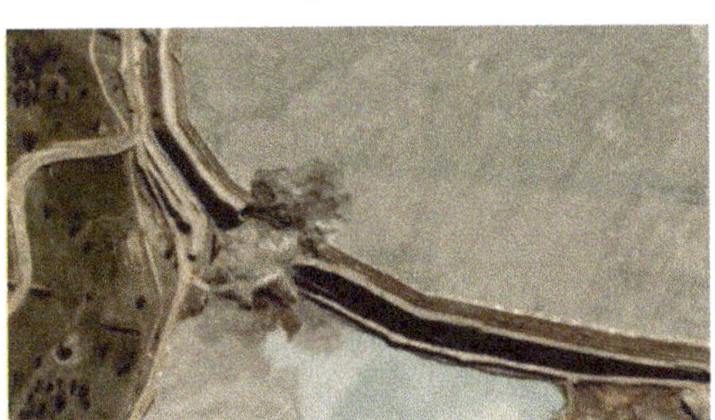

(**c**)

Figure 7. Failure of tailings dams due to weak foundation layers: (**a**) Aznalcóllar [6], (**b**) Mount Polley [7], and (**c**) Cadia [8].

3.7. Rapid Flow Liquefaction of Upstream Tailings Dams

In Minas Gerais, Brazil, the Fundão tailings dam failure in November 2015 caused 19 fatalities and extensive infrastructure and environmental damage (Figure 8) [9], and the Feijão tailings Dam I failure near Brumadinho in January 2019 caused 270 fatalities and infrastructure and environmental damage (Figure 9) [10]. Both incidents were rapid flow liquefaction failures.

Figure 8. Failure of the Fundão tailings dam due to rapid flow liquefaction [9].

Figure 9. Failure of Feijão tailings dam I due to rapid flow liquefaction [10].

Based on the findings in [11], the technical causes of the Fundão tailings dam failure may be summarized as follows. Regarding the question of why flow liquefaction occurred, damage to the original starter dam resulted in increased saturation of the tailings, and slimes were deposited very rapidly and closer to the dam than the designer intended. The failure of a concrete drain and a setback of the left abutment caused the dam to be raised over the slimes, and compression and lateral extrusion of the slimes resulted in the loosening of sand layers above. Regarding the question of why the flow liquefaction occurred where it did, the setback left abutment, on softer, wetter slimes, was critical to stability. Regarding the question of why the flow liquefaction occurred when it did, a series of three small seismic shocks occurred about 90 min prior to the failure, resulting in a small deformation that was sufficient to initiate liquefaction of the marginally stable setback.

Based on [12], the tailings in the Feijão tailings Dam I had high iron content, and oxidation of the iron on exposure to the air led to bonding, rendering the tailings stiff and brittle, with the potential for significant and rapid strength loss with ongoing strain on undrained loading. It is noted that a limit equilibrium stability analysis assumes all points on a slip surface to be at the same state, which is not necessarily correct for a marginally stable dam, in which some material may be pre-peak strength, some at peak strength, and some post-peak.

Dam I was marginally stable. The upstream slope was over-steep. A setback pushed the upper raises of the dam over weaker fine tailings. The water level in the dam remained persistently high, due to a lack of sufficient internal drainage. The bonded tailings had the potential for very brittle behavior if triggered to become undrained. The dam was subjected to high and intense wet season rainfall that can result in a significant loss of suction in the unsaturated tailings above the water level, producing a small loss of strength.

It was concluded in [12] *"that the sudden strength loss and resulting failure of this marginally stable dam was due to a critical combination of ongoing internal strains due to creep, and a strength reduction due to loss of suction in the unsaturated zone caused by intense rainfall towards the end of 2018. This followed a number of years of increasing rainfall and intensity after tailings deposition ceased in July 2016. Calculated pre-failure strains from this combination of triggers matched well the small deformations of Dam I in the year prior to the failure."*

A schematic of the path to failure of Dam I is shown in Figure 10, which shows a plot of undrained strength to vertical effective stress ratio against vertical strain. Different points within the dam would be at different starting points on the stress/strain plot, shown in Figure 10.

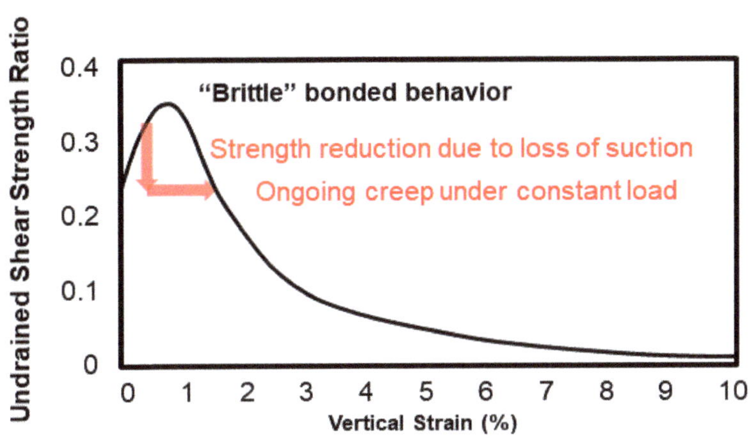

Figure 10. Schematic of the path to failure of Dam I.

4. As Low as Reasonably Practicable

The "as low as reasonably practicable" (ALARP) principle arose out of the UK Health and Safety at Work Act 1974 [14], and this is increasingly applied to managing tailings dam risk, among others. A typical schematic of the ALARP principle is shown in Figure 11 [14], which includes the probabilities of fatalities for workers and the public who are exposed to the risk (1×10^{-3} and 1×10^{-4}, respectively) and for the public generally (1×10^{-6}).

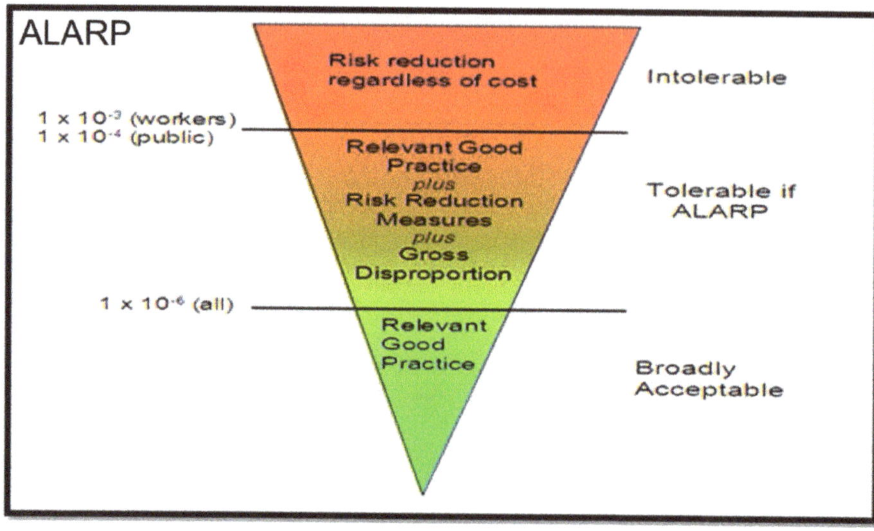

Figure 11. ALARP principle, including the probabilities of fatalities [14].

The ALARP principle can be applied in the risk assessment process, involving the frequency of occurrence versus the magnitude of consequence, as described in [15] and shown schematically in Figure 12. The aim is to carry out risk mitigations that either reduce the frequency of occurrence (F) or reduce the magnitude of consequence (N), or both, to shift the risk from the "intolerable" region into the "tolerable" region.

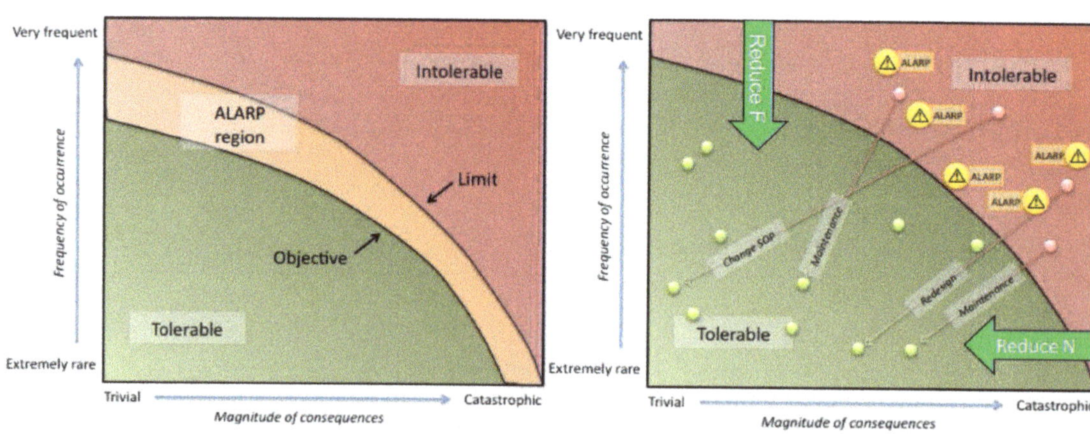

Figure 12. Schematic of ALARP principle applied in risk assessment [15].

In engineering risk assessment, the plot showing the frequency of occurrence versus the magnitude of consequence becomes a plot of annual probability of up to N fatalities (F) versus the number of fatalities (N). Some examples of actual F–N data for different engineering infrastructures are shown in Figure 13, after [16]. Superimposed on Figure 13 are the ALARP limits recommended by the Canadian Dam Association [17], shaded in red for additional risk, orange for tolerable risk, and green for broadly acceptable risk.

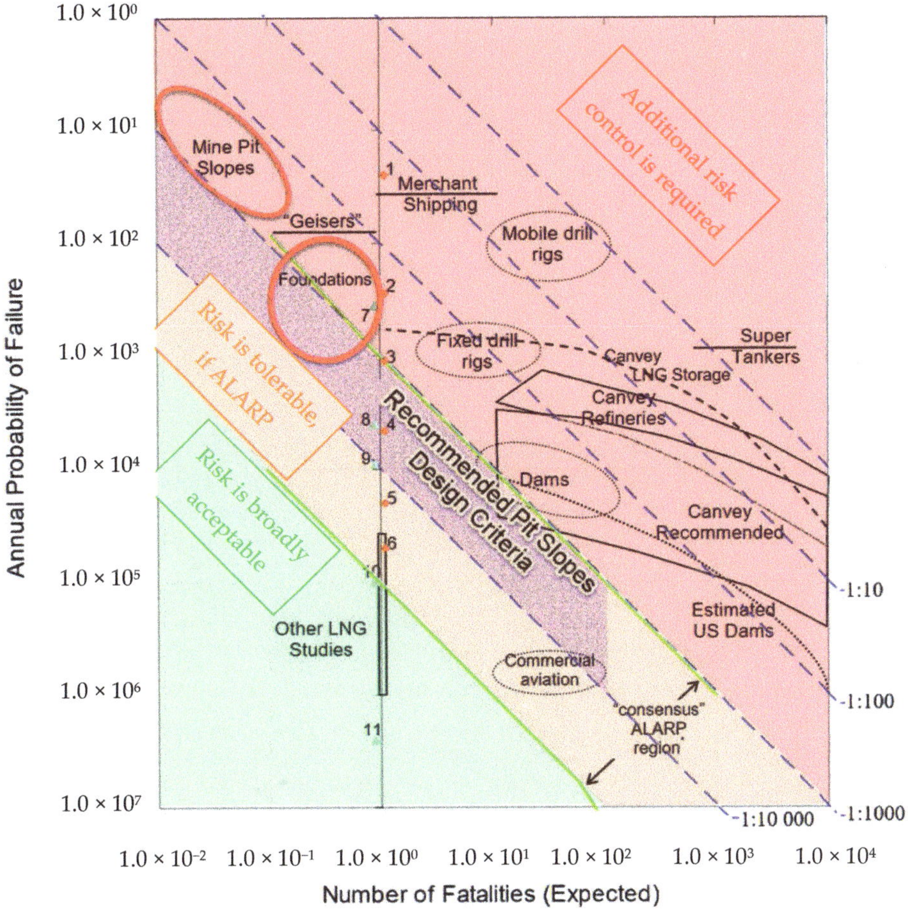

* region defined by the upper most and lower most ALARP boundaries from: Hong Kong planning department, ANCOLD, U.S. Bureau of reclamation and U.K. HS executive guidelines.

Individual Fatality statistics U.S.A - Voluntary

1. Space Shuttle program (per flight) 2. Cigarette smoking (1 pack per day) 3. Average individual voluntary sporting risk 4. U.S. Police killed in line of duty (total) 5. Frequent flying profession 6. Alcohol (light drinking)

Individual Fatality statistics U.S.A - Involuntary

7. Cancer 8. Motor vehicle 9. Home accidents 10. Air Travel 11. Hurricanes / Lightning / Tornadoes

Figure 13. Examples of actual F–N data for different engineering infrastructure [16,17].

There is a mismatch between actual F–N data and the aspirational ALARP target range, let alone the "acceptable" range. It could be argued that the actual F–N data are "accepted", since they represent the status quo that we live with. "Acceptable" is a subjective term, with different people having different perceptions depending on their exposure to risk, whether this exposure is "involuntary" or "voluntary", their level of awareness of risk, and their sense of social responsibility.

It is clear from Figure 13 that water dams plot up to an order of magnitude above the ALARP region, warranting additional risk control. Tailings dams plot on average two orders of magnitude above water dams, warranting considerably more risk control than water dams. The *ICMM Tailings Management Good Practice Guide* [18] provides the ALARP schematic shown in Figure 14, with the "sweet spot" shown by the green circle representing the optimal balance between reduced risk for the resources and effort input.

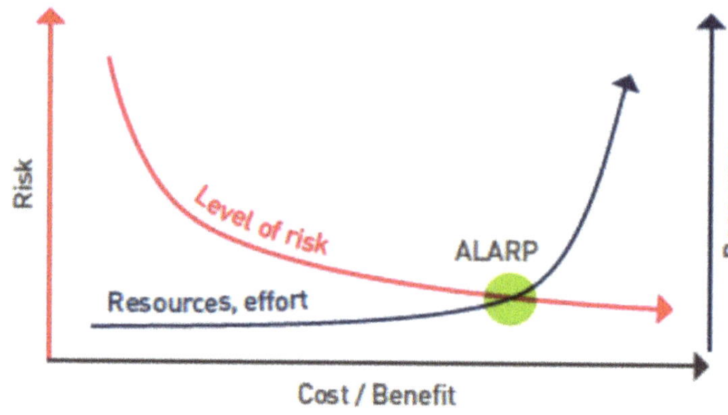

Figure 14. Schematic of the ALARP "sweet spot" [18].

5. Global Industry Standard on Tailings Management

The failure of the Feijão tailings Dam I led to the launch of the Investor Mining and Tailings Safety Initiative [19], which sought reassurance from the minerals industry about the safety of their tailings dams. The International Council of Mining and Metals (ICMM), the United Nations Environment Programme (UNEP), and the Principles for Responsible Investment (PRI), under the "Global Tailings Review", co-convened the development of a Global Tailings Standard under an independent chair and expert panel. The Global Industry Standard on Tailings Management (GISTM) was launched in August 2020 [20].

5.1. What Does the GISTM Require of Tailings Facility Operators?

The GISTM has the aspirational goal of *"zero harm to people and the environment"* from tailings facilities [20]. It elevates accountability to the highest organizational levels, with new requirements for independent oversight. It expects global transparency and disclosure to improve stakeholder understanding. It sets as the design basis an *"Extreme Consequence Classification external loading criteria or the current Classification, with upgrade to Extreme maintained throughout the tailings facility lifecycle"*. Many active tailings facilities are not classified as having an "Extreme" consequence classification and may need to be re-classified and/or upgraded to meet "Extreme".

The post-closure classification is "Extreme" since tailings facilities must remain safe, stable, and non-polluting in perpetuity. This criterion is in line with current tailings dam guidelines, such as in [17,21]. The extreme consequence return interval is 1 in 10,000 years, which is presumed to be equivalent to "in perpetuity". This long return interval is not applied to other engineered structures, which can be attributed to the loss of confidence and trust in the capability of the industry, their consultants, and regulators to ensure safe and stable tailings dams.

The purpose of the GISTM is to provide *"a framework for safe tailings facility management, while affording operators flexibility as to how best to achieve this goal"*, in other words, to be self-regulating. The design, construction, operation, monitoring, and closure of tailings facilities are required to be robust in order to minimize the risk of harm to people and the environment. The GISTM calls for emergency response and long-term recovery of the community affected, in the event of a tailings facility failure. Public disclosure and access to information are required to support public accountability, requiring clear communication. The GISTM also specifies governance requirements for the management of tailings facilities.

5.2. What Are the Next Steps in the Implementation of the GISTM?

Since the tailings dam failure near Brumadinho in Brazil, some regions of the world, now including Brazil, have been refused insurance for tailings dams. Other regions have had their insurance coverage reduced (by about one-third, from coverage that was typically USD 300 million to USD 200 million, which is clearly insufficient to cover a catastrophic failure), and their premiums have increased (about two-fold). Some major mining companies are self-insuring their tailings facilities. However, mid-tier and small mining companies with few mines cannot afford to not be insured. Access to finance for new tailings facilities could be withheld from companies that do not comply with the GISTM.

The Global Tailings Review estimated that there are approximately 18,000 tailings facilities worldwide, of which approximately 3500 are active. The GISTM does not cover the thousands of inactive and abandoned tailings facilities worldwide.

ICMM members, comprising 27 of the world's major mining and metals companies and operating about 1200 tailings facilities (about one-third of the world's active facilities), have committed that all tailings facilities with "Extreme" and "Very high" potential consequences will be in conformance with the GISTM by August 2023, and all other facilities in conformance by August 2025 [20]. Non-ICMM members companies are encouraged to comply with the GISTM.

5.3. Management of Tailings Facilities

The ICMM and its members are adapting the management structure and interactions of their "Extreme" and "Very high" consequence classification tailings facilities to accommodate the governance requirements of the GISTM. The new structure and interactions are along the lines of those shown schematically in Figure 15. The GISTM defines the roles and responsibilities of all personnel in this structure. The Accountable Executive is directly answerable to the Chief Executive Officer (CEO) on matters related to the GISTM, and will communicate with the Board of Directors (Board). The Accountable Executive is accountable for the safety of the tailings facility, and for minimizing the social and environmental consequences of a potential tailings facility failure. The Accountable Executive interacts closely with the Tailings Facility Operator, who in turn interacts closely with the Responsible Tailings Facility Engineer (RTFE). The Accountable Executive should not have Key Performance Indicators (KPIs) that are based on production improvements.

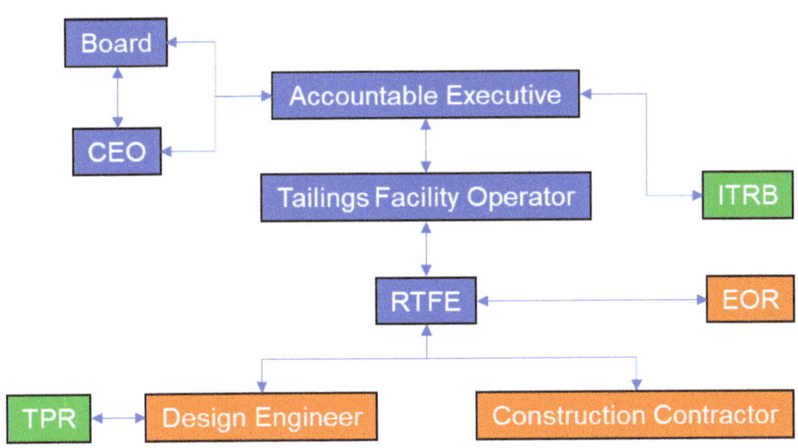

Figure 15. Schematic tailings facility management structure and interaction.

The RTFE works closely with the (usually) external Engineer of Record (EOR), typically interacting monthly, meeting on-site perhaps quarterly, and with the two having the authority between them to make recommendations to the Accountable Executive about the ongoing operation of the tailings facility. The EOR has responsibility for preparing and updating the Design Basis Report that provides a basis for the monitoring and risk management of all design phases of the tailings facility.

The RTFE and EOR also work closely with the Design Engineer and Construction Contractor for the tailings facility. The design of the facility and sequential raises are reviewed by a third-party reviewer (TPR). According to the GISTM [20], dam safety reviews are periodically and systematically carried out by a TPR *"to assess and evaluate the safety of a tailings facility against* ["credible"] *failure modes, in order to make a statement on the safety of the facility. A safe tailings facility is one that performs its intended function under both normal and unusual conditions, does not impose an unacceptable risk to people, property or the environment, and meets applicable safety criteria."*

Credible failure modes are *"technically feasible failure mechanisms given the tailings and foundation, their parameters, geometry, drainage conditions and surface water control, throughout the lifecycle of the tailings facility. They typically vary during the lifecycle of the facility as the conditions vary, and the tailings facility requires sufficient resilience against each credible failure mode. Different failure modes will result in different failure scenarios. Credible catastrophic failure modes do not exist for all tailings facilities, and are not associated with likelihood and not a reflection of facility safety."*

According to the GISTM [20], the Independent Tailings Review Board (ITRB) *"provides independent technical review of the design, construction, operation, closure and management of the tailings facility. ITRB members are third parties who are not, and have not been, directly involved with the design or operation of the particular tailings facility. The expertise of the ITRB members should reflect the range of issues relevant to the facility, and its context and the complexity of these issues."*

An ITRB would typically include a geotechnical expert, a hydrology or hydrogeology expert, as appropriate, a seismology expert for highly seismic locations, and possibly a geochemistry or geomorphology expert, as appropriate. The ITRB would typically meet prior to key changes in the life of the tailings facility, such as a dam raise, interacting most closely with the Tailings Facility Operator and the RTFE, and also with the EOR and Design Engineer, and possibly also with the TPR. The ITRB should report its findings through the Accountable Executive to inform the company's CEO and board.

A qualified EOR, TPR, or ITRB member would typically be able to handle perhaps four tailings facilities. An estimated 70 ITRBs would be required for all operating tailings facilities globally, requiring over 200 tailings experts, with perhaps 10 times this number of EORs and TPRs required. There are insufficient tailings experts available worldwide to meet this demand, particularly in the current COVID-19 climate that restricts travel to sites. The increasing availability of online training and certification in tailings management will assist in meeting this demand in the future, although the experience required for these roles will take time to develop. Senior tailings professionals will need to mentor talented early-career professionals.

5.4. ICMM Documents Accompanying the GISTM

With the release by the ICMM of the *Tailings Management Good Practice Guide* [18], and conformance protocols [22], comes a structure for tailings management, as shown in Figure 16. At its base are technical guidelines from several institutions: the International Committee on Large Dams (ICOLD, which is currently working toward a global tailings guideline); the Australian National Committee on Large Dams (ANCOLD) [21]; the Canadian Dam Association (CDA) [17]; the Japanese Commission on Large Dams (JCOLD); and the South African National Committee on Large Dams (SANCOLD). The United States Society on Dams is drafting a national tailings dam guideline to replace the individual state guidelines, many of which were drafted in the 1990s and lack consistency. See [23] for a review of mine tailings guidelines, initiatives and standards.

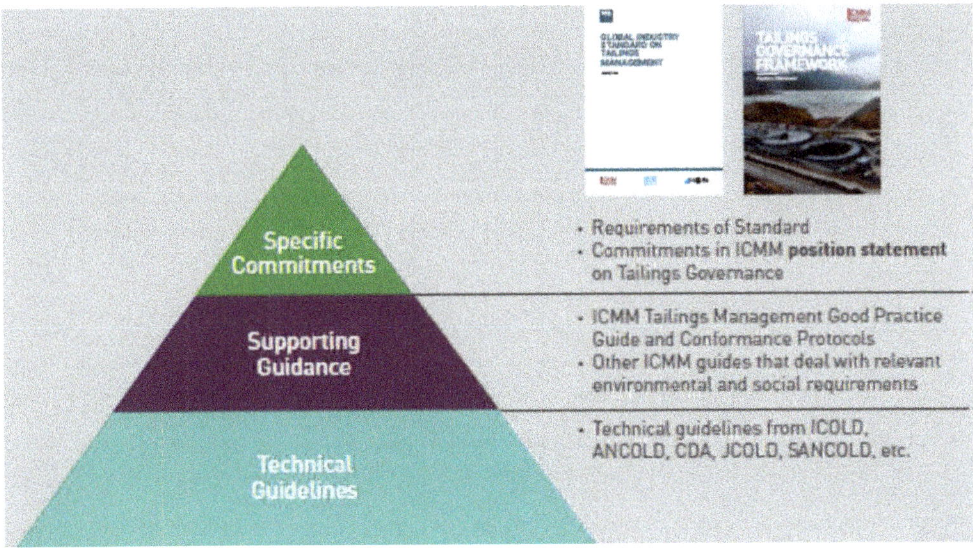

Figure 16. Structure for tailings management [18].

At the apex of the structure shown in Figure 15 are the requirements of the GISTM and commitments in the ICMM position statement on tailings governance [24]. Supporting guidance linking the technical guidelines and the specific commitments includes the *ICMM Tailings Management Good Practice Guide* and conformance protocols, plus other ICMM guides that deal with relevant environmental and social requirements.

It remains to be seen how this structure will be accommodated within the regulation of tailings facilities, both regionally and globally. None of the structure shown in Figure 15 is specifically included in the legislation, although in the jurisdictions where they are applied, the ANCOLD [21] and CDA [17] guidelines are taken by mining companies, regulators, and the courts to be de facto "standards" for tailings dam design.

6. Alternative Tailings Management Options

"In the wake of recent tailings dam failures, improved Management Resilience (that is, Governance) is leading over improved Engineering Resilience, which would reduce Liability", according to Dr. Andy Robertson [25].

The GISTM [20] does not specify particular tailings management options that should be allowed or disallowed, leaving this choice to the Operator. However, the GISTM does require that tailings management options be reviewed and analyzed. For existing tailings facilities, the Operator is required *"to periodically review and refine the tailings technologies and design, and management strategies, to minimize risk and improve environmental outcomes. An exception applies to facilities that are demonstrated to be in a state of safe closure."* For new tailings facilities, the Operator is required *"to use the knowledge base and undertake a multi-criteria alternatives analysis of all feasible sites, technologies and strategies for tailings management. The goal of this analysis is to: (i) select an alternative that minimizes risks to people and the environment throughout the tailings facility lifecycle; and (ii) minimize the volume of tailings and water placed in external tailings facilities."*

The GISTM also requires the development and documentation of a *"breach (dam break and runout) analysis for the tailings facility using a methodology that considers credible failure modes, site conditions, and the parameters of the tailings. When flowable materials (water and liquefiable solids) are present at tailings facilities with a Consequence Classification of 'High', 'Very High' or 'Extreme', the breach analysis results should include estimates of the physical area impacted by a potential failure, flow arrival times, depth and velocities, and depth of material deposition."*

The results of the breach analysis should be used to identify and document the potential human exposure and vulnerability to credible failure scenarios regarding a particular tailings facility. The breach analysis should be updated whenever there is a material change either to the tailings facility or the physical area impacted.

According to Williams [26], *"while the conventional disposal of tailings slurry can be the optimal NPV and life-cycle choice for a given operation, there is often a divergence when a whole-of-life approach is fully considered."* Alternative approaches to tailings management are described in the following sections.

6.1. Geotechnical versus Geochemical Stability of Stored Tailings

Fundamentally, the geotechnical stability of tailings is enhanced by dewatering and densification, while the geochemical stability of potentially contaminating tailings (such as those containing sulfides) is ensured by maintaining the tailings at near-saturation, to limit oxygen ingress. For the storage of tailings slurry in a surface facility, geotechnical stability and the potential for upstream construction are enhanced by cycling tailings deposition in thin layers, and by leaving time for consolidation and desiccation before the next layer is deposited.

Geochemical stability is ensured by maintaining the tailings at a level that is at least 85% saturated, to limit the diffusion of oxygen into a desaturating surface. This can be achieved by depositing fresh tailings slurry before the previous layer has been allowed to desiccate to less than this saturation limit. Geochemical stability is also enhanced for fine-grained and clay mineral-rich tailings that tend to hold onto water due to their low permeability. Perhaps counterintuitively, sulfidic tailings that tend to hardpan result in a virtually impermeable crust that limits moisture and hence contaminant flow, either as net downward percolation or as evapotranspirative uptake into an overlying growth medium. Hardpans can have a permeability that is 10 to 100 times lower than that of uncemented tailings, and an oxygen diffusion rate up to 1000 times lower, causing them to maintain a high degree of saturation and to not pass water or oxygen [27]. However, hardpans may not be continuous across the tailings' surface.

For potentially contaminating tailings that are deposited in-pit, any pit lake should not be allowed to become a "source" of contaminated water for surface or ground waters. It is preferable that final pit lakes containing water of diminishing quality should be avoided by completely back-filling the pit, even if the pit were to remain a surface- and groundwater "sink".

6.2. In-Plant Dewatering of Tailings

The in-plant recovery of process water is the most effective means of maximizing water return for recycling and for recovering any residual process chemicals, assuming that the return water is suitable for reuse in the plant and that the tailings can be dewatered. The choice of the appropriate amount of in-plant dewatering is related to the tailings dewatering continuum adapted from [28], as shown in Figure 17, which ranges from the initial slurry-like state to a potentially soil-like state. The cost of mechanical dewatering of tailings increases exponentially with increasing dewatering, which is balanced by benefits including the greater recovery of water and process chemicals and reduced tailings storage volume.

Clay mineral-rich tailings, such as oil shale tailings, some coal tailings, and caustic processing residues, such as red mud from bauxite refining, nickel laterite tailings, and oil sands tailings, are difficult to thicken, let alone filter. Slurry and thickened tailings may readily be pumped using robust, inexpensive centrifugal pumps. Paste tailings require positive displacement or diaphragm pumps, which are an order of magnitude more expensive than centrifugal pumps and are more sensitive to variations in the input tailings particle size distribution and chemistry. Furthermore, a true paste will discharge like "toothpaste" and require continuous management of the deposition. To avoid the need to pump paste tailings, they may be discharged through gravity to a completed underground mine, possibly as cemented paste tailings backfill, or to a completed pit.

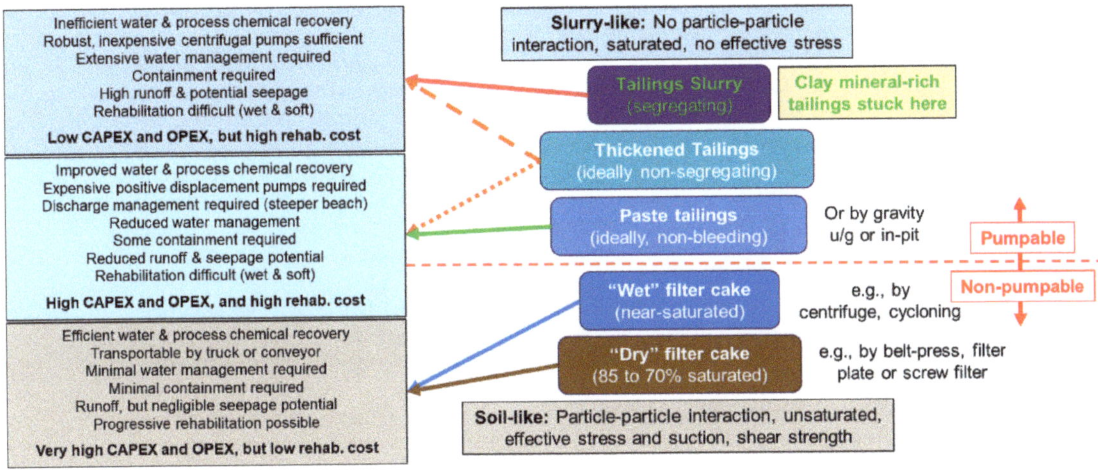

Figure 17. The tailings dewatering continuum [28].

"Wet" tailings filter cake, which is potentially produced by centrifuging or cycloning, is not pumpable, can potentially be conveyed or trucked, and typically flows on deposition. "Dry" tailings filter cake, which is potentially produced by a belt-press or filter plate, and possibly by screw filter (used in sewage treatment, although it has not been applied to tailings), can be conveyed, trucked, and potentially mixed with coarse-grained wastes, and will not flow on deposition.

The optimal in-plant dewatering of tailings for disposal to a surface tailings storage facility is likely to be of tailings thickened as much as possible but that is still able to be pumped by robust, inexpensive centrifugal pumps. Paste tailings are best reserved for gravity disposal underground or in-pit. Filtered tailings are potentially suited to "dry" stacking, possibly with the compaction of at least the outer zone to prevent potential liquefaction.

6.3. In-Facility Dewatering of Tailings

Water recovery from a surface tailings facility is generally limited to the recovery of supernatant water (the water that pools at the end of the tailings beach). Tailings water is lost to evaporation from the wet tailings and from the decant pond, from entrainment within the tailings, and from seepage into the foundation and through the dam.

While the majority of the water from a tailings facility may be lost through evaporation off freshly deposited wet tailings, minimizing the size of the decant pond and the rapid return of supernatant water will minimize additional evaporation losses. Tailings densification, due to self-weight consolidation, and desiccation on exposure, is far more robust than in-plant thickening or filtration, being less affected by variations in the nature of the tailings.

Water recovery from an in-pit tailings facility is more difficult than from a surface tailings facility. Firstly, the rate of rising in-pit is generally much faster, due to the steep geometry of the pit, compared with more shallow surface tailings facilities. Secondly, it is difficult to recover water from a pit as it fills. Thirdly, the opportunity for desiccation of in-pit tailings is limited by the usual water cover, and by shadows thrown from the pit walls onto any exposed tailings. The faster rate of rising in-pit, only partial consolidation, and limited or absent desiccation, result in limited dewatering and densification of the tailings, and much higher water entrainment.

6.4. Coagulation, Flocculation and Secondary Flocculation

Coagulation concerns the initial supernatant water clarification through the settling of colloids. Coagulation is both a chemical (charge neutralization of colloids) and a physical (consequent aggregation of colloids) process. Coagulants are mostly cheap iron or aluminum salts, forming a highly ionic polymer of low molecular weight.

Flocculation concerns the agglomeration of dispersed particles. Flocculants are normally high molecular weight, long-chain polymers that achieve flocculation relatively slowly, primarily by bridging, but also by wrapping water. Polyacrylamide is a common flocculant, either as negatively charged anionic for mineral particles or positively charged cationic for organic particles, with divalent cations such as Ca^{2+} and Mg^{2+} aiding bridge formation. The effectiveness of flocculation is dominated by the clay mineral content and type of the tailings, and the salinity of the process water.

If conventional thickening and slurry tailings disposal fail to achieve adequate settling and consolidation and supernatant water recovery, secondary (inline) flocculation can be applied just prior to the discharge of the tailings, to re-flocculate conventionally thickened tailings that have been shear-thinned by pumping.

It is important that the appropriate type and dosage of coagulants and flocculants applied to tailings be carefully selected, which generally involves trialing them both in the laboratory and on-site. Flocculants, in particular, can be expensive, and over-dosing can be counter-productive, leading to excessive "wrapping" of both the tailings particles and entrained water. With variable tailings feeds, online monitoring of pH and electrical conductivity may be required, and the flocculant type and dosage should be adjusted to suit.

6.5. On-Off Tailings Cells

As described in Williams [26], the desiccation and harvesting of black coal tailings in "on-off" tailings cells have been employed at a number of mines, including at Charbon Coal Mine in New South Wales, Australia, since 1990, as shown in Figure 18. This method involves the disposal of slurry tailings in cells in thin layers, preferably no more than 600 mm thick, and leaving sufficient time for the tailings to consolidate and desiccate on exposure to the sun and the wind in dry weather, prior to the disposal of the next layer. The limited layer thickness is dictated by desiccation by solar and wind action, which drops off exponentially with depth. The time required for the consolidation and desiccation of each layer will be of the order of several weeks, depending on the nature of the tailings and the ambient weather conditions. The cells are filled to a depth that can readily and safely be excavated by the available equipment, such as an excavator (as shown in Figure 18a for poorly desiccated tailings) or a front-end loader (as shown in Figure 18b for well-desiccated tailings). A typical depth is about 3 m. The full depth of dried tailings is then harvested and can be co-deposited with coarse wastes (as shown in Figure 18c). This method of dewatering tailings requires a large number of shallow cells covering a large footprint, although probably no larger a footprint than would ultimately be needed for a conventional surface slurry tailings facility. Reliance on evaporation means that a considerable volume of water is lost. Ultimately, there could potentially be no tailings facility remaining. The cost of operating on-off tailings cells is comparable to the cost of tailings filtration and dry stacking.

Figure 18. On-off coal tailings cells, showing: (**a**) insufficient desiccation, (**b**) effective, full-depth desiccation, and (**c**) dumping harvested tailings with the coarse reject.

6.6. Farming of Tailings

Some forms of wet and soft tailings, particularly clay mineral-rich tailings and caustic process residues, may benefit from "farming" by the use of equipment, such as an amphirol or scroller (being the more generic term), and/or later compaction by a D6 Swamp Dozer. Farming is widely applied to red mud in Australia and has been trialed on other tailings, including on coal tailings and fly ash in Australia, and oil sands tailings in Canada.

A scroller, as shown in Figure 19a, has a very low bearing pressure of 3 to 5 kPa and is used first. The principles of tailings or residue farming by scroller are as follows:

- The tailings or residue is poured to a thickness of 700 to 900 mm, up to three times the thickness to which surface desiccation would extend.
- The first pass of the scroller is applied after some initial drying and strengthening of the tailings or residue surface, to allow safe and efficient operation.
- If the bearing pressure from the scroller and/or the tailings or residue surface is too soft, bogging of the scroller can occur.
- A scroller will only achieve minimal consolidation or compaction of the tailings or residue since its bearing pressure is low.
- A scroller should essentially "float" over the tailings or residue surface, creating trenches down the tailings or residue beach to facilitate the drainage of surface water, maximizing the tailings or residue surface area that is exposed to evaporation and strengthening, and exposing undesiccated tailings or residue after further farming.
- A scroller should not over-shear the tailings or residue by excessive or repeated farming, with about four scroller passes being optimal.
- The last scroller pass may be perpendicular to the beach, to smooth the surface prior to subsequent dozing since there should no longer be any surface water to drain.

(a) (b)

Figure 19. Farming red mud, by (**a**) a scroller, and (**b**) bulldozing with a D6 Swamp Dozer.

A D6 Swamp Dozer has a bearing pressure of about 35 kPa and can be applied once farming of the tailings or residue by scrolling has achieved sufficient shear strength and bearing capacity to safely support it, as shown in Figure 19b. The initial dozer pass, with the blade not in contact, serves to further remove any near-surface moisture. Dozing improves the already desiccated tailings or residue by compaction, leading to a further increase in dry density and shear strength. Typically, four to six dozer passes are applied.

Red mud typically has a specific gravity of about 3.0 and is difficult to densify, due to its forming, on slurry deposition, a loose "house of cards" structure of low permeability. Without farming, the dry density achieved is typically limited to about 0.7 $t.m^{-3}$, and desiccation is limited to a depth of about 300 mm. Scrolling can increase the dry density to about 0.9 $t.m^{-3}$, and dozing can increase it further to 1.3 to 1.4 $t.m^{-3}$. The cost of red mud farming is offset by the increased utilization of the available storage, by facilitating upstream raising (potentially using compacted red mud), and rehabilitation on closure.

6.7. Paste Tailings

Paste thickeners can raise the percentage of solids by mass to between 45% (for red mud) and 75% for metalliferous tailings. Paste tailings can readily be delivered under gravity as underground backfill (usually with cement added), or in-pit if the dewatering facility is located close to the discharge point, possibly on a mobile skid.

Underground tailings paste backfill will generally reach its intended destination under gravity, provided that the angle between the discharge and final points is steeper than 45°. Paste tailings disposal to a surface facility has had a relatively limited take-up due to the high cost of pumping and discharge management, with Bulyanhulu in Tanzania, Africa the most well-known application.

The overall tailings water recovery, as a percentage of the total water used in processing, increases to about 80% for tailings disposal as a high slump paste, and to 85 to 90% for a low slump paste, although the cost of paste production is high.

6.8. Filtration and Dry Stacking

The difference in consistency between wet and dry tailings filter cake is illustrated in Figure 20. The wet tailings filter cake is near-saturated and has the potential to flow on disposal, while dry tailings filter cake has a stress-induced "structure" and tends to remain intact on disposal.

(**a**) (**b**)

Figure 20. Consistency of (**a**) centrifuged wet tailings filter cake, and (**b**) filtered dry tailings filter cake.

Tailings filtration is best achieved under high pressure (1600 to 2100 kPa), with a cycle time of up to about 20 min. Although termed "dry", filtered tailings retain moisture and are more correctly described as "unsaturated". Dry tailings filter cake may be deposited in a stack, although compaction may be required for geotechnical stability, to prevent potential liquefaction, and to limit oxygen ingress and rainfall infiltration into potentially contaminating tailings, in order to minimize contaminated seepage.

Dry stacking has found most favor in dry climates such as in northern Chile, at La Coipa (Figure 21) [29] and other sites, and in southern Peru, to maximize water recovery in this region of limited freshwater supply. The cost of tailings filtration in the Atacama Desert region is comparable to the cost of seawater desalination. However, the dry stacking of filtered tailings has also been applied in wetter and colder climates, including at Greens Creek Metal Mine in Alaska [30] and at CSN's Casa de Pedra Iron Ore Mine at Congonhas, Minas Gerais, Brazil. The high cost of filtration has, to date, limited its application to tailings production rates of less than 20,000 dry tpd, although this limit is rising.

(**a**) (**b**)

Figure 21. Dry stacking of filtered tailings at La Coipa, Atacama region, northern Chile, showing (**a**) stacker, and (**b**) overall stack [29].

With increasing the mechanical dewatering of tailings, costs increase dramatically. Thickening is about 1.5 times more costly than conventional slurry tailings, high-density thickening about 3 times more costly, paste about 8 times more costly, and filtration about 27 times more costly than slurry tailings [31]. However, this cost comparison ignores cost reductions due to increased process water recovery, the reduced storage volume required with increasing tailings dewatering, the consequently reduced containment required, and the reduced cost of rehabilitation and enhanced potential post-closure land use.

A more comprehensive cost comparison between slurry tailings deposition behind a dam, and tailings filtration and dry stacking with compaction, is provided in [32]. Tailings filtration and dry stacking with compaction indicated initial and average capital expenditure savings of 38% and 64%, respectively (almost 4 times the operating expenditure), with an average total cost saving of 33% over slurry tailings deposition behind a dam. The main cost of slurry tailings deposition behind a dam is the cost of dam raising, exacerbated by the low settled density and hence the high storage volume required for slurry tailings. In addition to the 80% saving in storage volume, the filtered tailings would also: recover about twice as much water by mass, plus metals and process chemicals; reduce seepage dramatically; be stable under seismic loading; be readily rehabilitated progressively and to a higher level of future land use. This cost comparison did not include closure costs.

6.9. Co-Disposal of Tailings and Coarse-Grained Waste

In British Columbia, Canada, coal tailings and coarse reject have been trucked and compacted in stable stacks that resemble a conventional dump rather than a tailings facility.

Since it was first trialed and introduced at Jeebropilly Coal Mine in the Ipswich Coalfields of South-East Queensland, Australia in 1990 [33], the combined coal tailings and coarse coal reject have been cost-effectively co-disposed by pumping at numerous coal mines in Australia and Indonesia. Pumped co-disposal in-pit at Jeebropilly Coal Mine is shown in Figure 22. In order to avoid pipeline blockages, the combined washery wastes are pumped at a low solids concentration of 25 to 30% by mass and at high velocity (up to 4 $m.s^{-1}$). While a steep upper coarse-grained beach (with about a 1-in-10 slope) is formed, the low solids concentration and high velocity result in the segregation of most of the fines and the generation of an undesirable flat (at about a 1-in-100 slope) fines beach (mostly tailings) beyond the upper beach. In addition, the inclusion of the coarse reject in a low-density medium results in high pump and pipeline wear.

Figure 22. Pumped co-disposal in-pit at Jeebropilly Coal Mine.

The co-deposition in-pit of waste rock by end-dumping and thickening tailings by gravity has been practiced very cost-effectively at a number of sites, where completed pits have become available. At Kidston Gold Mines in north Queensland, Australia, the tailings and waste rock streams from the second pit were co-deposited entirely within the completed first pit [34], as shown in Figure 23. The pit was 240 m deep, and the waste rock was end-dumped from one side of the pit, while thickened tailings were deposited by gravity from a mobile thickener on a skid from the other side of the pit. The initial angle of repose of the waste rock slopes of 40° was flattened slightly to 38° as the waste rock settled, about 1.5% of its height or about 4 m, and extended at the "toe" by about 10 m.

Figure 23. In-pit co-deposition of waste rock and thickened tailings (left-hand pit) at Kidston Gold Mines in north Queensland, Australia.

Dewatered tailings are also co-deposited in spoil dumps at coal mines in the Queensland Bowen Basin and New South Wales Hunter Valley Coalfields in Australia. The coal tailings are dewatered by belt-press filtration or centrifuging and are deposited from trucks in the coarse waste dumps, either in dedicated cells where the tailings remain flowable, or by end-dumping over a tip-head where the tailings are more soil-like. The main determinate as to whether the dewatered tailings are flowable is the clay mineral content and type of the coal seams being washed. A small percentage of smectite clay can render the tailings very difficult to dewater by any means.

6.10. Integrated Waste Landforms

Integrated waste landforms are increasingly being planned and employed in Australia, particularly at coal and iron ore projects, and elsewhere worldwide. This involves either the construction of a robust containment for thickened tailings using waste rock (including a low permeability core and/or drainage layers, as required) or the co-disposal of mixtures of filtered tailings and waste rock or coarse-grained processing wastes. The mixtures are delivered by combined pumping for coal washery wastes, or by haul truck or conveyor. This approach has also been employed in the wet tropics to encapsulate tailings and waste rock that are potentially acid-forming, placed behind a robust containment of more benign waste rock, created by paddock-dumping and compacted in layers.

6.11. Tailings Reprocessing and Reuse and Reduced Tailings Production

There is a long history of reprocessing gold tailings, sometimes more than once, such as in Johannesburg in South Africa and Kalgoorlie in Western Australia. What is relatively new is the water monitoring of base metal tailings for reprocessing, such as at Century Zinc Mine in North Queensland, Australia, as shown in Figure 24. Reprocessing tailings at Century has made it the world's tenth-largest zinc producer (formerly the third largest). The residual tailings are deposited into the completed pit, with the potential to completely remove the environmental liability of the surface tailings facility.

Figure 24. Water monitoring of tailings for reprocessing at Century Zinc Mine, showing: (**a**) water monitor, (**b**) re-slurried tailings, and (**c**) flowing slurry tailings toward a sump for recovery.

Examples of tailings reuse include for bricks and other building products, the use of pozzolanic power-station fly ash as a partial replacement for cement, and for the purposes of geopolymerization, as an environmentally friendly and low carbon replacement for natural sources [35,36].

In response to the ever-increasing production of tailings because of decreasing ore grades and increasing demand for minerals, attention is now being paid to finding ways of reducing tailings production. Another driver has been the rising cost of energy and other mining and processing costs. The primary focus has been on coarse particle and/or dry processing.

6.12. PasteRock™ and GeoWaste™

The co-disposal of filtered tailings and waste rock can be achieved by mechanical mixing to form "PasteRock™", patented by Golder Associates, which has been trialed in Papua New Guinea, and in Canada for mine waste covers [37]. More recently, Goldcorp patented "GeoWaste™", which incorporates filtered tailings, combined with screened or crushed waste rock [38].

The practical and economic challenges that must be overcome to promote the combination of filtered tailings and waste rock include:

- Minimizing the extent to which the tailings must be dewatered, to save costs, while not compromising the stability of the combined tailings and waste rock.

- Minimizing the crushing or screening of the waste rock to allow mixing with the filtered tailings and transportation. The largest size of the waste rock for conveying is about 200 to 300 mm, while run-of-mine waste rock can be trucked.
- Achieving adequate mixing of the filtered tailings and waste rock. This is unlikely to occur on a conveyor since the two waste streams tend to remain separate. It is also unlikely to occur upon dumping from a haul truck since the coarse fraction tends to ravel further. However, mixing may be achieved by placing a number of drop points into hoppers along a conveyor line. It may also be achieved by placing the coarse-grained waste rock in the base of a truck, with the filtered tailings over the top to the designed ratio, to avoid the filtered tailings sticking to the truck tray and to force mixing on dumping.
- Compaction of the mixture may be required to produce a stable deposit, although compaction could be restricted to the perimeter of the emplacement.

The benefits of combining filtered tailings and waste rock can include an increased shear strength, reduced compressibility, and permeability that is lower than that of the waste rock alone but is higher than that of tailings alone.

6.13. Briquetting of Tailings

Briquetting involves forcing a slurry between two rollers under very high stress (of the order of 50 MPa), and has been shown to be very effective in dewatering ultra-fine black product coal [39]. In this study, the ultra-fine product coal that was initially at 40 to 45% total moisture content (mass of water/total mass, expressed as a percentage) was dewatered to briquettes of about 15% total moisture content (85% solids by mass). The very high stresses imposed over a very limited time duration resulted in further dewatering of the briquettes in a dry atmosphere, to about 2 to 5% total moisture content (95% solids). The air-dried briquettes can re-wet in a humid atmosphere, but only to about 15% total moisture content, and they retain their "briquette" structure. However, the very high initial capital and operating costs of briquetting, and scale-up issues, would discourage its application to tailings.

6.14. Barriers to the Implementation of Innovative Tailings Management

The conventional disposal of slurry tailings remains the preferred choice for many mine sites. Barriers to the implementation of the wide range of innovative tailings management options available include:

- The continued use of NPV accounting with a high discount factor (typically 6 to 10%, which is three to five times the consumer price index). This approach favors tailings management options that are less costly (particularly in capital expenditure terms) in the short term, delaying long-term expenditure and rehabilitation. Favoring low capital expenditure can come at the expense of increasing operating expenditure, and is likely to exacerbate undesirable outcomes and blow-out rehabilitation costs.
- The mechanical dewatering of tailings, and the co-disposal of tailings and waste rock, are seen as too costly in a narrow comparison with conventional tailings transport and disposal as a thickened slurry, a view that is reinforced by NPV accounting.
- There are perceived and real technical difficulties associated with the mechanical dewatering of tailings and the co-disposal of tailings and waste rock. Very fine-grained tailings and tailings with a high clay mineral content are difficult to thicken and filter, and it is difficult to mix tailings and coarse-grained waste.
- There is an underlying inherent resistance to change and doing something other than what has always been done, often disguised as unsubstantiated claims about perceived high costs, perceived technical obstacles, and perceived uncertainty.
- The uncertainty, and hence the perceived higher risk, attached to new approaches also serve to discourage innovation.

7. Closure and Rehabilitation of Tailings Facilities

The closure and rehabilitation of tailings facilities should be considered early on in their planning, and continuously during operation, since options decrease and costs increase during the mine's life, as highlighted schematically in Figure 25 (taken from the *Global Acid Rock Drainage* (GARD) *Guide* [40]).

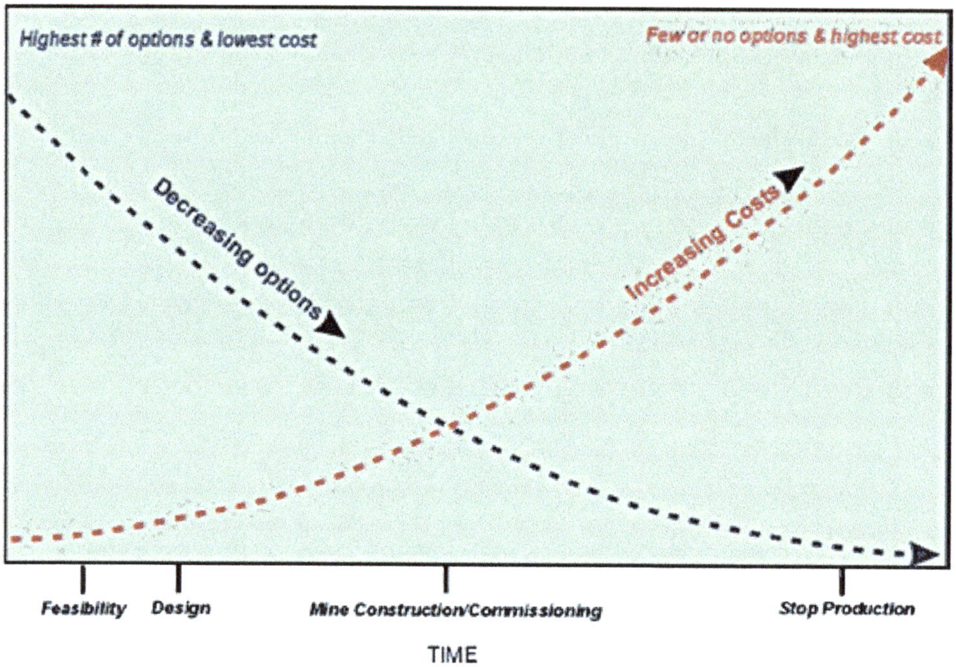

Figure 25. Choice of cover type, related to climate [40].

The aim of the closure of tailings facilities is to leave safe, stable and non-polluting structures in perpetuity, with stakeholder input to address the agreed post-mining land use and/or ecological function post-closure. Irrespective of the post-closure use(s) that are agreed upon, considerations for tailings facilities entering their closure phase should include:

- Geotechnical stability: the aim should be to convert tailings dams to be stable landforms, removing credible failure modes. This may or may not require slope flattening or buttressing. With the cessation of tailings deposition, the daily water input to the tailings facility will generally be dramatically reduced, and the tailings would generally be expected to drain down. This would render the dam more geotechnically stable. However, the tailings may be recharged by high rainfall (in the absence of a spillway), and may not drain down in a wet climate, leading to no improvement in geotechnical stability.
- Erosional stability: erosion may be exacerbated by a lack of revegetation on the tailings cover and dam slope and/or a fine-grained soil surface texture.
- Differential tailings settlement: differential settlement will arise due to variable tailings depth and consistency, potentially affecting the slope profile and drainage.
- Geochemical stability: Poor water quality (such as saline, and/or acidic, or alkaline water) may develop, after a lag:
    - In ponded water on a surface of the tailings facility;
    - In seepage emerging at low points around the toe of a surface facility;

- ○ On seepage infiltrating to any groundwater resource beneath a tailings facility; or
- ○ In a pit backfilled with potentially contaminating tailings, particularly if it could become a source of contaminated seepage or runoff.

Wet and soft tailings resulting from conventional slurry tailings deposition are difficult and expensive to physically cover and rehabilitate, particularly at the end of the mine life, when the mine is no longer producing revenue and construction equipment is being demobilized. Furthermore, such tailings will limit the future land-use potential of the facility. On the other hand, their high degree of saturation will limit the oxidation of any sulfides present in the tailings, reducing the potential for contaminated seepage.

7.1. Covering Tailings

The rehabilitation of tailings can vary, including benign water covers in wet climates, the direct revegetation of benign tailings, and soil covers, particularly in dry climates. As shown in Figure 26, the GARD guide [40] recommends that the choice between cover types is based on climatic conditions, guided by:

- Water covers: appropriate in wet climates as effective oxygen barriers, provided that they are a nominal 2-m deep to cater for wave action and potential mobilization of tailings, are retained by stable dams, and have an adequate supply of rainfall-runoff to maintain them.
- Water-shedding soil covers: appropriate in moist climates to promote revegetation for erosion protection, and rainfall runoff to limit net percolation of rainfall.
- Store and release soil covers that store wet season rainfall, releasing it through evapotranspiration during the dry season; appropriate for dry or seasonally dry climates to sustain revegetation, limit net percolation of rainfall, and prevent erosion through preventing rainfall-runoff during heavy rainfall events.

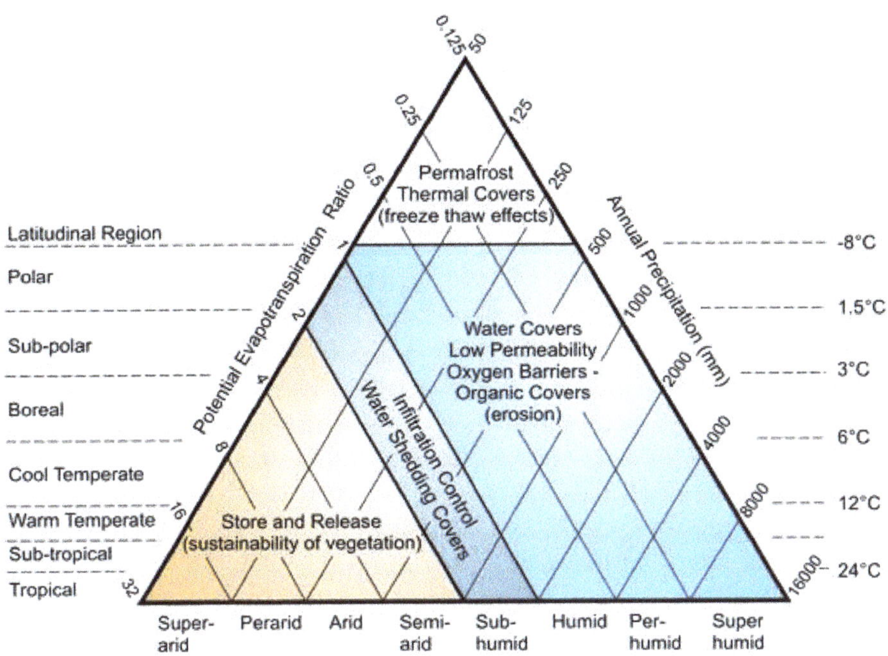

Figure 26. Choice of cover type related to climate [40].

In seasonally dry climates, store and release covers are more robust than rainfall-shedding covers since they better sustain revegetation, due to their greater rainfall infiltration and water storage capacity, and since they limit erosion by preventing rainfall runoff. Store and release covers require a base sealing layer (or a desiccated, hardpan, or compacted tailings surface, if suitable) to limit the breakthrough of rainfall infiltration into the underlying tailings, and may take advantage of the natural tailings beach slope to direct clean excess rainfall infiltration toward a collection point, limiting breakthrough into the underlying tailings.

7.2. Soil Cover Designs

The GARD guide [40] also provides schematics of soil cover designs, as shown in Figure 27, which increase in complexity, construction difficulties, potential performance and cost from left to right. The implication from Figure 27 is that the more complex and thick the cover, the more effective it will be. However, covers should be appropriate to the site conditions, in particular the climate, and more complex and thicker covers may not be the most effective choice for a given site, just more expensive. The schematics have generated much confusion, particularly in Australia. The Base Case is a single growth medium layer. Cover I, indicating a thicker single layer of growth medium than the Base Case, is inferred to be "better" than the base case. However, thicker is not necessarily better. A thick growth medium can lead to the infiltration of rainfall to a depth that makes it inaccessible to revegetation. This can also lead to increased net percolation into the underlying tailings, producing a result that may be worse than having no cover at all, since the cover would constitute a "sponge" that would increase rainfall infiltration, compared with a desiccated, hardpan, or compacted tailings surface.

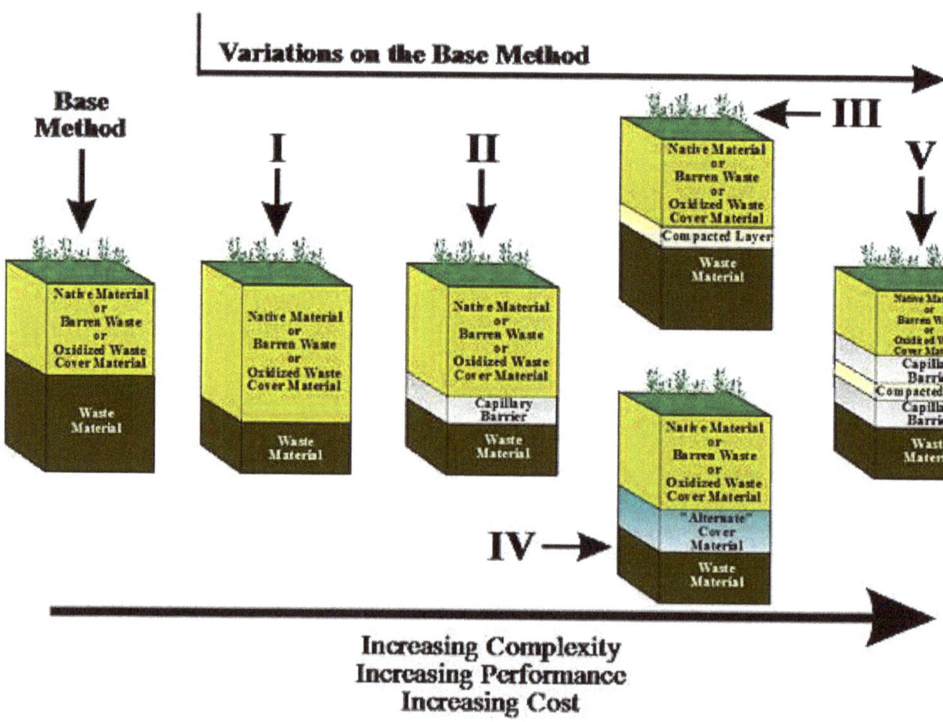

Figure 27. Various soil cover designs [40].

Cover II adds a capillary barrier (or break) beneath the growth medium, which may be desirable for limiting the uptake of any salts from the underlying tailings into the growth medium. A capillary barrier relies on a material that limits saturation and flow driven by capillary action and must be carefully selected and sized to ensure that it is effective and will remain so. One suitable capillary barrier material is clean gravel. Run-of-mine waste rock would likely not be suitable for use as a capillary barrier without the crushing of coarse-grained particles and screening to remove fines. A capillary barrier is a "drain", and so it should be overlain with a sealing layer to limit the net percolation of rainfall into the underlying tailings.

Clean gravel would require a thickness of greater than 200 mm to remain effective as a barrier. The capillary barrier thickness must allow for the possible infiltration of fines from the overlying growth medium, which would render it less effective over time. The particle size of the capillary barrier must be matched to that of the overlying growth medium, using filter criteria, to ensure that the infiltration of fines into the capillary barrier is limited by "arching" between particles. Finding suitable material for an effective capillary break, and ensuring its sustainability, are problematic.

Cover III adds a compacted (clayey) sealing layer beneath the growth medium, which is desirable, particularly for a store and release cover, to "hold up" rainfall infiltration within the overlying "rocky soil mulch" layer. Cover IV is a variation on Cover III, in which the compacted layer is replaced by an "alternative" sealing layer, such as a geomembrane, bituminous geomembrane, or geosynthetic clay liner (GCL). A sealing layer could also potentially be achieved by a desiccated, hardpan, or compacted tailings surface.

Cover V incorporates three layers separating the growth medium from the tailings, comprising a compacted fine-grained layer sandwiched between two capillary barriers. This cover was apparently based on the cover with capillary barrier effects (CCBE) applied at the Les Terrains Aurifères (LTA) mine site sulfide tailings impoundment, near Malartic, Abitibi, in Québec, Canada, in a net positive precipitation climate [41].

The CCBE, constructed on the LTA tailings in 1998, comprises 500 mm of sand (a capillary barrier) placed on the sulfidic tailings, overlain by 800 mm of fine-grained, non-acid-generating tailings (a moisture-retaining layer) that is in turn overlain by more than 300 mm of sand and gravel (protection and drainage layer). The design objective was to maintain a minimum degree of saturation of 85% in the moisture-retaining layer to effectively reduce the oxygen flux from the atmosphere to the underlying acid-generating tailings. Near-saturation of the moisture-retaining layer was to be maintained by a combination of enhanced rainfall infiltration and suction in the lower capillary barrier. The intention of the sand and gravel surface layer was to limit revegetation, so as not to reduce rainfall infiltration. The same cover was applied to the side slopes of the tailings impoundment (also comprising sulfidic tailings).

The cover initially functioned as intended on the top of the impoundment, which initially remained un-vegetated, but it was not so successful on the side slopes, due to gravity drainage. However, volunteer revegetation was established on the LTA CCBE, commencing the year after construction [42]. Eight functional groups of plants were identified, with herbaceous plants being the most abundant. Of the 11 tree species identified, the four most abundant were poplar, paper birch, black spruce and willow. Root excavation showed that tree roots penetrated the moisture-retaining layer, with an average root depth of 400 mm and a maximum root depth of 1.7 m.

After 10 years, although the LTA CCBE was effective in reducing the oxygen flux from the atmosphere to the acid-generating tailings, the quality of the seepage from the tailings impoundment still did not meet the Québec water quality standards, and dolomitic drains were constructed as a passive treatment [43].

The CCBE was designed for a specific purpose, in a net positive precipitation climate, and appears to be the only Cover Type V double capillary barrier cover applied in practice. Double capillary barrier (or break) covers have been promoted by some regulators in

Australia, notably in Queensland. This is perplexing, since the majority of mine sites in Queensland, and elsewhere in Australia, are in semi-arid to arid climates, for which the GARD guide recommends a store and release cover. Attention is best paid to the design, materials selection, and construction of such store and release covers.

7.3. Store and Release Cover Design

The key elements of a store and release cover [44,45], which was developed for seasonal, dry climates, are: (i) a thick, loose, rocky soil with a mulch growth medium layer that has an undulating surface to store the wet season rainfall without inducing runoff; (ii) an effective sealing layer at the base of the cover to hold-up rainfall infiltration; (iii) the appropriate choice of sustainable revegetation to release the stored rainfall during the wet season, through evapotranspiration.

The required thickness of rocky soil mulch growth medium will depend on the wet season rainfall pattern and the rooting depth of the vegetative cover applied and is typically up to 2-m thick. It is sized so as to accommodate within its available voids the majority of the wet season rainfall. The typical porosity of the loose, rocky soil mulch is about 0.25, providing up to 250 mm of storage volume per meter thickness of mulch.

Too thin a growth medium will not accommodate the wet season rainfall infiltration and will not support revegetation during the dry season. Too thick a growth medium could lead to rainfall infiltration beyond the reach of the revegetation. In a dry climate, store and release covers are more robust than rainfall-shedding covers, which have a more limited store and release capability and promote runoff and the potential for erosion.

The sealing layer should achieve a saturated hydraulic conductivity of less than 10^{-8} m/s (equivalent to a potential percolation rate of less than 300 mm/year, when water is available), so that in its usual unsaturated state in a dry climate its hydraulic conductivity will be less than perhaps 10^{-10} m/s (a potential percolation rate of less than 3 mm/year).

In the typically dry climate found over much of Australia, and in other dry regions, rainfall occurs on only perhaps 30 days/year, so that water may be available on top of the sealing layer for perhaps 10% of the time, reducing the potential percolation rate to less than 30 mm/year, or less than 5% of the typical average annual rainfall, similar to the typical natural percolation rate. High net percolation will be associated mainly with extreme rainfall events. A store and release cover should cycle annually between wet and dry states without a net wetting-up (which would lead to net percolation) or drying out (which would cause revegetation dieback and subsequent rainfall-induced erosion). An extreme rainfall event may wet up the cover, potentially leading to net percolation into the underlying tailings. In Australia's generally arid to semi-arid climate, a mixed eucalypt tree cover represents the only sustainable means of achieving the required evapotranspiration rates from a store and release cover that can handle extreme rainfall events and be sustainable in the long term. Australian eucalypts are quite adaptable to the climatic variability in the country, and their water uptake can vary by an order of magnitude, according to the availability of water.

There is ongoing concern and debate about the effect of tree roots on cover integrity on the tops of tailings facilities, from a number of standpoints. Tree roots can potentially penetrate through the cover thickness toward the underlying tailings. However, a desiccated, hardpan or compacted tailings surface is essentially a root barrier. This gives tree roots no encouragement to penetrate to depth in search of moisture or nutrients. Instead, the tree roots would grow laterally through the overlying cover. Since the rocky soil mulch is loose and granular, trees are unlikely to promote cracking and the development of preferred seepage pathways through the cover. The height of the trees will be limited by the climate, and by the thickness, water-holding and nutrient capacity of the rocky soil mulch, while their root patterns will not penetrate the sealing layer but will instead grow laterally. Sufficient rocky soil mulch thickness and limited tree height will lessen the possibility of wind blow-down and any possible threat to the integrity of the cover.

Should wind blow-down of shrubs and trees occur, the limited rooting depth will limit the impact on the cover, and the coarse-grained rocky soil mulch will tend to self-heal.

Although store and release covers have gained popularity for mine wastes in dry climates, they have not always been well designed and constructed. Failings include the lack of suitable and sufficient cover materials and poor construction control. A further failing is that cover placement is necessarily delayed until the completion of filling of the tailings facility, by which time tailings that have the potential to generate poor-quality seepage are already doing so. Placing any cover on wet and soft conventional tailings is also a challenge. In response to the perceived poor performance of the store and release covers, a composite cover is seen by regulators to be "better", while operators see it as being more costly.

7.4. Treatment of the Side Slopes of a Surface Tailings Facility

The erosion of the side slopes of a surface tailings facility is a function of the surface texture and revegetation (Figure 28) [46]. A grass cover needs to extend over the majority of the surface of a slope to be effective in limiting erosion and, during the long dry season over much of Australia, and other dry regions, this cannot be relied upon. Further, extended droughts cause revegetation dieback. This has implications for the use of erodible, fine-grained topsoil on tailings facility side slopes. Erosion resistance under such climatic conditions requires a rocky surface texture on the slopes.

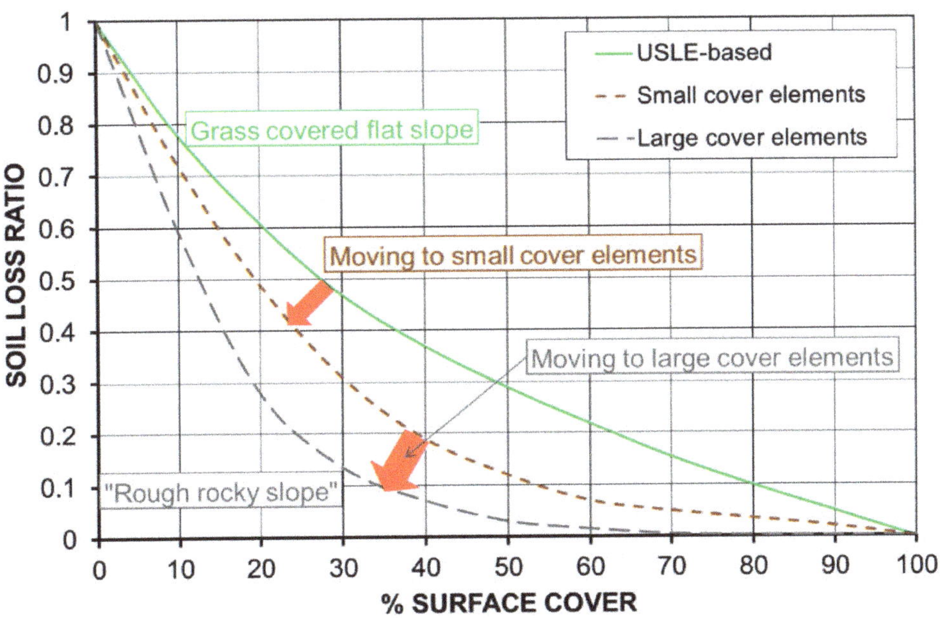

Figure 28. Soil loss as a function of the percentage of surface cover and surface texture [46].

7.5. Conventional Cost-Focused Rehabilitation versus Value-Added Rehabilitation

The conventional cost-focused approach by both operators and regulators to the rehabilitation of surface tailings storage is often at odds with the potential for value-added rehabilitation, as described in Table 1 [47]. A focus on the cost discourages and delays rehabilitation activities post-closure, which in turn is likely to lead to increased impacts over time, exacerbating the situation. In contrast, identifying and realizing potential opportunities for value-added rehabilitation and post-closure land use sets the rehabilitation budget and is a potential win for all stakeholders, including the operator, future land users and the government.

Table 1. Conventional cost-based rehabilitation versus value-added rehabilitation.

Conventional Cost-Based Rehabilitation
Production rules Rehabilitation is seen by the operator and regulator as a "cost" The operator discounts cost over time, discouraging rehabilitation Infrastructure such as power lines and buildings are stripped for little financial gain Rehabilitation is limited to "smoothing" and "greening" (where sustainable) Post-closure land use and function are limited The operator is threatened with loss of financial and social licenses to operate
Value-Added Rehabilitation
Post-closure "value" is identified up-front Examples of post-closure value include: • Reprocessing of tailings to extract metals of value, depositing the residual tailings in-pit and reducing the rehabilitation liability; • Industrial land use; • Renewable energy–solar, wind and pumped storage, delivered to the grid via mine transmission lines; • Agriculture and/or fishery impoundment; • Tourism and heritage (the older the better). Value sets the rehabilitation budget Potential wins for the operator, future land user and the government

8. Discussion and Conclusions

The commonly held perception that transporting tailings as a slurry to a surface dam is the most economical solution has driven conventional slurry tailings disposal and needs to be challenged. The way in which tailings management is costed needs to move beyond relying largely on an NPV approach, with a high discount factor, to a more whole-of-life cost approach. Tailings management must also consider the nature of the tailings and how they may change through the life of the mine and, most importantly, the climatic, topographic and seismic settings of the mine. The site settings are not under the control of the operator, although they can and must be accommodated during operations. Post-closure, the site settings will dominate the long-term performance of a tailings facility.

The ongoing rate of tailings facility failures is unacceptable to both industry and society, and there is a need to restore lost confidence and trust in the industry's ability to safely manage tailings, the ability of their consultants and contractors to design and construct safe facilities, and the capability of regulators to oversee this.

Past tailings dam failures have highlighted that tailings dams that fail have marginal stability, and the technical causes of failures are reasonably well known. They include the inappropriate use in the past of upstream construction in highly seismic regions such as Chile, and in wet climates such as Brazil. The industry in Chile responded to this, following the fatal 1965 El Cobre tailings dam failure, and turned around tailings practice there to be among the safest worldwide. Since the 2019 Brumadinho tailings dam failure in Brazil's wet climate, the upstream construction of tailings dams is no longer allowed there, and a global tailings standard (GISTM) has been developed. In more suitable climatic, topographic and seismic settings, where required controls are in place, upstream construction continues to be successfully employed. Weak foundation layers have been identified as a potential risk to tailings dam stability, particularly since they are raised progressively.

Risk assessment and the ALARP approach are now routinely applied to tailings management, and past tailings dam failures have raised the ALARP bar to an extremely high level, comparable to that applied to nuclear power stations. This is well above that achieved by water dams and is much higher than that achieved by conventional slurry tailings facilities in the past.

The GISTM has the aspirational goal of zero harm to people and the environment from tailings facilities, and elevates accountability to the highest organizational levels, with new requirements for independent oversight. It expects global transparency and disclosure to improve stakeholder understanding. It sets as the design basis an extreme consequence classification or the current classification, with an upgrade to "Extreme" maintained throughout the tailings facility's life-cycle, including at closure. Many active tailings facilities are not classified as having an "Extreme" consequence and may need to be re-classified and/or upgraded to meet "Extreme". A plethora of guides accompany and complement the GISTM. Compliance with the GISTM is required of ICMM members, and is being taken up across the industry. This is challenging in the face of a shortage of qualified and experienced tailings practitioners at all levels.

There is much scope for the further development and implementation of alternative tailings management technologies and innovations, which will change the overall mine plan, moving away from a siloed approach to mining, processing and waste management. The natural resistance to doing things differently from the way they are usually done, yet expecting a different outcome, needs to be challenged.

Tailings facilities can be built to a similar margin of safety to that of water dams, at a probability of failure of about 10^{-4}. This would prevent many tailings facility failures, with the associated loss of life, damage to infrastructure, and environmental harm. It would also restore the industry's financial and social licenses to operate, earning and restoring greater control of their operations. The implementation of existing and new approaches to technologies in tailings management could help to eliminate the risks posed by some conventional tailings facilities, possibly removing the risks altogether. Such approaches and technologies include:

- Addressing both the geotechnical and geochemical stability of tailings, which can be in conflict;
- Optimizing in-plant dewatering of tailings, particularly by thickening or filtration, and also considering paste tailings, particularly for gravity disposal in underground mines or in completed pits;
- Optimizing in-facility dewatering of tailings, possibly including on-off cells, and farming of deposited tailings that settle and consolidate poorly;
- Optimizing coagulation, flocculation and, where necessary, secondary flocculation, particularly for very fine-grained and clay mineral-rich tailings;
- Dry stacking of filtered tailings;
- Co-disposal of tailings and coarse-grained wastes, including mixing filtered tailings and waste rock;
- In-pit tailings disposal, particularly if final pit lakes containing water of diminishing quality can be avoided by complete back-filling;
- Integrated waste landforms that recombine tailings and coarse-grained wastes;
- Tailings reprocessing and reduced tailings production through coarse or dry processing;
- Possible briquetting of tailings, although this is unlikely to be economic; and
- Value-added tailings rehabilitation post-closure.

As discussed in the paper, there are several barriers to the implementation of innovative tailings management where they are indicated by site-specific conditions, particularly where existing facilities are concerned. Change will be more readily achieved in new mining projects and, hence, change in tailings management for the minerals industry as a whole will necessarily be generational. Case studies of the successful implementation of alternative tailings management and closure approaches should be published to inform the industry, their consultants and contractors, regulators, and other stakeholders. By this means, confidence and trust in the industry's management and closure of tailings facilities can be restored.

Funding: This review received no external funding.

Conflicts of Interest: The author declares no conflict of interest.

Nomenclature of Abbreviations and Terms

ALARP	As low as reasonably practicable
ANCOLD	Australian National Committee on Large Dams
Board	Board of Directors
CCBE	Cover with capillary barrier effects
CDA	Canadian Dam Association
CEO	Chief Executive Officer
EOR	Engineer of Record
F	Frequency of occurrence or annual probability of up to N fatalities
GCL	Geosynthetic clay liner
GISTM	Global Industry Standard on Tailings Management
ICMM	International Council of Mining and Metals
ICOD	International Committee on Large Dams
ITRB	Independent Tailings Review Board
JCOLD	Japanese Commission on Large Dams
KPI	Key Performance Indicators
N	Magnitude of occurrence or number of fatalities
NPV	Net present value
Pre-consolidation pressure	Highest historical stress
PRI	Principles for Responsible Investment
RTFE	Responsible Tailings Facility Engineer
SANCOLD	South African National Committee on Large Dams
TPR	Third-Party Reviewer
UNEP	United Nations Environment Programme

References

1. Troncoso, J.H.; Avendaño, A.; Vergara, A. The Seismic Failure of Barahona Tailings Dam. In Proceedings of the Third International Conference on Case Histories in Geotechnical Engineering, St. Louis, MO, USA, 1–4 June 1993; Paper No. 2.56. pp. 1473–1479.
2. Dobry, R.; Alvarez, L. Seismic failures of Chilean tailings dams. *ASCE J. Soil Mech. Found. Div.* **1967**, *93*, 237–260. [CrossRef]
3. Valenzuela, L. The 2015 Casagrande Lecture: Tailings Dams and Hydraulic Fills. In Proceedings of the Panamerican Conference, Buenos Aires, Argentina, 17–22 September 2015; p. 45.
4. Fourie, A.B.; Blight, G.E.; Papageorgiou, G. Static liquefaction as a possible explanation for the Merriespruit tailings dam failure. *Can. Geotech. J.* **2001**, *38*, 707–719. [CrossRef]
5. Merriespruit Tailings Dam Failure, Virginia, South Africa. Available online: https://www.tailings.info/casestudies/merriespruit.htm (accessed on 30 June 2021).
6. Remembering Aznacóllar: 20 Years after the Mining Disaster. Available online: https://www.birdlife.org/europe-and-central-asia/news/remembering-aznac%C3%B3llar-20-years-after-mining-disaster (accessed on 30 June 2021).
7. Six Years after Mount Polley Disaster, Expert Recommendations not Fully Implemented: Report. Available online: https://thenarwhal.ca/mount-polley-mine-expert-recommendations-not-implemented-report/ (accessed on 30 June 2021).
8. Cadia Gold Mine: Planet Labs Imagery of the Sequence of Events of the Tailings Dam Failure in March 2018. Available online: https://blogs.agu.org/landslideblog/2018/07/05/cadia-gold-mine/ (accessed on 30 June 2021).
9. Another Two Brazil Dams 'at Risk of Collapsing'. Available online: https://www.bbc.com/news/world-latin-america-34851895 (accessed on 30 June 2021).
10. The Speed of the Brumadinho Tailings Dam Landslide. Available online: https://blogs.agu.org/landslideblog/2019/02/04/brumadinho-tailings-dam-landslide/ (accessed on 30 June 2021).
11. Fundão Tailings Dam Review Panel Report on the Immediate Causes of the Failure of the Fundão Dam. Available online: http://fundaoinvestigation.com/wp-content/uploads/general/PR/en/FinalReport.pdf (accessed on 30 June 2021).
12. Report of the Expert Panel on the Technical Causes of the Failure of Feijão Dam I. Available online: https://bdrb1investigationstacc.z15.web.core.windows.net/assets/Feijao-Dam-I-Expert-Panel-Report-ENG.pdf (accessed on 30 June 2021).
13. Idriss, I. (University of California, Davis, CA, USA). Personal communication, 2020.

14. Guidance on ALARP Decisions in COMAH. SPC/Permissioning/37. Available online: https://www.hse.gov.uk/foi/internalops/hid_circs/permissioning/spc_perm_37/ (accessed on 30 June 2021).
15. ALARP: A Lazy and Risky Policy? Mark Horton, Director at MRO Analytics. Available online: https://www.linkedin.com/pulse/alarp-lazy-risky-policy-mark-horton/ (accessed on 30 June 2021).
16. Amoushahi, S.; Grenon, M.; Locat, J.; Turmel, D. Deterministic and probabilistic stability analysis of a mining rock slope in the vicinity of a major public road—case study of the LAB Chrysotile mine in Canada. *Can. Geotech. J.* **2018**, *55*, 1391–1404. [CrossRef]
17. CDA. *Dam Safety Guidelines 2007 (Revised 2013)*; Canadian Dam Association: Ottawa, ON, Canada, 2013.
18. ICMM Tailings Management Good Practice Guide. Available online: https://www.icmm.com/website/publications/pdfs/environmental-stewardship/2021/guidance_tailings-management.pdf (accessed on 30 June 2021).
19. The Investor Mining and Tailings Safety Initiative. Available online: https://www.churchofengland.org/about/leadership-and-governance/church-england-pensions-board/pensions-board-investments/investor-1 (accessed on 30 June 2021).
20. Global Industry Standard on Tailings Management. 2020. Available online: https://globaltailingsreview.org/wp-content/uploads/2020/08/global-industry-standard_EN.pdf (accessed on 30 June 2021).
21. ANCOLD. *Guidelines on Tailings Dams—Planning, Design, Construction, Operation and Closure*; Revision 1; Australian National Committee on Large Dams: Hobart, Australia, 2019.
22. ICMM Tailings Management Conformance Protocols. Available online: https://www.icmm.com/website/publications/pdfs/environmental-stewardship/2021/tailings_conformance_protocols.pdf (accessed on 30 June 2021).
23. Mine Tailings: A Review of Guidelines, Initiatives and Standards. Available online: https://gmggroup.org/tailings/ (accessed on 30 June 2021).
24. ICMM Tailings Governance Framework Position Statement. Available online: https://www.icmm.com/website/publications/pdfs/mining-principles/position-statements_tailings-governance.pdf (accessed on 30 June 2021).
25. Robertson, A.M. Tailings Governance Workshop. In Proceedings of the Mine Waste and Tailings 2018 Conference, Brisbane, QLD, Australia, 22 July 2018.
26. Williams, D.J. Chapter VI: The Role of Technology and Innovation in Improving Tailings Management. In towards Zero Harm—A Compendium of Papers Prepared for the Global Tailings Review. Available online: https://globaltailingsreview.org/wp-content/uploads/2020/09/Ch-VI-The-Role-of-Technology-and-Innovation-in-Improving-Tailings-Management.pdf (accessed on 30 June 2021).
27. Agnew, M.K. The Formation of Hardpans within Tailings as Possible Inhibitors of Acid Mine Drainage, Contaminant Release and Dusting. Ph.D. Thesis, The University of Adelaide, Adelaide, Australia, 1998.
28. Davies, M.P.; Rice, S. An Alternative to Conventional Tailing Management—"Dry Stack" Filtered Tailings. In Proceedings of the Tailings and Mine Waste 2004, Vail, CO, USA, 10–13 October 2004; pp. 411–422.
29. La Coipa, Atacama Region, Northern Chile. Available online: https://tailings.info/casestudies/lacoipa.htm (accessed on 30 June 2021).
30. Condon, P.D.; Lear, K.G. Geochemical and Geotechnical Characteristics of Filter-Pressed Tailings at the Greens Creek Mine, Admiralty Island, Alaska. In Proceedings of the 7th International Conference on Acid Rock Drainage, St. Louis, MO, USA, 26–30 March 2006; pp. 350–363.
31. Sahi, A.; El Mahboub, K.; Belem, T.; Maqsoud, A.; Mbonimpa, M. Dewatering of Mine Tailings Slurries Using Superabsorbent Polymers (SAPs) Reclaimed from Industrial Reject of Baby Diapers: A Preliminary Study. *Minerals* **2019**, *9*, 785. [CrossRef]
32. Stockman, C.; Soto, J.; Gonzalez, D. Peruvian Gold-Cyanide Tailings Filtration—A Detailed Cost Comparison. In Proceedings of the 17th International Seminar on Paste and Thickened Tailings, Vancouver, BC, Canada, 8–12 June 2014; p. 4.
33. Morris, P.H.; Williams, D.J. Co-disposal of washery wastes at Jeebropilly Colliery, Queensland, Australia. *Trans. Inst. Min. Metall. Sect. A Min. Ind.* **1997**, *106*, A25–A29.
34. Williams, D.J.; Currey, N.A. Innovative Mine Site Rehabilitation in a Semi-Arid Climate. In Proceedings of the Tailings and Mine Waste 2004, Vail, CO, USA, 10–13 October 2004; pp. 383–391.
35. Krishna, R.S.; Shaikh, F.; Mishra, I.; Lazorenko, G.; Kasprzhitskii, A. Mine tailings-based geopolymers: Properties, applications and industrial prospects. *Ceram. Int.* **2021**, *47*, 17826–17843. [CrossRef]
36. Lazorenko, G.; Kasprzhitskii, A.; Shaikh, F.; Krishna, R.S.; Mishra, J. Utilization potential of mine tailings in geopolymers: Physicochemical and environmental aspects. *Process. Saf. Environ. Prot.* **2021**, *147*, 559–577. [CrossRef]
37. Wilson, G.W.; Wickland, B.; Miskolczi, J. Design and Performance of Paste Rock Systems for Improved Mine Waste Management. In Proceedings of the Fifth International Seminar on the Management of Rock Dumps, Stockpiles and Heap Leach Pads, Perth, WA, Australia, 5–6 March 2008; pp. 107–116.
38. Burden, R.; Wilson, G.W.; Williams, D.J.; Jacobs, M. The Shear Strength of Filtered Tailings and Waste Rock Blends. In Proceedings of the Tailings and Mine Waste 2018, Keystone, CO, USA, 29 September–1 October 2018; pp. 347–355.
39. Williams, D.J. Some mining applications of unsaturated soil mechanics. *Geotech. Eng.* **2012**, *43*, 83–98.
40. International Network for Acid Prevention. Global Acid Rock Drainage Guide. Available online: http://www.gardguide.com/index.php?title=Main_Page (accessed on 30 June 2021).
41. Bussière, B.; Maqsoud, A.; Aubertin, M.; Martschuk, J.; McMullen, J.; Julien, M. Performance of the oxygen limiting cover at the LTA site, Malartic, Québec. *CIM Bull.* **2006**, *1*, 1–11.

42. Smirnova, E.; Bussière, B.; Tremblay, F.; Bergeron, Y. Vegetation succession and impacts of bio-intrusion on covers used to limit acid mine drainage. *J. Environ. Qual.* **2011**, *40*, 133–143. [CrossRef] [PubMed]
43. Bussière, B.; Potvin, R.; Dagenais, A.-M.; Aubertin, M.; Maqsoud, A.; Cyr, J. Restauration du site minier Lorraine, Latulipe, Québec: Résultats de 10 ans de suivi. *Déchets Sci. Tech.* **2009**, *54*, 49–64. [CrossRef]
44. Williams, D.J.; Wilson, G.W.; Currey, N.A. A Cover System for a Potentially Acid Forming Waste Rock Dump in a Dry Climate. In Proceedings of the Tailings and Mine Waste 1997, Fort Collins, CO, USA, 13–17 January 1997; pp. 231–235.
45. Williams, D.J. Mine Closure as a Driver for Waste Rock Dump Construction. In Proceedings of the Mine Closure Seminar, Perth, WA, Australia, 13–15 September 2006; pp. 697–706.
46. So, H.B.; Sheridan, G.J.; Loch, R.J.; Carroll, C.; Willgoose, G.; Short, M.; Grabski, A. Post-Mining Landscape Parameters for Erosion and Water Quality Control. Final Report on ACARP Projects C1629 and C4011. Australian Coal Association Research Project. 1998. Available online: https://www.acarp.com.au/ (accessed on 5 August 2021).
47. Williams, D.J. Obstacles to Effective Mine Closure, Rehabilitation and Relinquishment. In Proceedings of the Tailings and Mine Waste 2019, Vancouver, BC, Canada, 17–20 November 2019; p. 16.

minerals

MDPI

Article

A Failure Modes and Effects Analysis Framework for Assessing Geotechnical Risks of Tailings Dam Closure

Haley L. Schafer [1,*], Nicholas A. Beier [1] and Renato Macciotta [1,2]

1 Department of Civil and Environmental Engineering, University of Alberta, 9211 116 Street NW, Edmonton, AB T6G1H9, Canada; nabeier@ualberta.ca (N.A.B.); macciott@ualberta.ca (R.M.)
2 School of Engineering Safety and Risk Management, University of Alberta, 9211 116 Street NW, Edmonton, AB T6G1H9, Canada
* Correspondence: hvandalf@ualberta.ca

Abstract: Tailings dams remain on site following mine closures and must be designed and reclaimed to meet long-term goals, which may include walk-away closure or long-term care and maintenance. The underperformance of these structures can result in significant risks to public and environmental safety, as well as impacts on the future land use and economic activities near the structure. In Alberta, Canada, the expectation is for a tailings dam to be reclaimed and closed so that it can undergo deregistration. To aid in assessing the risks of underperformance during and after closure, a Generalized Failure Modes and Effects Analysis (G-FMEA) framework was developed to assess the long-term geotechnical risks for tailings dams in Alberta, with the goal of assessing the potential success of a tailings dam closure strategy. The G-FMEA is part of an initiative to enhance closure evaluations in Alberta in a collaborative effort between industry, the regulator, and academia. The G-FMEA incorporates the element of time to account for the evolution of the system, which should be applied at the planning stage and updated continually throughout the life of the facility. This paper presents the developed G-FMEA framework for tailings dams in Alberta, including the developed risk matrix framework.

Keywords: external tailings facilities; tailings dams; closure; Alberta; failure modes and effects analysis; geotechnical risk

1. Introduction

Historically, tailings dams were designed with the consideration of geotechnical failure modes that could occur during the active life of the mine. Little consideration was given to closure of the facility and the long period afterwards. This approach has resulted in many tailings dams that are in a poor position for closure or have been orphaned all together. This is problematic as the failure of these structures can still occur after the closure of a mine site, such as at the Matachewan Consolidated Mine in Ontario [1]; or when the mine is in a period of care and maintenance, such as at Obed Mountain Coal Mine in Alberta [2]. Regardless of whether the mine is active or not, tailings dams can pose a significant risk to the public and the environment if not properly managed or decommissioned. In the medium-to-long term, after decommissioning, uncertainties associated with the performance of these structures increase, as there is limited information regarding how a tailings dam ages in perpetuity. This knowledge gap poses a significant environmental, public, and financial risk, especially when combined with the serious consequences associated with the failure of tailings dams. As a result, it is vital that risk management practices be developed to assess the risks associated with the long-term behavior. Mitigation strategies should be adopted early in the life of the structure, or as soon as they are practical for active or closed facilities. There is a transition towards integrating sustainable mining practices at the outset of mining. Of note, the Landform Design Institute (LDI) advocates for mining with an end in mind, noting that there will always be residual risks present following the closure of a mine [3]. Managing the long-term residual risks requires a transparent mechanism that includes collaborative design, construction, reclamation, and aftercare [3].

In Alberta, Canada, the mining industry consists primarily of coal and oil sands extraction. The external tailings facilities (ETFs) at these mines are at various points in their lifecycles. With some facilities reclaimed or in the process of being reclaimed, it is important that systems are in place to aid operators in understanding the regulator's requirements for the process of deregistration. Deregistration is the process where the governmental body in charge of regulating these structures (i.e., Alberta Energy Regulator, AER) assesses a dam to determine if it can be removed from the dam and pond registry and no longer be regulated as a dam; Alberta is making ongoing progress towards this goal. This is evident in the release of the Alberta Dam and Canal Safety Directive (the Directive) and Manual 019: Decommissioning, Closure, and Abandonment of Dams at Energy Projects (Manual 019) [4,5]. Manual 019 indicates that "in closure, a dam operator provides active care and may transition to passive care, depending on site-specific circumstances. As a result, the Directive adopts a number of formal risk-management principles to ensure the long-term care and maintenance of tailings dams after the construction and operation phase has ended" [5].

With the impending closure of a number of external tailings facilities in Alberta, a Generalized Failure Modes Effects and Analysis (G-FMEA) framework was developed to assess the landform design for closure. This assessment should ideally be conducted before mining begins (during permitting) and updated as the project progresses. Ultimately, the goal of the G-FMEA framework is to aid in managing the residual risks associated with ETFs in both a practical and economical way. The residual risks must be acceptable to stakeholders.

In order to complete a comprehensive risk assessment, it may be necessary to conduct various levels of modelling to evaluate the behaviour and/or assess consequences. For example, in operational risk assessments, runout modelling and inundation mapping are important elements of the risk assessment, as noted by Ghahramani et al. [6]. The utility of such assessments for conducting a long-term closure risk assessment for the purposes of deregistration may need to be evaluated in the context of the regulator. For example, the AER (Manual 019) requires flowable tailings to be removed or mitigated for a tailings facility to be deregistered, such that they do not pose an unacceptable risk to dam safety [5]. Similar criteria regarding flowable tailings is outlined by Al-Mamun and Small [7]. While further advancements and an increased confidence in the results of the runout modelling and inundation mapping are useful for understanding the risks associated with flowable tailings, they may not be useful tools within the current deregistration framework. However, they remain useful tools for assessing the long-term risks of tailings facilities where it is not possible to remove the flowable tailings or for orphaned facilities.

The G-FMEA will fit most structures and failure modes but may highlight failure modes that require additional analyses, such as quantitative risk assessments. As such, the G-FMEA is intended to be used as a screening tool for the closure phase of the life cycle of an external tailings facility, where the risks assessed as acceptable require no further analyses and higher risks (or multiplicity of relatively low risk) can trigger more detailed and/or quantitative approaches. The G-FMEA is, therefore, meant to investigate the potential of the facility to be deregistered as a dam by investigating the risk of geotechnical failure. This paper presents the developed G-FMEA framework, including a developed risk matrix framework. The G-FMEA charts developed as part of the framework present the failure modes that are applicable to closure and should not be used to assess dam safety during construction or operation.

2. FMEA Background

FMEA is a top-down risk assessment tool that allows for the systematic identification and analysis of the different failure modes and their associated consequences [8,9]. FMEA aims to identify all potential failure modes, the consequences of these failure modes, the mechanisms of failure, and how the risks associated with the failures can be avoided or controlled [10]. The entire process is based on determining what happens if a specific component or element of the system fails [11]. When extended and applied to a dam, FMEA focuses on developing a clear picture of the dam, including the various components and how they interact in a systematic way [11]. Using this basis, FMEA can be used to evaluate how component failures can lead to overall system failures, the consequences of component and system failures, and the criticality of various components for risk control [11].

FMEA is generally site-specific and considers the site data, construction method, construction materials, and past behaviour of the structure [12]. As a result, it can be a time-consuming, complex process (occasionally even requiring field or laboratory investigations). Despite this, it is proven to be an extremely useful tool as it allows risks to be assessed and managed. As shown by Santos et al., FMEA is effective in demonstrating the potential failure modes of all components of a tailings dam system which then allows for the development of a comprehensive dam monitoring and surveillance system [13]. To conduct an FMEA, the following structure can be used [11]:

1. Define the system, including all components.
2. Based on component interaction, de-aggregate the system into functional sub-systems.
3. Break the sub-systems down into key elements and functions.
4. Analyze the failure modes of the different elements.
5. Assess the failure effects and consequences of the various elements.
6. Summarize the findings.
7. Repeat as necessary.

Steps 4 and 5 involve analyzing the failure mode, effects, and consequences of the different elements. This is an important step as it allows the effect of component failure modes on other components of the sub-system and the overall system to be evaluated.

Applying an FMEA to a tailings dam is not a new or revolutionary task; however, the application of an FMEA to assess the risk of a tailings dam over post closure time frames is challenging as a major limitation of an FMEA is the inclusion of time dependence and the depreciation of performance of system components [13]. It is expected that the dam will evolve significantly over time (i.e., clogging of drains, ageing of sands, etc.) [14,15]. Ultimately, the likelihood and consequences of failure are not static and may increase or decrease over time [14–16]. Robertson and Shaw note that some risks have a different likelihood or consequences if they occur during operations or post closure [9]. Risk matrices aim to include time through defining the time frame in which the probability of failure (PoF) is valid. This is typically set as an annual PoF, for the duration of the life of the structure. This concept is complex for evaluating closure as PoF is expected to change over long periods of time.

Incorporating time into an FMEA requires an evaluation and consideration of how risk profiles may change over time due to system changes, as shown in Figure 1. Point A represents the state of a system (dam structure) in terms of its PoF. Point A has an associated probability of failure that is above the acceptable limit at time zero. To move from point A to A′ so that the risk meets the acceptable limit would require risk control measures to be implemented. While this satisfies the conditions at time zero, the question remains: What will happen as time progresses and the dam undergoes evolutionary processes? As shown in Figure 1, the risk profile may follow several different trajectories that range from decreasing over time to increasing over time.

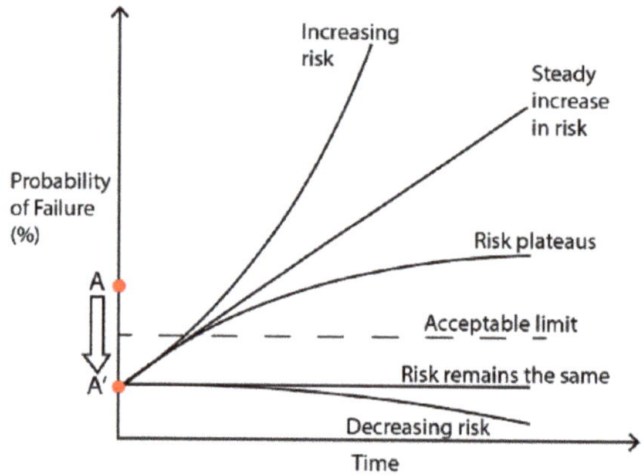

Figure 1. Changing probability of failure with time due to system changes.

The changing risk profiles over time are influenced by the closure plan and considerations of the custodial transfer scenario of the local jurisdiction. In cases where the closure risks are intended to be managed with long-term maintenance, consideration may be required for scenarios where the maintenance may be disrupted for a period of time (for example, by war or another pandemic). The developed G-FMEA considers different temporal scales as a means of accounting for the time dependence, depreciation of system elements, and the associated changing risk profiles over time, as shown in Figure 1.

3. Risk Matrix Background

FMEAs may be qualitative or quantitative in nature and are commonly combined with qualitative risk matrices to estimate the likelihood and consequences of different failure modes. Risk matrices combine the estimates of the likelihood of a negative outcome with estimates of the magnitude of consequences to determine a risk level [9,16,17]. The risk level then determines the level and timing of the required mitigative measures and critical controls to be implemented to reduce the risk level [18]. Risk matrices are often perceived as being simple to interpret and use. However, according to Porter et al. [16], "*guidance notes that explain the details of how the matrices were developed and how they were meant to be used for risk assessment, risk communication and risk management are often brief and cryptic*". Ultimately, this lack of guidance can lead to problems with the application of the risk matrix, which is already commonly criticized for being subjective. Risk matrices have been critiqued by a number of authors, as shown in Table 1.

Table 1. Risk matrix issues, references, and associated comments related to the G-FMEA.

Issue	References	Comments related to the G-FMEA
Subjective nature (not as simple and transparent as it would seem).	[8,19–23]	Quantitative data should be used as often as possible with supporting verbal descriptions and quantitative descriptors (ranges or anchor points to the definition of categories). The risk assessment should be accompanied by a description of the risk assessor, including their risk tolerances.
Can lead to incorrect risk prioritization.	[21,22,24,25]	The risk matrix does not focus on risk prioritization, but is intended as a screening tool to assess the closure design.
Can shut down conversations about risk instead of opening them up due to the pre-defined colour coding scheme determining risk prioritization.	[24]	The risk matrix is designed to determine if a closure design is adequate and is a screening tool. It involves input from stakeholders to determine the consequence ratings and is intended to open up dialogue.
Ranking reversal: where quantitatively smaller risks are assigned qualitatively higher rating levels than some quantitatively larger risks due to incorrect risk prioritization.	[19,20,26,27]	The risk matrix does not focus on risk prioritization, but is intended as a screening tool to assess the closure design.
Does not account for different risk tolerances of the individual conducting the risk assessment.	[22,28]	The risk assessor's background should be listed with the completed risk assessment.
Range compression where risks with very different likelihoods and consequences are grouped together. Risk matrices with too many categories may give false resolutions.	[19,26,28,29]	The number of risk categories should be developed with consideration of range compression.
Centering bias can be an issue where individuals have the tendency to avoid extreme values or statements when presented with a choice. This can exacerbate range compression.	[26,30]	An extra category can be added to both sides of the expected range for the consequences and likelihoods, as suggested by Duijm [31].
Category definition bias where different definitions exist for a given likelihood or consequence descriptor.	[26,31]	Clear definitions must be provided for the consequences and likelihoods categories.
Risk matrix can be misleading as it implies that risk is categorical as opposed to a position on a risk continuum.	[20]	The risk matrix should be accompanied by a clear definition of risk and associated discussion on the risk continuum.
Ambiguity of the consequence definition. There are different definitions used in practice for the consequence which can lead to issues if not clearly defined (worst case, most likely, a number of alternate discrete outcomes).	[31]	The consequence category must be clearly defined.
Risk matrices cannot provide aggregate measures of risk (i.e., total risk).	[23,27,32]	The goal of this risk matrix is to serve as a screening tool and is not intended to provide an indication of total risk.
Risk matrices are unable to aggregate risk from multiple consequence dimensions. This means that different types of consequences should not be directly compared (i.e., impact on the environment, human life). In practice, a hazard is often assigned a risk level based on the most severe consequence. This is misleading.	[23,31,33]	A risk rating should be assigned for each consequence category for a hazard.
Corporate-wide risk matrices are intended to be used as a way of standardizing risk assessment and risk acceptance criteria across a company. This is problematic as risk tolerance may vary throughout a company.	[31]	Corporate-wide risk matrices should not be used with the Generalized FMEA.

To aid in mitigating the known issues with risk matrices, a number of authors have suggested various improvements, including logarithmic scales [19,27]; probability-consequence diagrams with continuous scales [31]; and following the Cox axioms of weak consistency, betweenness, and consistent colouring [19]. Despite the issues associated with risk matrices, some authors suggest that an important element of using risk matrices is applying them in the appropriate scenarios. Bao et al. suggests that subjectivity is a vital characteristic of risk matrices as they are effective for assessments where data are insufficient and quantitative tools cannot be applied [32]. Baybutt notes that risk matrices should be used to provide " ... *initial decision guidance, which should be used with caution and the application of common sense. Risk matrices should not be used in isolation to make decisions*" [27]. Considering all of this, the risk matrix developed as part of this research is intended to be used as a screening tool, where the risks assessed as acceptable require no further analyses and higher risks (or multiplicity of relatively low risk) could inform where/if quantitative risk assessment techniques need to be used in the further decision-making stages for the closure of a facility.

4. Generalized FMEA Framework

A Generalized FMEA framework was developed, based on the current practice with FMEA, for use by regulators and the industry to assess the long-term risk of the failure of a tailings dam following closure. In order to conduct an FMEA, the definition of tailings dam failure must be clearly laid out. For the purposes of this work, a dual definition will be used for failure, similar to the Ultimate Limit States and Serviceability Limit States used in foundation engineering, as defined in the Canadian Foundation Engineering Manual [34]. Here, failure is defined as:

- Ultimate failure: the collapse of a tailings dam leading to catastrophic failure as defined by the Global Tailings Standard [35]:
    - *"A tailings facility failure that results in material disruption to social, environmental and local economic systems. Such failures are a function of the interaction between hazard exposure, vulnerability, and the capacity of people and systems to respond. Catastrophic events typically involve numerous adverse impacts, at different scales and over different timeframes, including loss of life, damage to physical infrastructure or natural assets, and disruption to lives, livelihoods, and social order. Operators may be affected by damage to assets, disruption to operations, financial loss, or negative impact to reputation. Catastrophic failures exceed the capacity of affected people to cope using their own resources, triggering the need for outside assistance in emergency response, restoration and recovery efforts."*
- Serviceability failure: failure to perform as intended.

The goal of performing the G-FMEA is to guide practitioners and regulators through the process of continuing to ask questions in a disciplined manner to prevent ultimate failure. The risk management tool is not intended to be a static tool and should be revisited at specific intervals and triggered by changes in the facility or events of underperformance. A clear trigger protocol needs to be developed for revisiting the G-FMEA. The completed G-FMEA should undergo a robust scrutiny, just like the designs, be accompanied by peer reviews, and assessed by external reviewers or review boards, where applicable. The development of the G-FMEA is assisted by the following:

1. The interviews presented in Schafer et al. [14,15];
2. Literature reviews related to:
    a. Tailings dam characteristics in the oil sands and coal industries;
    b. Tailings dam failure modes;
    c. Risk assessment tools.
3. The input from the regulator and industry stakeholders.

Schafer et al. [14,15] and a detailed literature review were used to develop the overall system definition, which included dam characteristics that impact dam behaviour.

The system definition serves as the foundation of the development of the G-FMEA and Ayyub notes that the first step to conducting a system definition is to establish the objectives [8]. This step is essential to providing context and ensuring that the users of the risk framework are asking the same questions when applying the G-FMEA to their sites. The objective question of the G-FMEA is: How should geotechnical risks associated with an external tailings facility in Alberta be managed in the long-term to achieve an acceptable closure plan, such that the facility is able to be deregistered as a dam? All users of the G-FMEA must have an understanding of the objective and intended use of the FMEA.

Following this, the system boundaries can be defined. The system definition is summarized in Figure 2 and shows the characteristics of the dam that should be carefully considered and defined prior to starting the FMEA. The general categories are summarized in the flow chart on the left with details on the general categories in the table on the right.

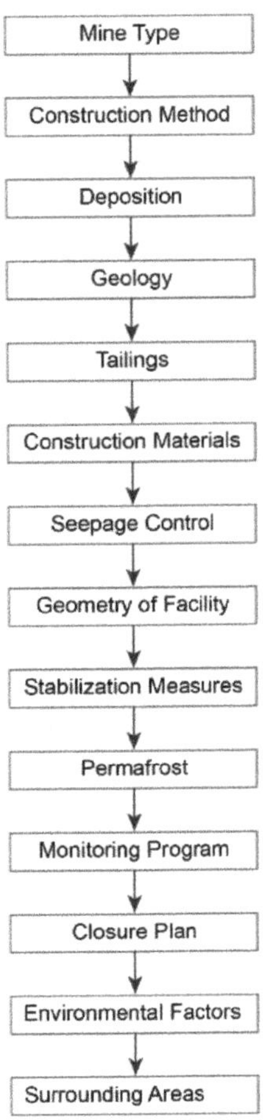

Mine Type	a) Oil sands b) Coal
Construction Method	Starter dyke (Y/N) a) Upstream b) Downstream c) Centreline d) Modified centreline e) Combination f) Unknown
Deposition	a) Hydraulic fill construction i) Spigotting ii) Cycloning with direct deposition iii) Cell construction iv) Combination b) Overburden construction c) Combination d) Unknown
Geology	Assess geology, including determination of liquefiable or brittle behaviour
Tailings	Characterize tailings, including state parameter
Construction Materials	a) Oil Sands i) Tailings ii) Interburden/overburden (i.e. lean oil sand, clay shale, till, etc.) iii) Coke b) Coal i) Interburden/overburden ii) Coarse coal refuse
Seepage Control	a) None b) Seepage reduction measures (low permeability core, grout curtain, upstream impervious blanket) c) Drains (natural materials, tailings, coke, geosynthetic, non-perforated pipe, perforated pipe, etc.) i) Chimney drain ii) Toe drain iii) Strip drain iv) Relief well v) Horizontal blanket drain vi) Seepage collector and outlet pipes
Geometry of Facility	a) Height b) Volume contained c) Slopes
Stabilization Measures	Define stabilization measures (i.e. berms, soil improvement, shear keys etc.)
Permafrost	a) Yes b) No
Monitoring	a) Yes b) No
Closure Plan	a) Plan for tailings (i.e. infilling, dewatering, natural processes, cover, etc.) b) Reclamation surface and cover c) Outlet (Y/N) d) Surface water management
Environmental Factors	a) Determine site climate (consider climate change) b) Precipitation and earthquake events c) Other
Surrounding Areas	Describe upstream/downstream areas (inhabitable/habitable areas, slopes, valleys, water bodies, upstream watershed size, culturally/ecologically significant areas, etc.)

Figure 2. System definition of oil sand and coal mines in Alberta, Canada.

The system definition serves as the minimum of what should be defined for a tailings dam prior to conducting the FMEA. However, these parameters should be defined comprehensively and may take the form of a fact sheet. This step aids in evaluating and conducting the FMEA.

Following the development of the system definition, a hazard mapping exercise was conducted, which focused on identifying potential failure modes, mechanisms, triggers, and hazards. A failure mode is defined as the overarching cause of failure. The failure mechanism is the mechanism by which a hazard leads to a failure mode. This may consist of a series of steps. A trigger is defined as that which initially leads to the failure mechanism. A hazard is defined as a condition with the potential to cause an undesirable consequence. Combined, the system definition and hazard mapping contributes greatly to the development of the G-FMEA. The steps required to use the G-FMEA in practice are outlined in Figure 3. If risk mitigation is required, or the closure plan is altered, the process may need to be repeated. This may require the risk assessor to return to Step 1 to update the system definition if substantial changes are made.

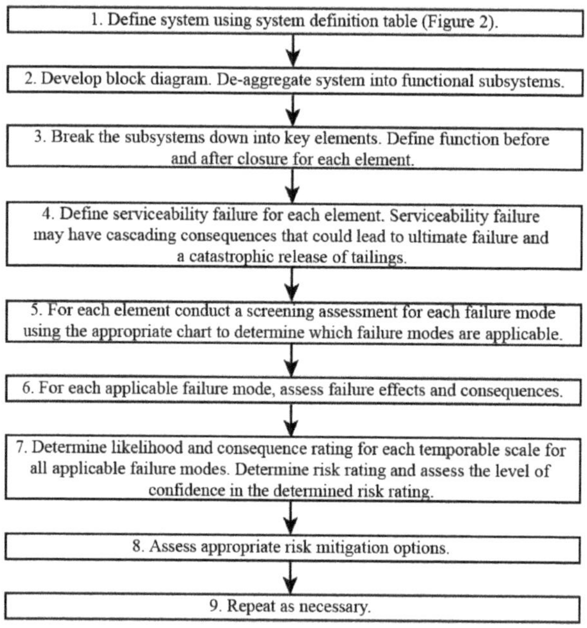

Figure 3. G-FMEA steps.

The G-FMEA includes four individual charts for the drainage system, foundation, dam body, and landform, which are provided in Appendix A. An excerpt of the drainage system chart is provided in Table 2. The drainage system and landform chart consider failure modes for a variety of different elements that could be present within a dam. The foundation chart and dam body chart do not provide failure modes for individual elements, since most failure modes are applicable to multiple elements. Each identified foundation element should be evaluated for every failure mode on the foundation list. Similarly, each dam body element must be evaluated for every failure mode on the dam body list.

Failure modes that are not applicable should be screened out prior to completing the risk assessment, as discussed in Step 5. A justification must be recorded for screening out a failure mode and re-evaluated each time the FMEA is revisited. This may include a justification of whether a failure mode is credible or non-credible as defined by the Global Industry Standard on Tailings Management [35]. It is important to note that failure modes that may be credible in the short-term may not be credible in the long-term and vice versa.

While the list of failure modes in Appendix A for the various elements is extensive and reasonably comprehensive, it is recommended as a pre-list of failure modes as no list can be completely comprehensive as there are always additional site-specific hazards. The G-FMEA charts should be applied to a facility using the complete worksheet provided in Appendix B (Table A5).

Table 2. Excerpt from drainage element G-FMEA chart.

Element	Failure Mode Description	Potential Trigger/Cause	Screening Assessment of Failure Mode	Failure Effects
Perimeter Ditch	Blockage (partial or full)	Sedimentation, sloughing/slope failure of walls, beaver dam, continuous build up of ice (icing)	Is there erosion protection in place? What is the slope of the ditch? Is it sufficient to keep particles suspended? What is the slope of the side slopes? What is the strength of material? Have there been failures in this material before? Are there beavers in the area? Is the mine located in an area that could experience icing?	Rise in phreatic surface, increase in seepage, pond on reclamation surface, internal erosion, global instability, flooding, blockage of drain outlets, toe erosion, discharge of process affected water to the environment
	Reduction in cross-sectional area	Sloughing/slope failure of walls, excessive vegetation	What is the slope of the side slopes? What is the strength of material? Have there been failures in this material before? Will the ditch regularly have water running through it or will it stay dry for a portion of the year? Are there deterrents in place to prevent the growth of vegetation?	Rise in phreatic surface, increase in seepage, pond on reclamation surface, internal erosion, global instability, reduced capacity, erosion, potential flooding, blockage of drain outlets, toe erosion, discharge of process affected water to the environment
	Change in slope	Erosion, differential settlement	Is there erosion protection in place? Does the material have the potential to consolidate or settle over time? Is it a cut into natural ground or is the material placed?	Change to water discharge velocity, creation of secondary channels, localized areas of erosion, instability of dam

4.1. Temporal Scales

It is essential that the FMEA for a closed tailings dam considers the element of time (i.e., how the dam evolves). Representing this is not straightforward and can be approached using different methods. The first method would be to complete a series of FMEAs for different points in time considering the different evolutionary processes that may occur. Ultimately, this adds a lot of complexity and may be prohibitively time-intensive. Another route, which is adopted here, is to complete immediate-term, short-term, medium-term, and long-term risk assessments using a risk matrix for each. The determination of the time frames requires consideration of what the overall time frame is for closure. An evaluation of this was carried using the following points: geological context, nuclear repository industry, Uranium Mill Tailings Remedial Action (UMTRA) surface project, the current databases that are relied on in the industry (often for less than 100 years), our ability to forecast the future, and transitions in periods of management.

For the long-term time period, time frames in excess of 10,000 years were deemed to be inappropriate in the mining industry as it is expected that glacial periods in time frames in excess of 10,000 years would change the geology and effectively bury the tailings facilities. The use of these time frames is perhaps better suited for systems such as nuclear waste repositories in tectonically inactive areas. With the consideration of this and the expectation for a tailings dam closure lasting 1000 years [5,36], the maximum time frame selected for the G-FMEA is 1000 years. This is also reflected by the UMTRA project where the facilities are designed to be effective for 1000 years where reasonably achievable with a minimum required performance of 200 years [37–40]. All risk assessments conducted for the 1000-year time frame should include a careful consideration of the uncertainties associated with the current databases, climate change predictions, and our ability to forecast the future.

The selection of appropriate time frames for the immediate-term, short-term, and medium-term risk assessment was completed with consideration of the phases of the tailings facility following closure. Initially, consideration was given to aligning the different risk assessments with the 'active care' and 'passive care' phases commonly used in industry. This was ultimately discarded as different organizations describe these phases in different ways, which may ultimately lead to confusion [41–43]. To mitigate this, the performance monitoring assessment periods recommended by the International Network for Acid Prevention (INAP) were selected. INAP outlines three specific assessment periods, including the adaptive management period, proactive management period, and reactive management period, which are outlined in Table 3 [44]. The primary difference between the assessment periods is the degree of direct management employed [44]. The focus on management, as opposed to the actions taken during the assessment period, is attractive from a safety management system perspective.

Table 3. Assessment periods [44].

Assessment Period	Definition
Adaptive Management	Occurs following closure activities when the closed facility may be at its greatest risk of failure prior to reaching equilibrium. The operator has the greatest capacity to respond. This period may be defined explicitly by the regulator, using accumulated knowledge, or using a site-specific scientific basis.
Proactive Management	Occurs as personnel and equipment are reduced. Involves a regular fixed frequency monitoring and maintenance schedule to confirm that the landform is trending along the designed trajectory. It is expected that the frequency will be less than during the adaptive management period.
Reactive Management	Issues are rectified strictly on a reaction basis, once a trigger event occurs. There should be a clear plan in place that outlines what the trigger events are and how they will be managed. Monitoring may occur in response to events such as fires, floods, earthquakes, and other extreme events.

Based on the characteristics associated with the adaptive management period, the immediate-term risk assessment should occur during this phase. The short-term risk assessment may occur during the adaptive management period or the proactive management period, depending on site-specific characteristics.

The selection of a time frame for the medium-term risk assessment involved the consideration of the reactive management period and the time frames outlined by the UMTRA project. The medium-term risk assessment should fall within the reactive management period where direct management efforts (and monitoring) decrease substantially and the facility largely shows that it meets closure objectives. The recommendations for the time frames for the various levels of risk assessment are provided in Table 4. The approximate assessment periods provided are intended as a guideline only and should be based on site-specific considerations. It is expected that the immediate-term and short-term risk assessments would have the most active forms of risk mitigation employed with risk measures, slowly transitioning to more passive methods as the risk assessment moves to medium- and long-term temporal scales. It is also expected that the level of confidence will decrease as the risk assessment moves from immediate-term to long-term scales for some failure modes.

Table 4. Time frames for various levels of risk assessment.

Risk Assessment	Conditions	Approximate Assessment Period [1]
Immediate-term	Should fall within the projected adaptive management period	0–10 years
Short-term	Should fall within the projected adaptive or proactive management period	10–50 years
Medium-term	Should fall within the projected reactive management period	50–200 years
Long-term	-	1000 years

[1] Assessment period time frames depend on site-specific considerations. Intended as a guideline.

4.2. Risk Matrix Framework

The development of the risk matrix for the G-FMEA involves a consideration of the known pitfalls with risk matrices, as outlined in Table 1; recommendations for improvement from Baybutt and Duijm [25,31]; and the evaluation and consideration of other published risk matrices, including those provided by Brown, Hadjigeorgiou, the Guidelines for Mine Waste Dump and Stockpile Design, Mine Environment Neutral Drainage (MEND), and the Oil Sands Tailings Dam Committee (OSTDC) [17,18,43,45,46]. A number of key observations can be taken from the reviewed published risk matrices, including:

- Clear descriptions of likelihood and consequence ratings are essential. as provided by Griffin and MEND [17,46]. It is common for different risk matrices to use the terms 'almost certain', 'likely', and 'possible' but it is equally common for these to not be accompanied by clear definitions as shown in Brown and Hadjigeorgiou [18,45]. This is problematic as terms such as these are not universally understood to have the same meaning amongst practitioners [47]. This aligns with the critiques in Table 1.
- A level-of-confidence scale (or similar) is recommended to go along with the risk matrix and selected risk rating [17].
- MEND and OSTDC were both developed for closure scenarios and provide useful analogues for the development of the risk matrix [17,43].
- OSTDC determines the consequence rating based on the loss of function of the structure, the degree of human intervention required on the structure after an event occurs, population at risk, and environmental economics [43]. In contrast, a risk matrix designed to assess the risk for an operational facility may consider elements such as health and safety, environment, community, reputation, and legal aspects when assessing the consequences of failure.

While the risk matrix developed by OSTDC provides the best analogue and a useful basis for this work, it cannot be applied directly to the G-FMEA, for the following reasons [43]:

- The G-FMEA presented in this research assesses the individual elements of a dam, not the dam as a whole, which is what the OSTDC [43] risk matrix was developed for. The developed risk matrix must be fit for purpose.
- The OSTDC risk matrix was developed to assess if a facility could be deregistered as a dam [43]. The consequence and likelihood ratings reflect this. In contrast, the risk matrix is used here to assess the risk of failure over time and if a closure plan is adequate to prevent failure to support a facility being deregistered as a dam.

The risk matrix framework is intended as an example of how a qualitative risk assessment method could be combined with the G-FMEA charts to assess a closure plan and includes a likelihood rating table, consequence rating table, and a risk matrix (including details on how to colour code the risk matrix and an example of a colour-coded matrix). The risk matrix to be used for a site should be defined by the industry, regulators, and other stakeholders with considerations of the technical, social, and economic aspects.

4.2.1. Likelihood Rating

The developed likelihood rating shown in Table 5 has seven different categories for likelihood from 'Close to non-credible' to 'Almost certain', which are defined with considerations of annual probabilities. Qualitative descriptors were developed to provide further context on the annual probabilities. The inclusion of seven likelihood categories is more extensive than many risk matrices, which commonly have five likelihood ratings. A more extensive likelihood rating list was selected due to the time frames associated with the closure and to aid in mitigating centering bias (Table 1). It also serves to cover the full range of likelihoods that may be encountered, which is an essential step to developing a likelihood table [25]. The risk matrix is intended to be applicable from the time that closure works have been completed to 1000 years following the completion of closure works. This means that the risk matrix is not intended to capture failure modes that may occur during the construction of the closure landscape. The risk matrix is also intended to be applied to orphaned facilities, which are defined as facilities with "owners that cannot be found or for which the owner is financially unable or unwilling to carry out the clean up" [48]. These facilities should not be confused with those that undergo abandonment as per the Alberta Dam and Canal Safety Directive [4]. It could be argued that the categories of 'Almost certain' and 'Likely' should not be applied to a closure scenario as the assumption is that the facility is designed to a standard such that the failure of an element will not occur within a year. While this may be true for many facilities, it is unlikely to be true for all facilities (i.e., orphaned facilities). Furthermore, some failure modes may become more likely when conducting a long-term risk assessment as factors such as degradation and climate change occur. The likelihood of occurrence is often described in terms of the probability of failure. While this is an important way of evaluating the likelihood, it is valuable to describe the likelihood categories using qualitative descriptors to anchor the probabilities [25]. Further, humans are notoriously poor at estimating the probability of extreme events [49]. As such, providing qualitative guidance for assessing the likelihood of a failure mode would be useful. Table 5 describes the likelihood categories in terms of the annualized probability and provides a qualitative interpretation. The quantitative interpretation guidance for the example risk matrix considers that each rating is a range of probabilities. Anchor values can also be adopted, if preferred, as long as it is clear to the users that the anchor values are a point approximation of this range of probabilities.

Table 5. Likelihood rating.

Likelihood Rating	Qualitative Interpretation Guidance [1]	Quantitative Interpretation Guidance	Annualized Probability of Occurrence
Almost certain	Almost certain that an incident will occur given the circumstances. Very high probability of one or more occurrences per year.	Higher than 10% probability in a year	$p \geq 0.1$
Likely	High likelihood. Commonly observed at similar facilities.	Higher than 10% probability in 10 years	$p \geq 0.01$
Possible	Has occurred a number of times within the industry and at least once at the site (or similar facilities in the region).	Higher than 1% probability in 10 years	$p \geq 0.001$
Unlikely	Has occurred before within the industry, but not at the site.	Less than a 1% probability in 10 years	$p < 0.001$
Rare	Low likelihood of occurrence, but not impossible. Has not occurred at the site but has occurred in industry.	Less than a 1% probability in 100 years	$p < 0.0001$
Very rare	Very low likelihood of occurrence, but not impossible. Occurrence cannot be deemed non-credible	Less than a 1% probability in 1000 years	$p < 0.00001$
Close to non-credible	Extremely remote likelihood of occurrence. Although the mechanisms are technically plausible for the occurrence, it is seen as near non-credible.	Less than a 1% probability in 10,000 years	$p < 0.000001$

[1] Industry encompasses the mining industry as a whole.

4.2.2. Consequence Rating

The consequence rating shown in Table 6 has five categories that range from 'Slight' to 'Severe'. The consequence rating is selected based on:

- The degree of the consequence of failure of an element on the rest of the system.
    - This involves an assessment regarding if the failure will result in cascading consequences to other elements.

- The degree of human intervention (post failure) required.
    - It is important to note the degree of human intervention required in response to a failure as the ultimate goal is to reach a state of minimal or no human intervention. In cases where this is not possible, this may suggest a careful evaluation of the custodial transfer scenario for that facility.
- Community.
    - The impact on the community should be assessed using input from the relevant stakeholders. This is difficult to evaluate in light of the time frames associated with closure. Our ability to predict downstream and upstream populations in the future is limited by our short-term knowledge; however, a consideration can be made of the likelihood that the affected area could be inhabited in the future (i.e., is the mine located in inhospitable terrain or in an area that is desirable for development?). This would allow baseline assumptions to be made about the future community impact. A key step to defining the impact on a community is identifying the community. As defined by the Impact and Benefit Agreement (IBA) Community Toolkit [50], this involves answering the questions: "Who is the community? How is the geographic, ethnic, or scope of community defined? Who legitimately represents the community? Is it simply representatives from local community organizations, or is it necessary to reach out to more diverse groups to ensure all elements are consulted? The definition of "community" should be inclusive enough to promote equity and avoid future conflict resulting from lack of inclusion" [50]. The community consequence rating provided in Table 6 is provided as a guideline only and requires input from the relevant stakeholders to explicitly define the community impact. This should be accomplished using a meaningful engagement as defined by the Global Tailings Standard [35]. Defining the consequence rating for the community may involve a consideration of health (including fatalities), loss of access/destruction of traditional lands, housing, destruction/damage of farmland, harm to livestock, damage to water or soil resources, impacts to trapping and fishing, the loss of animals, overall cultural impact, and employment. It is critical that the impact on all valued components (and their condition following a failure) to the relevant stakeholders are considered. A valued component is that which is considered important by the community [50]. It should be noted that Impact and Benefit Agreements developed between the mining companies and Aboriginal communities may include agreement provisions to account for catastrophic failures and losses and should be consulted when assessing the consequences of a dam failure to the community [50].
- Environment.
    - The environmental impact must be assessed with a consideration of the impacted land (both surrounding the facility and in the facility itself, with respect to its post mining land use) and waterbodies and the toxicity of the tailings. The Canadian Dam Association (CDA) developed a Working Group to revamp their environmental classification system, noting that the existing system lacked a clear scientific basis and used vague criteria that was open to interpretation based on personal beliefs and principles [51]. Nikl et al. provides a summary of the draft environmental consequence classification system [51]. The framework considers three variables: ecological impact, the intrinsic hazards of contents, and the duration of the impact, to determine the consequence category (low, significant, high, and very high) using a matrix and dial combination method approach [51]. The consequence classification is intended to assess the environmental consequences from a global tailings dam failure (i.e., ultimate failure in this case). In contrast, the risk matrix

and G-FMEA is intended to assess the failure modes of individual elements; however, the principles and concepts from Nikl et al. can still be used to support the environmental consequence category [51]. As such, the methodology used by Nikl et al. is adapted to fit within the consequence rating framework [51]. The goal of a closure plan is for the facility to remain safe and sustain a particular land use. Consequently, the environment category should be assessed with a consideration of the impact on the post mining land use, in addition to the surrounding environment. It is recognized that post mining land uses may change over time as the closure plan develops [52]. As the agreed post-mining land use changes, hazards should be re-evaluated. The most serious consequence associated with the post mining land use is when the promised land use is destroyed leaving the land sterilized. In such a situation, the facility may be fenced off to prevent all access to the site. This situation may also lead to downstream effects where additional land is sterilized. Ultimately, this would yield a severe consequence rating.

Fatalities were included in the consequence rating table in the 'Severe' consequence rating in the community column. The inclusion of fatalities in risk decisions can be a contentious issue as various stakeholders may have different risk tolerances for fatalities, especially with regard to voluntary versus involuntary risks. It could be argued that fatalities should be included throughout the risk matrix. For example, one fatality could classify a moderate risk rating, 10–100 fatalities could classify a major risk rating, and greater than 100 fatalities could classify a severe risk rating. Including fatalities only in the 'Severe' consequence rating is out of step with the risk tolerances for other industries and fails to reflect the reality that multiple-fatality events are more consequential than single-fatality events (e.g., road vs. air accidents). However, this goes against the Global Industry Standard on Tailings Management, which has a goal of zero harm to people and the environment with zero tolerance for human fatalities [35]. Considering this and the overall goals of the closure design, the current consequence rating table classifies one fatality as 'Severe'. This may be adapted through a meaningful engagement with stakeholders, if appropriate.

The approximate time frames for the environment and community categories are provided: short-term (<5 years), medium-term (5–25 years), and long-term (>25 years). These time frames are intended as guidance and may be amended with the input of stakeholders. Depending on the risk tolerance of the stakeholders, it may be necessary to reduce these time frames.

Table 6. Consequence rating.

Consequence Rating [1]	Consequence of Failure of Element on the Rest of the System	Degree of Human Intervention Required	Environment	Community
Slight	Failure of element does not have cascading consequences.	Structural integrity maintained. No intervention or maintenance required.	No movement of tailings beyond the structure footprint.	No impact on local community.
Minor	Failure of element may have cascading consequences that do not result in global failure.	Structural integrity maintained. Minor or localized intervention or maintenance required.	Released tailings are not toxic [2], and/or minimal loss of habitat (<5%) of species of special interest [3], and/or acceptable restoration of water bodies and environment feasible in a short time frame (<5 years).	Impact [4] on local community for less than 1 year.
Moderate	Failure of element has cascading consequences that do not result in global failure.	Intervention or maintenance required to limit impact of cascading consequences.	Released tailings are not toxic [2], and/or moderate loss of habitat (5–20%) of species of special interest [3], and/or acceptable restoration of water bodies and environment feasible in a short time frame (<5 years).	Short-term (<5 years) impact [4] on local community.
Major	Global failure of tailings dam with minor release of tailings.	Intervention or maintenance required to maintain function of structure as a whole.	Released tailings are toxic [2], and/or significant loss of habitat (20–50%) of species of special interest [3], and/or acceptable restoration of water bodies and environment feasible in a moderate time frame (5–25 years).	Medium-term (5–25 years) impact [4] on local community.
Severe	Global failure of tailings dam with catastrophic release of tailings.	Structural repair not possible.	Released tailings are toxic [2], and/or very significant loss of habitat (>50%) of species of special interest [3], and/or acceptable restoration of water bodies and environment unlikely within an extended time frame (>25 years).	Long-term (>25 years) impact [4] on local community. Fatalities.

Notes: [1] Assigned consequence should reflect the most likely outcome. If assigning consequence with consideration of the worst case or a combination of discrete outcomes, this must be declared. [2] Toxicity assessment of tailings should consider an assessment of the fluids and solids (leaching potential, acidity, radioactivity). [3] Species of special interest is defined as a species that lives in the inundation area that would be greatly impacted by habitat loss (preferable to select a species that is provincially or federally listed). [4] Community impacts must be determined through meaningful engagement with stakeholders and may include a consideration of health, loss of access/destruction of traditional lands, housing, destruction/damage of farmland, harm to livestock, damage to water or soil resources, impacts to trapping and fishing, loss of animals, overall cultural impact, and employment. Reputation, legal aspects, and economics are not considered in this consequence table as they are considered site- and corporation-specific. It may be necessary to assess these aspects on a site-specific basis.

4.2.3. Risk Matrix and Rating

The seven likelihood ratings and five consequence ratings can be combined to form a risk matrix, as shown in Figure 4. The challenging part of finalizing the risk matrix lies in determining the assigned risk category for a given likelihood and consequence rating. The risk categories for the G-FMEA are presented in Table 7. The risk categories were developed with the consideration of the suitability of the closure design to prevent failure such that a facility could be deregistered as a dam. This resulted in four risk categories from 'Low' to 'Extreme'. As the risk category increases, the level of the required risk mitigation increases as the closure plan is assessed as being inappropriate in preventing serviceability failure of a particular element.

It is common practice for risks to be managed using the ALARP principle: As Low As Reasonably Practicable (ALARP). In ALARP, all risk reduction measures should be employed as long as the cost of implementing them is reasonably practicable with a consideration of cost effectiveness [53]. In Table 7, the high-risk category is defined with consideration of the principle of ALARP. In the high-risk category, the risks are undesirable and must be reduced using ALARP. If the risk category cannot be reduced using ALARP, the closure plan should be altered to accommodate risk mitigation.

Likelihood Rating	**Consequence Rating[1]**				
	Slight	**Minor**	**Moderate**	**Major**	**Severe**
Almost Certain					
Likely					
Possible					
Unlikely					
Rare					
Very Rare					
Close to Non-Credible					

[1]Assign risk rating for each consequence category.

Figure 4. Risk matrix.

Table 7. Risk category.

Risk Category	Description of Risk Category
Low	Risk minimal. Monitor risks. Acceptable closure plan.
Moderate	Risk tolerable with controls. Assess risk mitigation options and monitor these risks. Minor re-design of closure plan may be required to accommodate risk mitigation.
High	Risk undesirable. Risk mitigation should be employed to ALARP to reduce risk category. Closure plan may require alteration to accommodate risk mitigation.
Extreme	Risk intolerable. Risk mitigation required immediately to reduce risk category. Requires more detailed risk analysis. Closure plan requires alteration.

Once the risk categories are developed, they can be applied to the risk matrix to develop a colour-coded 'heat map' based on the combination of the likelihood rating and consequence rating. The risk matrix is not colour-coded or populated with risk ratings as this defines risk tolerance. This step should be completed with the input of all relevant stakeholders (i.e., industry, regulator, the public). However, an example of the colour-coded risk matrix is provided in Figure 5. The colour-coded matrix presented in Figure 5 demonstrates major hazard aversion and presents a threshold line that is suggested to be used to trigger a more detailed quantitative analysis. The framework for attaining the colour-coded matrix is provided in this section.

Minimal guidance is provided in the literature on how to colour code a risk matrix. Ayyub indicates that colour coding a risk matrix involves shading each box depending on a "subjectively assessed risk level" [8]. IEC provides more guidance on assigning risk categories to the risk matrix, indicating that assigning risk categories should depend on the definitions of the likelihood and consequence tables [23]. Using these definitions, the risk matrix can have risk categories assigned to provide extra weight to the consequences or likelihoods, or it may be symmetrical [23]. Oboni suggests that a risk matrix cannot be symmetrical as this typically indicates an incorrect prioritization of risks [54]. Risk matrices should be coloured with consideration of what the cells tell the user.

The developed risk matrix for this research helps to assess if the risks associated with decommissioning an ETF are acceptable or where/if a more detailed analysis is required, and could be applied throughout a tailings facility's life. In light of this, an alternative way to use the developed risk matrix is to work with a "threshold". Any hazard categorized with a risk higher than the defined threshold (orange in Figure 5) requires a more detailed quantitative risk assessment method, which may require a more in-depth investigation, design, or modelling to be completed. This is an important step to aid in mitigating issues associated with upper severity limits [25].

The framework for developing the example risk matrix was developed with a consideration of the tool being used as a screening method; guidance provided by Baybutt [25], Duijm [31], and Levine [29]; and a consideration of the known pitfalls that can be associated with risk matrices, as discussed in Table 1. The development of the example risk matrix in Figure 5 was accomplished using the following steps, and may be used as guidance for colour coding the risk matrix for a specific site (see Figure 4). Appendix C shows how Figure 5 was developed using the following steps.

1. Evaluate the likelihood ratings (Table 5), consequence ratings (Table 6), and risk ratings (Table 7).
2. Develop iso-contours of equal risk based on the estimated quantitative consequence measure and the provided likelihoods (Likelihood = Risk/Consequence). Assuming that the consequence quantitative measure is an order of magnitude between categories, the consequence and likelihood iso-contours should be plotted in log-log space. Plot the iso-contours on the risk matrix. Assign the risk categories to the iso-contours of equal risk and use these to develop an initial colour-coded risk matrix. This step serves as a starting point for colour coding the risk matrix. Any available known acceptance criteria can be used as an initial starting point for this exercise. Decisions will need to be made about the cells that have an iso-contour cutting through them (i.e., do you assign the higher or lower risk category?). This exercise demonstrates a known issue with risk matrices where risk is represented as categorical, rather than on a continuum, and hazards are binned into these categories.
    a. A note from Baybutt [25]: When consequences and likelihood categories differ by orders of magnitude, they are represented in log-log space. Practically, this means that high consequence-low likelihood events (negatively correlated) have the most uncertain risks in these type of rating schemes. This is especially problematic as these events already have a lot of inherent uncertainty. Ultimately, it is not possible to increase the precision of these events on a risk matrix. This concept of uncertainty related to these types of events supports the idea of using a threshold value on the risk matrix to trigger more extensive risk analysis.
3. Assess the initial colour-coded risk matrix against the Cox axioms for a well-defined risk matrix [19]. This step is simply a logical 'check'. Duijm recommends that a key to using the risk matrix is to recognize that the colouring of the risk matrix is a risk definition in its own right (and cannot be separately and stringently defined as Risk = probability × consequence) as it expresses a subjective risk perception (i.e., major hazard aversion), which is an important element of risk decisions [31]. Duijm notes that, when no reference is made to an external risk definition, then "the colouring of the matrix itself is the only relevant risk definition, then the axioms of weak consistency and consistent colouring are trivial" [31]. If a quantitative definition of risk is desired by stakeholders, then the Cox axioms should be satisfied [19]. Cox axioms can be summarized as [19]:

   a. Weak consistency, where each hazard in the red category represents a higher risk than the hazards in the green category.
   b. Betweenness, where every positively sloped line segment that starts in a green cell and ends in a red cell must pass through an intermediate risk category.
   c. Consistent colouring, where hazards with an approximately equal risk have the same colour.
4. Assess if the Levine lettering scheme is more appropriate for your risk analysis goals [29]. In the study by Levine, logarithmic axes are used for the consequence and likelihood axes [29]. Straight line iso-contours of equal risk are drawn. Instead of using colours, the different areas are labelled A, B, C, D, E, F, and G (each line represents a new letter). This method results in a matrix that is somewhat unintuitive, but this prevents risk matrix users from making assumptions about the risks based on the colouring scheme. When a risk matrix is constructed in this manner, the following conclusions can be drawn:
   a. Risks in one letter category can only be distinguished from risks in another category if they are more than one letter apart (i.e., C > A, D > B).
   b. Risks in categories that are zero or one letter apart are not able to be distinguished from one another (i.e., it is not known if C > B or B > C).
5. As noted by Duijm, another way to develop risk scores and colouring is by using basic arithmetic (multiplication and addition) based on ordinal numbers assigned to each consequence and probability category [31]. When the categories are essentially linearly spaced, then the multiplication of the ordinal numbers is an appropriate way of defining the risk score. When the categories are essentially logarithmically spaced, the addition of the ordinal numbers is desired. Apply the appropriate mathematical operations and compare to the risk matrix developed in Step 3.
6. Assess if major hazard aversion is required and apply as necessary. Hazard aversion is the concept of low-probability–high-consequence events being assigned a higher risk than a high-probability–low-consequence event, even if the expected loss is mathematically the same [31]. This concept is used in scenarios where low-probability–high-consequence events are of greater concern and may require different decisions [25].
7. Conduct "logic checks" by stress testing the risk matrix with different scenarios. Focus on what the risk matrix is telling the user. Adjust the colour scheme as needed and repeat. This is an important step as the risk matrix may be developed using one quantitative measure. As such, it must be assessed to determine if it is applicable across the different consequence categories.

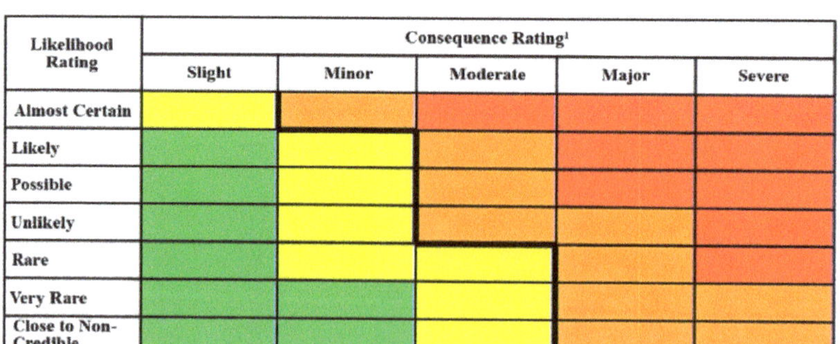

[1]Assign risk rating for each consequence category.
[2]This matrix is provided for illustration purposes. Site specific risk matrices should be developed using Steps 1-7 and include meaningful engagement with stakeholders to ensure that risk tolerance of these stakeholders is represented.

Figure 5. Example risk matrix considering major hazard aversion and threshold for quantitative analysis.

4.2.4. Limitations of the Risk Matrix Framework

As noted in Table 1, a common issue related to risk matrices is their inability to aggregate risks, both in general and from multiple consequence categories. With regard to aggregating risks, as noted by IEC "one cannot define that a particular number of low risks or a low risk identified a particular number of times is equivalent to a medium risk" [23]. This is an important limitation of risk matrices, and one that risk users must be cognizant of. Therefore, sound engineering judgement is required for evaluating the results of the matrix where multiple low-risk failure modes are present. Once the assessment is complete, a careful evaluation of the results must be completed. All hazards that receive a risk rating of red or orange need to undergo further analyses and may need additional controls. If a scenario occurs where all the hazards are green (or green and yellow), the organization should evaluate if the aggregation of them represents an intolerable risk. In that case, the quantified descriptions in the likelihood matrix can help with the aggregation of likelihood.

Next, it is common practice for a hazard to be assigned a risk level based on the most severe consequence. Caution should be employed with this practice, and if practitioners choose to use this methodology, it must be clearly declared. Instead, it is recommended that risk ratings are assigned for each consequence category (i.e., consequence of failure of an element on the rest of the system, degree of human intervention required, environment, and community). This is largely due to the inability of risk matrices to aggregate the risk from multiple consequence dimensions, meaning that the consequences should not be directly compared (as noted in Table 1). Therefore, a careful consideration is needed when determining a final risk rating for the hazard.

One final issue that must be addressed is related to the correlated risk. The G-FMEA involves breaking the system down into its individual components. It is possible that one hazard may become a trigger for another hazard. Risk matrix users must be aware of the correlated risks and note them explicitly in the reporting of results, along with how they were considered in the study.

Given the time frames associated with closure, it is recommended that a level of confidence regarding the risk estimates is provided. There are uncertainties associated with assessing the likelihood and consequence ratings, which may be associated with a lack of data, lack of system understanding, uncertain future operating conditions or uncertain maintenance, and regional development post closure [9]. As the time frames associated with the G-FMEA may extend to 1000 years, it is expected that the level of confidence may decrease as the temporal scale increases. A level of confidence framework is currently being developed at the University of Alberta to accompany the G-FMEA.

5. Summary and Conclusions

Tailings dams can pose a significant risk to the public and the environment following closure. To prevent the failure of these large structures after mining, their long-term behavior must be understood and incorporated into the closure design, ideally from the conception of the project and initial design. Challenges lie in the limited information available regarding how these structures age over time and the forecasting, loading, and environmental scenarios over long-term periods (weather events, seismicity, human activity, etc.). A G-FMEA framework was developed to aid in assessing the potential success of a tailings dam closure strategy for external tailings facilities in Alberta, Canada. The goal of the G-FMEA was to reduce the future risk of geotechnical failure on the environment and the public to a degree that is both practical and economical. The G-FMEA can be used to assess the closure risks at the onset of mining if adopted early in the design process, which allows meaningful changes to be made to reduce the long-term risks from operational design decisions and ultimately support the goal of deregistration. In many cases, this is not possible as the tailings facility may be further into its life cycle (as is the case at many Alberta tailings facilities).

While this geotechnical risk management tool was developed with Alberta in mind, it does have global significance and could be easily adapted to other jurisdictions or organizations (i.e., to account for the different types of mines present in other parts of Canada or the world). Furthermore, the G-FMEA may be adapted for in-pit facilities with perimeter dykes (as desired) by updating the G-FMEA tables to account for failure modes applicable to these types of facilities. The current G-FMEA tables were developed with a consideration of external tailings facilities at oil sands and coal mines in Alberta. Differences in physiographic regions, seismic regions, expected climate change, etc., could lead to other failure modes that should be included in the G-FMEA tables.

The G-FMEA requires that the dam be broken down into the individual elements. Each element should be analyzed using the G-FMEA framework, which includes four individual charts for the drainage system, foundation, dam body, and landform. The G-FMEA is accompanied by a developed risk matrix, including a likelihood rating table, consequence rating table, and final risk matrix. The risk matrix framework is intended as an example of how a qualitative risk assessment method could be combined with the G-FMEA charts to assess a closure plan. It is recommended that the risk matrix to be used for a site should be defined by industry and regulators with considerations of technical, social, and economic aspects. Guidelines are provided on how to colour code the risk matrix. It is recommended that the risk matrix be accompanied by a level of confidence rating, which is currently being developed at the University of Alberta.

This risk management tool is timely and necessary in Alberta as many tailings facilities approach closure. It provides a systematic method for assessing the post closure risks associated with tailings facilities, so that they can be managed, and, ultimately, aims to develop actionable tools consistent with the Global Industry Standard on Tailings Management. The G-FMEA framework aids in satisfying the risk management requirements required by the Alberta Energy Regulator for the closure of a tailings dam to support these facilities being deregistered as dams.

Author Contributions: Development of G-FMEA framework, including the G-FMEA tables, and risk matrix methodology, H.L.S.; writing—original draft preparation, H.L.S.; writing—review and editing, N.A.B. and R.M.; project administration, N.A.B.; supervision, N.A.B. and R.M. All authors have read and agreed to the published version of the manuscript.

Funding: This research was funded by the Vanier Banting Secretariat and the Alberta Energy Regulator (Service Agreement: 17SA-OP010).

Data Availability Statement: Not applicable.

Acknowledgments: The authors would like to thank Tim Eaton (Alberta Energy Regulator), Scott Martens (Canadian Natural Resources Ltd.), and Gord McKenna (McKenna Geotechnical Inc.) for their ongoing support throughout this research.

Conflicts of Interest: The authors declare no conflict of interest. The funders had no role in the design of the study; in the collection, analyses, or interpretation of data; in the writing of the manuscript, or in the decision to publish the results.

Appendix A. G-FMEA Charts

Table A1. Drain FMEA.

Element	Failure Mode Description	Potential Trigger/Cause	Screening Assessment of Failure Mode	Failure Effects
Perimeter ditch	Blockage (partial or full)	Sedimentation, sloughing/slope failure of walls, beaver dam, continuous build up of ice (icing)	Is there erosion protection in place? What is the slope of the ditch? Is it sufficient to keep particles suspended? What is the slope of the side slopes? What is the strength of the material? Have there been failures in this material before? Are there beavers in the area? Is the mine located in an area that could experience icing?	Rise in phreatic surface, increase in seepage, pond on reclamation surface, internal erosion, global instability, flooding, blockage of drain outlets, toe erosion, discharge of process affected water to the environment
	Reduction in cross-sectional area	Sloughing/slope failure of walls, excessive vegetation	What is the slope of the side slopes? What is the strength of the material? Have there been failures in this material before? Will the ditch regularly have water running through it or will it stay dry for a portion of the year? Are there deterrents in place to prevent the growth of vegetation?	Rise in phreatic surface, increase in seepage, pond on reclamation surface, internal erosion, global instability, reduced capacity, erosion, potential flooding, blockage of drain outlets, toe erosion, discharge of process affected water to the environment
	Change in slope	Erosion, differential settlement	Is there erosion protection in place? Does the material have the potential to consolidate or settle over time? Is it a cut into natural ground or is the material placed?	Change to water discharge velocity, creation of secondary channels, localized areas of erosion, instability of dam
	Sand channel buoyancy	Freezing conditions in channels composed of sand	Are the drainage channels constructed of sand? Could the channel experience freezing?	Rise in phreatic surface, increase in seepage, pond on reclamation surface, internal erosion, global instability, reduced capacity, erosion, potential flooding, blockage of drain outlets, toe erosion, discharge of process affected water to the environment
Pipes (perforated and non perforated)	Breakage of pipe	Break in pipe, buckling, physical degradation	Is the pipe capable of breaking, buckling, and/or physically degrading over time?	Lack of control of phreatic surface (potential rise in phreatic surface), increase in seepage, pond on reclamation surface, internal erosion, global instability, release of water into downstream shell, erosion on downstream slope
	Pipe clogging	Biological, chemical, particulate clogging	Is chemical, biological, or sediment clogging possible?	Lack of control of phreatic surface (potential rise in phreatic surface), increase in seepage, pond on reclamation surface, internal erosion, global instability, release of water into downstream shell, erosion on downstream slope
	Clogging surround (woven sock or sand and gravel bed)	Biological, chemical, or particulate clogging	Is it possible for the material surrounding the pipe to become clogged?	Lack of control of phreatic surface (potential rise in phreatic surface), increase in seepage, pond on reclamation surface, internal erosion, global instability, release of water into downstream shell, erosion on downstream slope
	Blockage at outlet	Blockage from perimeter channel or other (snow, debris, etc.)	Is the pipe outlet close enough to the base of the perimeter channel that it could become blocked?	Lack of control of phreatic surface (potential rise in phreatic surface), increase in seepage, pond on reclamation surface, internal erosion, global instability
	Breakage of connection between a drain and outlet pipe	Overloading, degradation of connection, poor installation	How are the drain and outtake connected?	Lack of control of phreatic surface (potential rise in phreatic surface), increase in seepage, pond on reclamation surface, internal erosion, global instability, release of water into downstream shell, erosion on downstream slope

Table A1. *Cont.*

Element	Failure Mode Description	Potential Trigger/Cause	Screening Assessment of Failure Mode	Failure Effects
Granular materials	Erosion from flowing water	Increase in seepage gradients, development of preferential flow paths adjacent to drain (i.e., burrowing from animals, cracks)	Could cracks develop along drains? Is the site known for having issues with burrowing animals? Is the material erodible?	Decreased capacity of drain resulting in failure, lack of control of phreatic surface (potential rise in phreatic surface), increase in seepage, pond on reclamation surface, internal erosion, global instability, erosion on downstream slope
	Clogging	Biological, chemical, particulate clogging	Is chemical, biological, or sediment clogging possible? What is the grain size distribution?	Lack of control of phreatic surface (potential rise in phreatic surface), increase in seepage, pond on reclamation surface, internal erosion, global instability, release of water into downstream shell, erosion on downstream slope
	Failure to meet drain criteria	Change in material properties, including permeability due to aging, change in gradation due to movement with seepage, weathering/degradation	To what extent is aging and weathering/degradation of the material possible?	Lack of control of phreatic surface (potential rise in phreatic surface), increase in seepage, pond on reclamation surface, internal erosion, global instability, release of water into downstream shell, erosion on downstream slope
	Obstruction of drainage at outlet	Snow and ice blocking outlet	Will an obstruction prevent the drain from performing as intended? Is the outlet protected from a blockage?	Lack of control of phreatic surface (potential rise in phreatic surface), increase in seepage, pond on reclamation surface, internal erosion, global instability, release of water into downstream shell, erosion on downstream slope
	Deformation leading to reduction in drain capacity	Slow and continuous deformation under long-lasting shear and pressure forces from consolidation of overlying material	How much deformation is expected over time as the materials above the drain consolidate and settle? Has the impact on the drain been considered?	Lack of control of phreatic surface (potential rise in phreatic surface), increase in seepage, pond on reclamation surface, internal erosion, global instability, release of water into downstream shell, erosion on downstream slope
	Crushing/breakage of granular drain	Overloading drain, settlement of the dam	How much settlement is expected to occur over time? Was settlement accounted for in the design? Will the drain continued to be loaded?	Lack of control of phreatic surface (potential rise in phreatic surface), increase in seepage, pond on reclamation surface, internal erosion, global instability, release of water into downstream shell, erosion on downstream slope
Geosynthetics	Aging	Degradation of geosynthetic over time (temperature, oxidation, hydrolytic, chemical, biological, radioactive, etc.)	Is the geosynthetic capable of aging in the given time frame? Have sufficient tests been performed to investigate this?	Brittle rupture of geosynthetic, lack of control of phreatic surface (potential rise in phreatic surface), increase in seepage, pond on reclamation surface, internal erosion, global instability, release of water into downstream shell, erosion on downstream slope
	Creep deformation	Slow and continuous deformation under long-lasting shear and pressure forces from consolidation of overlying material	How much deformation is expected over time as the materials above the geotextile consolidate and settle? Has the impact on the geosynthetic been considered?	Reduction in thickness leading to reduction in drain capacity or shear failure of drain if in-place deformation reaches a critical value, lack of control of phreatic surface (potential rise in phreatic surface), increase in seepage, pond on reclamation surface, internal erosion, global instability, release of water into downstream shell, erosion on downstream slope

Table A1. *Cont.*

Element	Failure Mode Description	Potential Trigger/Cause	Screening Assessment of Failure Mode	Failure Effects
Geosynthetics	Clogging	Biological, chemical, particulate clogging	Are there dispersive soils present? Are there ferrous soils? Does the permeant contain oily waters or sludge? Is there turbid water with high suspended solids? Is there potential for chemical precipitation or biological growth? What is the end land use (does it involve agriculture or sewage systems that could result in clogging)? Is sediment capable of clogging the drain? What is the grain size distribution?	Lack of control of phreatic surface (potential rise in phreatic surface), increase in seepage, pond on reclamation surface, internal erosion, global instability, release of water into downstream shell, erosion on downstream slope
	Blockage	Intrusion of adjacent materials (i.e., geotextile), blockage of downstream or exit surface caused by sedimentation, vegetation, etc.	Does the downstream or exit surface of the geosynthetic have the potential to be blocked and prevent drainage from sediment, vegetation, ice, snow, etc.? Could adjacent materials impede movement of water to the geosynthetic?	Lack of control of phreatic surface (potential rise in phreatic surface), increase in seepage, pond on reclamation surface, internal erosion, global instability, release of water into downstream shell, erosion on downstream slope
	Blinding where fine-grained soils are prevented from entering the geotextile, which creates a filter cake	Formation of a filter cake at the interface of the geosynthetic from coarse particles being retained by the geotextile and intercepting fine particles migrating from the soil	Is the geosynthetic in intimate contact with the soil? Have all appropriate filter criterion been followed during design?	Lack of control of phreatic surface (potential rise in phreatic surface), increase in seepage, pond on reclamation surface, internal erosion, global instability, release of water into downstream shell, erosion on downstream slope
Low permeability cores	Hydraulic fracture	Decrease in total stress (i.e., differential settlement, arching in narrow cores), increase in porewater pressure	Is there a narrow core? Is there the potential for excessive differential settlement that could lead to a decrease in total stress? Is there the potential for an increase in porewater pressure? Is there an effective downstream filter to prevent internal erosion of the core?	Cracking, internal erosion, global instability
	Internal erosion in dam from suffusion	High hydraulic gradients, design/construction defect, presence of widely gap-graded or non-plastic gap-graded soils	Is the material widely gap-graded or gap-graded non plastic? Is there an effective downstream filter to prevent internal erosion of the core?	Global instability, seepage on the downstream slope, settlement of the crest, permeability may increase as erosion progresses or decrease if clogging occurs
	Internal erosion in dam from concentrated leak	Cracks from vertical deformation in foundation, starter dyke, or other tailings materials or differential settlement; tunnels created by burrowing animals; hydraulic fracture; high hydraulic gradient; design/construction defects	Is there a crack or gap that could allow for a concentrated leak to develop? Is there an effective downstream filter to prevent internal erosion of the core?	Global instability, development of a pipe
	Internal erosion in dam from contact erosion	Parallel flow in coarser layer to the interface between the coarse-grained and fine-grained soil, high hydraulic gradients, design/construction defects	Is there a contact between a coarse and fine-grained soil? Is there a filter in place? Is there an effective downstream filter to prevent internal erosion of the core? Is there an effective downstream filter to prevent internal erosion of the core?	Global instability, static liquefaction, settlement of the crest, loss of stability or unravelling, eroded material can clog the permeable layer and increase the porewater pressure (could result in hydraulic fracture and uplift of the downstream toe or a rise in the phreatic surface), development of a pipe

Table A1. *Cont.*

Element	Failure Mode Description	Potential Trigger/Cause	Screening Assessment of Failure Mode	Failure Effects
Low permeability cores	Shear failure from changing shear strength	Degradation/weathering, porewater pressure change, change in permeability over time, failure of drains, progressive failure of strain softening materials, brittle failure of contractive materials	Is there potential for weathering or degradation of materials? Does the porewater pressure rely on drain performance? Could drains fail over time? Are the materials strain softening or brittle? Is there an effective downstream filter to prevent internal erosion of the core?	Slumping of downstream slope, translational slide, rotational slide
	Shear failure from changing shear stress	Loading/unloading crest, toe, upstream, or downstream; surface erosion of downstream slope; excessive and uncontrolled seepage through foundation resulting in erosion of toe; subsurface stress changes (geothermal development, in situ oil or gas production, wastewater injection, etc.)	Is there potential for anthropogenic contributions (i.e., excavations or construction)? Erodibility of material? Is there an effective downstream filter to prevent internal erosion of the core?	Slumping of downstream slope, translational slide, rotational slide
	Vertical deformation (differential or otherwise) from consolidation	Consolidation/settlement	Does the material have the potential to consolidate? How much consolidation has occurred already? How much is expected to occur? Is there an effective downstream filter to prevent internal erosion of the core?	Release of pore water and loss of height (potential for pond to develop on reclamation surface), development of cracks above starter dyke, internal erosion, overtopping

Table A2. Foundation FMEA.

Failure Mode Description	Potential Trigger/Cause	Screening Assessment of Failure Mode	Failure Effects
Heave (seepage forces create zero effective stress condition)	Embankment loading, excessive rainfall, embankment seepage	What are the current hydraulic gradients and maximum possible due to geometry? What are the materials present? Are there cohesionless soils confined by an overlying lower permeability layer?	Global instability
Vertical deformation from collapse of karst formation	Collapse of karst formation	Is there karst present in the foundation?	Cracking (transverse cracks - perpendicular to dam crest are larger problems than longitudinal cracks) in dam, internal erosion in dam, crest subsidence
Vertical deformation caused by settlement of material	Consolidation	Will the materials in the foundation consolidate over time? How much consolidation has already occurred? Does the material have the potential to collapse?	Cracking (transverse cracks - perpendicular to dam crest are larger problems than longitudinal cracks) in dam, internal erosion in dam, crest subsidence
Excessive/uncontrolled seepage through foundation or foundation/dam contact	Excessive rainfall	Is there potential for seepage through the foundation? What is the permeability of the materials?	Erosion of downstream toe, increase in porewater pressure in dam, global instability

Table A2. *Cont.*

Failure Mode Description	Potential Trigger/Cause	Screening Assessment of Failure Mode	Failure Effects
Shear failure along pre-existing shear plane from changing shear stress	Loading/unloading of foundation, earthquake, subsurface stress changes (geothermal development, in situ oil or gas production, wastewater injection, etc.)	Are there pre-existing shear planes? Is there the potential for anthropogenic loading or unloading events? Is the material erodible?	Slumping of downstream slope, translational slide, rotational slide, static liquefaction
Shear failure along new shear plane from changing shear stress	Loading/unloading of foundation, earthquake, subsurface stress changes (geothermal development, in situ oil or gas production, wastewater injection, etc.)	Is there the potential for anthropogenic loading or unloading events? Is the material erodible?	Slumping of downstream slope, translational slide, rotational slide, static liquefaction
Shear failure along pre-existing shear plane from changing shear strength	Degradation/weathering, porewater pressure change, progressive failure of strain softening materials, brittle failure of contractive materials	Are there pre-existing shear planes? Is there the potential for degradation or weathering of the material? Is the material strain softening or brittle?	Slumping of downstream slope, translational slide, rotational slide, static liquefaction
Shear failure along new shear plane from changing shear strength	Degradation/weathering, porewater pressure change, progressive failure of strain softening materials, brittle failure of contractive materials	Is there the potential for degradation or weathering of the material? Is the material strain softening or brittle?	Slumping of downstream slope, translational slide, rotational slide, static liquefaction
Internal erosion in foundation or dam/foundation contact from global backward erosion	Failure of soil above or around a backward erosion pipe to hold a roof, heave, high hydraulic gradients, design/construction defect, presence of non-plastic soils in the foundation	Are there non-plastic soils in the foundation?	Static liquefaction, global instability, unravelling/sloughing of downstream face, sub vertical cavities
Internal erosion in foundation or dam/foundation contact from backward erosion piping	Heave, high hydraulic gradients, design/construction defect, presence of non-plastic soils that are capable of holding a roof	Are there non-plastic soils in the foundation and soils capable of 'holding a roof'?	Enlargement of pipe, global instability, static liquefaction
Internal erosion in foundation or dam/foundation contact from contact erosion	Parallel flow in coarser layer to the interface between the coarse-grained and fine-grained soil, high hydraulic gradients, design/construction defect	Is there a contact between a coarse-grained and a fine-grained soil? Is the geometrical and hydraulic condition for contact erosion met?	Global instability, static liquefaction, settlement of the crest, loss of stability or unravelling, eroded material can clog the permeable layer and increase the porewater pressure (could result in hydraulic fracture and uplift of the downstream toe or a rise in the phreatic surface), development of a pipe
Internal erosion in foundation or dam/foundation contact from suffusion	High hydraulic gradients, design/construction defect, presence of widely gap-graded or non-plastic gap-graded soils	Is the material widely gap-graded or gap-graded non plastic?	Global instability, seepage on the downstream slope, settlement of the crest, permeability may increase as erosion progresses or decrease if clogging occurs
Internal erosion in foundation or dam/foundation contact from concentrated leak	Fracture in foundation soil, hydraulic fracture, high hydraulic gradient, cracks at dam/foundation contact from vertical deformation in foundation or poor construction practices or differential settlement, design/construction defects	Is there a crack or gap that could allow for a concentrated leak to develop?	Global instability, development of a pipe
Thawing of foundation permafrost	Climate change	Is there permafrost in the foundation?	Cracking (transverse cracks - perpendicular to dam crest are larger problems than longitudinal cracks) in dam, piping in dam, crest subsidence

Table A3. Dam body FMEA.

Failure Mode Description	Potential Trigger/Cause	Screening Assessment of Failure Mode	Failure Effects
Destruction of vegetation	Suffocation by eroded material, forest fires, pests and disease, climate change, large storm event, anthropogenic contributions, surface erosion on downstream slope, evolution of vegetation over time due to climate change	Does the resistance to erosion rely on the vegetation? Is the area susceptible to forest fires? Are there pests/disease that could lead to vegetation destruction? What is the downstream slope? Is the area remote? Are there surrounding communities that could lead to destruction of vegetation (i.e., recreational vehicles)?	Increase in surface erosion, instability of downstream slope, global instability, change in overall evapotranspiration and water balance impacts (infiltration versus runoff)
Internal erosion in dam from suffusion	High hydraulic gradients, design/construction defect, presence of widely gap-graded or non-plastic, gap-graded soils	Is the material widely gap-graded or gap-graded non plastic?	Global instability, seepage on the downstream slope, settlement of the crest, permeability may increase as erosion progresses or decrease if clogging occurs
Internal erosion in dam from concentrated leak	Cracks from vertical deformation in foundation, starter dyke, or other tailings materials or differential settlement; tunnels created by burrowing animals; hydraulic fracture; high hydraulic gradient; design/construction defects	Is there a crack or gap that could allow for a concentrated leak to develop?	Global instability, development of a pipe
Internal erosion in dam from contact erosion	Parallel flow in coarser layer to the interface between the coarse-grained and fine-grained soil, high hydraulic gradients, design/construction defects	Is there a contact between a coarse and fine-grained soil? Is there a filter in place?	Global instability, static liquefaction, settlement of the crest, loss of stability or unravelling, eroded material can clog the permeable layer and increase the porewater pressure (could result in hydraulic fracture and uplift of the downstream toe or a rise in the phreatic surface), development of a pipe
Dynamic liquefaction	Earthquakes, induced seismicity, construction traffic, blasting	Seismic events in area? Induced seismicity? Density of material (contractive or dilative)? Saturated or unsaturated? Hydraulically placed or compacted?	Global instability, local slumps, crest drops
Shear failure from changing shear strength	Degradation/weathering, porewater pressure change, change in permeability over time, failure of drains, progressive failure of strain softening materials, brittle failure of contractive materials	Is there potential for weathering or degradation of materials? Does the porewater pressure rely on drain performance? Could drains fail over time? Are the materials strain softening or brittle?	Slumping of downstream slope, translational slide, rotational slide
Shear failure from changing shear stress	Loading/unloading crest, toe, upstream, or downstream; surface erosion of downstream slope; excessive and uncontrolled seepage through foundation resulting in erosion of toe; earthquake; subsurface stress changes (geothermal development, in situ oil or gas production, wastewater injection, etc.)	Is there potential for anthropogenic contributions (i.e., excavations or construction)? Erodibility of material?	Slumping of downstream slope, translational slide, rotational slide
Static liquefaction from changing mean effective stress	Change in pore pressures caused by a phreatic surface change (i.e., failure of drainage system)	Density of material (contractive or dilative)? Saturated or unsaturated? Hydraulically placed or compacted? Does control of the phreatic surface rely on drain function? Could drains become clogged or fail in the future?	Global instability

Table A3. *Cont.*

Failure Mode Description	Potential Trigger/Cause	Screening Assessment of Failure Mode	Failure Effects
Static liquefaction from changing shear stress	Loading/unloading; overloading, including increasing the load, construction activities at the crest, fill placement at toe; over steepening of downstream slope or toe (slumping of downstream slope from shear failure), including erosion or excavation of toe; foundation shear; shear in starter dyke; shear in other tailings materials; excessive and uncontrolled seepage through foundation resulting in erosion of toe; subsurface stress changes (geothermal development, in situ oil or gas production, wastewater injection, etc.)	Density of material (contractive or dilative)? Hydraulically placed or compacted? Saturated or unsaturated? Is there a likelihood for anthropogenic contributions in the future (i.e., unexpected construction)? Is the site remote? What is the material of the downstream slope? Is there a nearby river?	Global instability
Static liquefaction from changing shear stress and mean effective stress	Lateral extrusion	Density of material (contractive or dilative)? Hydraulically placed or compacted? Saturated or unsaturated? Weak layers interbedded in tailings? Could the tailings 'squish' out like toothpaste during loading?	Global instability
Static liquefaction from long-term change in material properties resulting in changing shear strength	Changing shear strength caused by degradation/weathering, progressive failure, porewater pressure change, failure of drains	Density of material (contractive or dilative)? Hydraulically placed or compacted? Saturated or unsaturated? To what extent could weathering or degradation of the material occur? Will it result in an increase or decrease in strength? Will a change in the phreatic surface impact the strength of the material?	Global instability
Surface erosion from spring sapping (headward erosion of gullies due to concentration of seepage forces at the locus of the gully which accentuates erosion)	Destruction of vegetation, increased seepage on downstream slope from failure of drainage system, increased seepage from internal erosion of starter dyke or tailings deposits)	Is the material susceptible to erosion? Is there a vegetative cover or erosion protection? Will seepage daylight on the downstream slope?	Slope failures (shallow surficial movement, slumps), change in downstream slope angle, blockage in perimeter channel with sediment, development of negative drainage, development of large erosion scarps
Surface erosion from wind and overland flow resulting in rills, gullies, or sheet erosion	Destruction of vegetation, rainfall, melting of snow, wind, increased seepage on downstream slope from failure of drainage system, increased seepage from internal erosion in embankment	Is the material susceptible to erosion? Is there a vegetative cover or erosion protection?	Slope failures (shallow surficial movement, slumps), change in downstream slope angle, blockage in perimeter channel with sediment, development of negative drainage, development of large erosion scarps
Toe erosion	Flow action from perimeter ditch or nearby river, release of a dam from a beaver, flood event, river changing course over time, destruction of vegetation, excessive and uncontrolled seepage through foundation, excessive erosion from internal erosion	Is the material susceptible to erosion? Is there a vegetative cover or erosion protection? Is there a nearby perimeter ditch or river? Is there known animal activity in the area?	Slope failures (shallow surficial movement, slumps), change in downstream slope angle, blockage in perimeter channel with sediment, development of negative drainage, development of large erosion scarps, beaver bafflers
Vertical deformation (differential or otherwise) from consolidation	Consolidation/settlement	Does the material have the potential to consolidate? How much consolidation has occurred already? How much is expected to occur?	Release of pore water and loss of height (potential for pond to develop on reclamation surface), development of cracks above starter dyke, internal erosion, overtopping

Table A4. Landform FMEA.

Item/Functional Identification	Failure Mode Description	Potential Trigger/Cause	Screening Assessment of Failure Mode	Failure Effects-End Effects
Cap	Excessive settlement	Consolidation	Does the material have the potential to settle over time?	Formation of ponds on reclamation surface, overtopping, piping (increase in seepage forces and gradients), infiltration of previously unsaturated tailings that could increase vulnerability to liquefaction for materials previously considered "not flowable"
	Differential settlement	Consolidation, poor construction practices	Does the material have the potential to settle over time? Are there areas that have the potential to settle more that others?	Formation of ponds on reclamation surface, failure to direct surface water runoff towards drainage channels, development of cracks, formation of preferential flow paths, localized depressions, infiltration of previously unsaturated tailings that could increase vulnerability to liquefaction for materials previously considered "not flowable"
Infilled material	Excessive settlement	Consolidation	Does the material have the potential to settle over time?	Formation of ponds on reclamation surface, overtopping, piping (increase in seepage forces and gradients)
	Differential settlement	Consolidation, poor construction practices	Are there areas that have the potential to settle more than others?	Formation of ponds on reclamation surface, overtopping, piping (increase in seepage forces and gradients), failure of drainage channels to behave as intended, localized depressions
Hummocks	Shear failure from changing shear stress	Loading/unloading crest, toe, slopes; surface erosion; failure of underlying underlying material to support hummock	Is there potential for anthropogenic contributions (i.e., excavations or construction)? Erodibility of material?	Slumping, translational slide, rotational slide, blockage of drainage channels
	Shear failure from changing shear strength	Degradation/weathering, porewater pressure change, change in permeability over time, failure of drains, progressive failure of strain softening materials, brittle failure of contractive materials	Is there potential for weathering or degradation of materials? Are the materials strain softening or brittle?	Slumping of slopes, translational slide, rotational slide, blockage of drainage channels
	Surface erosion from wind and overland flow resulting in rills, gullies, or sheet erosion	Destruction of vegetation, rainfall, melting of snow, wind	Is the material susceptible to erosion? Is there a vegetative cover or erosion protection?	Slope failures (shallow surficial movement, slumps), change in downstream slope angle, blockage in drainage channel with sediment, development of negative drainage, development of large erosion scarps
Drainage channels	Washout of erosion protection (riprap)	Precipitation event larger than design events (including extreme or repeat events)	What precipitation event are the channels designed for? What is the chance of exceedance over 1000 years? How susceptible are the underlying materials to erosion? Was the erosion protection properly designed and constructed?	Excessive erosion (erosion gullies, etc.), change in slope of drainage channels, erosion and release of materials underlying drainage channels
	Blockage (complete or partial)	Debris, beaver dam, icing, sedimentation, slumping from slope failure, ingress of vegetation, slope failure/excessive erosion from nearby hummock	Are there beavers in the area? Is there a chance for a slope failure? Could debris be carried downstream and deposited in the channels resulting in a complete or partial blockage? Is the mine located in an area that could experience icing?	Formation of a pond upstream of the drainage channel, blockage breakthrough resulting in flooding, overtopping from pond formation, revert back to a pond, piping through dam (increase in seepage forces and gradient)

Table A4. *Cont.*

Item/Functional Identification	Failure Mode Description	Potential Trigger/Cause	Screening Assessment of Failure Mode	Failure Effects-End Effects
Drainage channels	Sand channel buoyancy	Freezing conditions in channels composed of sand	Are the drainage channels constructed of sand? Could the channel experience freezing?	Flooding
	Erosion control failure	Improper design/construction, differential settlement	Do the drainage channels rely on erosion control for stability? Is there the chance for differential settlement of the channel?	Excessive erosion (erosion gullies, etc.), change in slope of drainage channel, erosion and release of materials underlying drainage channel, formation of secondary channel
Outlet	Washout of erosion protection (riprap)	Precipitation event larger than design event (including extreme or repeat events), flood following sand channel buoyancy event in drainage channel	What precipitation event is the outlet designed for? What is the chance of exceedance over 1000 years? How susceptible are the underlying materials to erosion? Was the erosion protection properly designed and constructed?	Excessive erosion (erosion gullies, etc.), change in slope of outlet, erosion and release of materials underlying outlet
	Blockage (complete or partial)	Debris, beaver dam, icing, sedimentation, slumping from slope failure, ingress of vegetation, increase in depositional material due to failure of erosion protection in drainage channels upstream	Are there beavers in the area? Is there a chance for a slope failure? Could debris be carried downstream and deposited in the outlet resulting in a complete or partial blockage? Is the mine located in an area that could experience icing?	Formation of a pond upstream of the outlet, blockage breakthrough resulting in flooding, overtopping from pond formation, revert back to a pond, piping through dam (increase in seepage forces and gradient)
	Sand channel buoyancy	Freezing conditions in channels composed of sand	Is the outlet constructed of sand? Could the channel experience freezing?	Flooding
	Erosion control failure	Improper design/construction, differential settlement	Does the outlet rely on erosion control for stability? Is there the chance for differential settlement of the channel?	Excessive erosion (erosion gullies, etc.), change in slope of outlet, erosion and release of materials underlying outlet, formation of secondary channel
Vegetative cover	Destruction of vegetation	Suffocation by eroded material, forest fires, pests and disease, climate change, large storm event, anthropogenic contributions	Does the resistance to erosion rely on the vegetation? Is the area susceptible to forest fires? Are there pests/disease that could lead to vegetation destruction? Is the area remote? Are there surrounding communities that could lead to destruction of vegetation (i.e., recreational vehicles)?	Increase in surface erosion, deposition of material in drainage channels that could lead to a blockage via sedimentation, development of negative drainage on reclamation surface, ponding of water near dam crest, internal erosion (increase in seepage forces and hydraulic gradient)
Tailings	Excessive settlement	Consolidation	Does the material have the potential to settle over time?	Formation of ponds on reclamation surface, overtopping, piping (increase in seepage forces and gradients)
	Differential settlement	Consolidation, different material properties/infilling techniques, etc.	Are there areas that have the potential to settle more than others?	Formation of ponds on reclamation surface, overtopping, piping (increase in seepage forces and gradients), failure of drainage channels to behave as intended, localized depressions

Appendix B. FMEA Worksheet

Table A5. G-FMEA worksheet.

								Immediate-term/Short-term/Medium-term/Long-term Assessment *														
									Consequences				Risk Rating				Level of Confidence					
Element	Failure mode identification	Failure mode description	Potential trigger/cause	Screening assessment of failure mode	Is this failure mode applicable?	If yes, is there sufficient data to evaluate the risk? List any resource gaps.	Failure effects	Likelihood	Element failure consequence	Human intervention	Environment	Community	Element failure consequence	Human intervention	Environment	Community	Element failure consequence	Human intervention	Environment	Community	Controls	Remarks

* Likelihood rating, consequence rating, risk rating, level of confidence, and controls must be determined for each failure mode for the short-term assessment, medium-term assessment, and long-term assessment.

Appendix C. Risk Matrix Development

The steps to colour code the risk matrix outlined in Section 4.2.3 are described here with the associated development of the example risk matrix in Figure 5 used for illustration purposes.

1. The likelihood ratings (Table 5), consequence ratings (Table 6), and risk ratings (Table 7) were evaluated.
2. In order to develop the iso-contours of equal risk, quantitative values from 0.01 to 10,000 were assigned to the consequence categories and assumed to have an order of magnitude increase between the categories. For individual projects, site-specific consequences could be considered here where there are known magnitudes of the consequences (i.e., financial impacts of environmental consequences). Iso-contours of equal risk were developed based on the estimated quantitative consequence measure and the provided likelihoods using the definition (Likelihood = Risk/Consequence). The iso-contours are shown in Figure A1, which show the annualized probability plotted against the consequences. It is important to remember that this is an estimation technique only and serves as a first-order step for colour coding the matrix. It is desirable to use quantitative measures of the consequences that extend across the full range of categories. Risk categories were then assigned to the iso-contours of equal risk (Table A6). This was used to colour code the initial risk matrix shown in Figure A2. Cells that had an iso-contour cutting through them were assigned to the higher risk category.

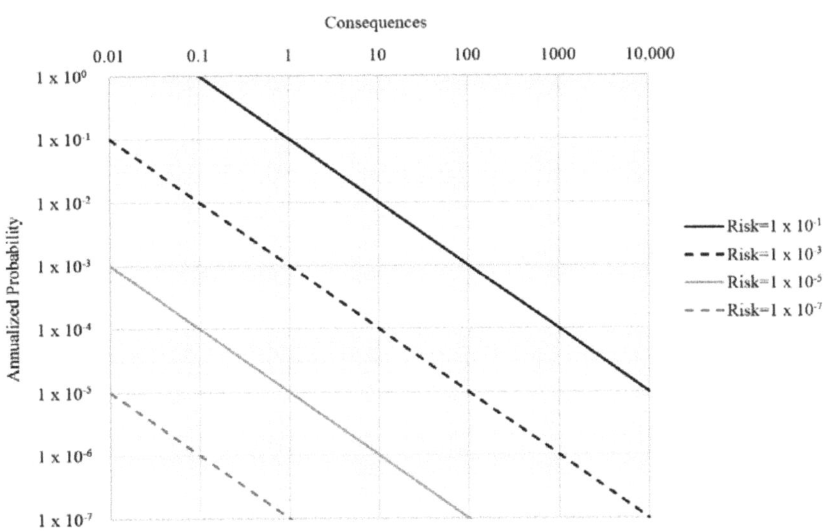

Figure A1. Iso-contours of equal risk.

Table A6. Assigned risk categories to iso-contour risk levels.

Risk Category	Risk Level
Red	$>1 \times 10^{-1}$
Orange	$>1 \times 10^{-3}$
Yellow	$>1 \times 10^{-5}$
Green	$>1 \times 10^{-7}$

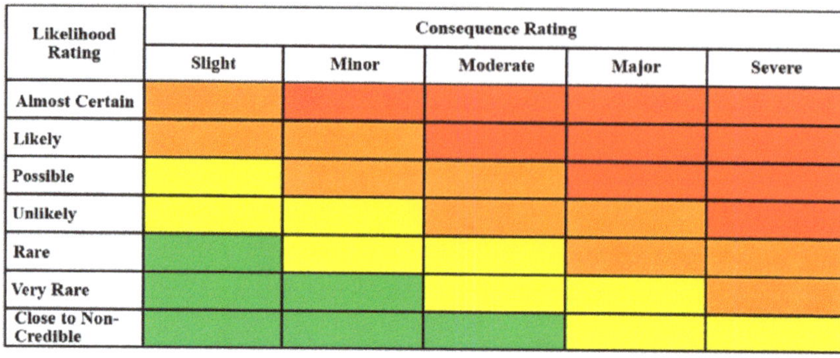

Figure A2. Initial colour-coded risk matrix with consideration of iso-contours.

3. The initial colour-coded risk matrix was assessed to determine if it satisfied the Cox axioms (weak consistency, betweenness, and consistent colouring).
4. The risk matrix was assessed using the Levine lettering scheme [29]. For this, logarithmic axes were used for the consequence and likelihood axes and straight-line iso-contours of equal risk were drawn (similar to Figure A1). Following this, each area was labelled with a letter as opposed to a colour, as shown in Figure A3 (each line represents a new letter). Levines method results in a matrix that is somewhat unintuitive, but this prevents risk matrix users from making assumptions about risks based on the colouring scheme. When a risk matrix is constructed in this manner, the following conclusions can be drawn:
    a. Risks in one letter category can only be distinguished from risks in another category if they are more than one letter apart (i.e., C > A, D > B).
    b. Risks in categories that are zero or one letter distant are not able to be distinguished from another (i.e., it is not known if C > B or B > C).

Likelihood Rating	**Consequence Rating**				
	Slight	**Minor**	**Moderate**	**Major**	**Severe**
Almost Certain	C	B	A	A	A
Likely	D	C	B	A	A
Possible	E	D	C	B	A
Unlikely	F	E	D	C	B
Rare	G	F	E	D	C
Very Rare	G	G	F	E	D
Close to Non-Credible	G	G	G	F	E

Figure A3. Risk matrix based on Levine [29].

5. As noted by Duijm, another way to develop risk scores and colouring is by using basic arithmetic (multiplication and addition) based on ordinal numbers assigned to each consequence and probability category [31]. As the categories were logarithmically spaced, the addition of the ordinal numbers was used, as shown in Figure A4. For this example, this results in a risk matrix that is colour-coded in the same way as Figure A2.

Likelihood Rating	Consequence Rating				
	1	2	3	4	5
7	8	9	10	11	12
6	7	8	9	10	11
5	6	7	8	9	10
4	5	6	7	8	9
3	4	5	6	7	8
2	3	4	5	6	7
1	2	3	4	5	6

Figure A4. Colour-coded risk matrix based on addition of ordinal pairs.

6. Major hazard aversion was applied to the risk matrix in Figure A2 to assign a higher risk rating to high-probability–low-consequence events, as these events were considered to be of greater concern, as shown in Figure A5.

Likelihood Rating	Consequence Rating				
	Slight	Minor	Moderate	Major	Severe
Almost Certain					
Likely					
Possible					
Unlikely					
Rare					
Very Rare					
Close to Non-Credible					

Figure A5. Colour-coded risk matrix following application of hazard aversion.

7. Following the application of major hazard aversion, the risk matrix was stress tested by evaluating its performance in different scenarios and evaluating what the risk matrix told the user. The evaluation showed that the yielded risk ratings from Figure A5 for the 'Slight' and 'Minor' consequence rating columns were too high. Some amendments were made to the risk matrix and it was stress tested again. This resulted in the final example risk matrix in Figure A6.

Likelihood Rating	Consequence Rating				
	Slight	Minor	Moderate	Major	Severe
Almost Certain					
Likely					
Possible					
Unlikely					
Rare					
Very Rare					
Close to Non-Credible					

Figure A6. Example colour-coded risk matrix.

References

1. Baker, F.; Cook, D.; Deem, J.; Letient, H.; Walsh, H. Matachewan Mines Tailings Dam from Failure to Rehabilitation. In Proceedings of the 1996 CDA Annual Conference, Niagra Falls, ON, Canada, 6–10 October 1996.
2. Provincial Court of Alberta. *Agreed Statement of Facts, Docket No, Agreed Statement of Facts, Docket No. 151258456P1 &160061354P1*; Provincial Court of Alberta, 2017; Available online: http://www1.aer.ca/compliancedashboard/enforcement/201706-04_AgreedStatementofFacts_PrairieMines_Obed_2013-006.pdf (accessed on 6 November 2021).
3. LDI. *Mining with the End in Mind: Landform Design for Sustainable Mining. Position Paper 2021-01*; Landform Design Institute (LDI): Delta, BC, Canada, 2021.
4. Government of Alberta. *Alberta Dam and Canal Safety Directive, AEP Water Conservation 2018 No.3*; Government of Alberta: Edmonton, AB, Canada, 2018.
5. AER. *Manual 019 Decommissioning, Closure, and Abandonment of Dams at Energy Projects*; Alberta Energy Regulator (AER): Calgary, AB, Canada, 2020.
6. Ghahramani, N.; Mitchell, A.; Rana, N.M.; McDougall, S.; Evans, S.G.; Take, W.A. Tailings-flow runout analysis: Examining the applicability of a semi-physical area–volume relationship using a novel database. *Nat. Hazards Earth Syst. Sci.* **2020**, *20*, 3425–3438. [CrossRef]
7. Al-Mamun, M.; Small, A. Revision of Guidance of Landforms in CDA Minings Dams Bulletin-A Companion Paper with Additional Details. In Proceedings of the 2018 CDA Annual Conference, Quebec City, QC, Canada, 13–19 October 2018.
8. Ayyub, B.M. *Risk Analysis in Engineering and Economics*, 2nd ed.; CRC Press: Boca Raton, FL, USA, 2014; ISBN 978-1-4665-1826-1.
9. Robertson, A.; Shaw, S. *Mine Closure*; Info Mine E-Book: Vancouver, BC, Canada, 2006.
10. Valis, D.; Koucky, M. Selected Overview of Risk Assessment Techniques. *Probl. Eksploat.* **2009**, *4*, 19–32.
11. Hartford, D.N.D.; Baecher, G.B. *Risk and Uncertainty in Dam Safety*; Thomas Telford Publishing: London, UK, 2004; ISBN 9780727740397.
12. Kupper, A.G.; Hurndall, B.; Morgenstern, N.; Sobkowicz, J.; Abeda, W.; Baldwin, G.; Cameron, R.; Eenkooren, N.; Grapel, C.; Martens, S.; et al. De-Licensing Oil Sands Tailings Dams-When Is a Dam No Longer a Dam? In Proceedings of the 17th International Conference on Tailings and Mine Waste, Banff, AB, Canada, 3–6 November 2013.
13. Dos Santos, R.N.C.; Caldeira, L.M.M.S.; Serra, J.P.B.; Dos Santos, R.C. FMEA of a tailings dam. *Georisk: Assess. Manag. Risk Eng. Syst. Geohazards* **2012**, *6*, 89–104. [CrossRef]
14. Schafer, H.L.; Macciotta, R.; Beier, N.A. Tailings dam closure scenarios, risk communication, monitoring, and surveillance in Alberta. *CIM J.* **2020**, *11*, 80–90. [CrossRef]
15. Schafer, H.L.; Beier, N.A.; Macciotta, R. Closure and the Long-Term Behaviour of Tailings Dams: Using Industry Experience to Fill in the Gaps. In Proceedings of the 72nd Canadian Geotechnical Conference, St. John's, NL, Canada, 29 September–2 October 2019.
16. Porter, M.; Lato, M.; Quinn, P.; Whittall, J. Challenges with Use of Risk Matrices for Geohazard Risk Management for Resource Development Projects. In Proceedings of the First International Conference on Mining Geomechanical Risk, Perth, WA, Australia, 9–11 April 2019; Australian Centre for Geomechanics: Crawley, WA, Australian; pp. 71–84.
17. MEND. Cold Regions Cover System Design Technical Guidance Document, MEND Report 1.61.5c; Mine Environment Neutral Drainage (MEND) Program: 2012. Available online: http://mend-nedem.org/wp-content/uploads/2013/01/1.61.5c.pdf (accessed on 6 November 2021).
18. Brown, B.S. What Are the Real Risks for Tailings Facilities? In Proceedings of the First International Conference on Mining Geomechanical Risk, Perth, WA, Australia, 9–11 April 2019; Australian Centre for Geomechanics: Crawley, WA, Australia; pp. 21–30.
19. (Tony)Cox, L.A. What's Wrong with Risk Matrices? *Risk Anal.* **2008**, *28*, 497–512. [CrossRef]
20. Pickering, A.; Cowley, S.P. Risk Matrices: Implied Accuracy and False Assumptions. *J. Health Saf. Res. Pract.* **2010**, *2*, 11–18.
21. Ball, D.J.; Watt, J. Further Thoughts on the Utility of Risk Matrices. *Risk Anal.* **2013**, *33*, 2068–2078. [CrossRef]
22. Monat, J.P.; Doremus, S. Alternatives to Heat Map Risk Matrices for Project Risk Prioritization. *J. Mod. Proj. Manag.* **2018**, *6*, 104–113. [CrossRef]
23. IEC. *IEC/ISO 31010 Risk Management-Risk Assessment Techniques*; International Electrotechnical Commission (IEC): Geneva, Switzerland, 2009.
24. Center for Security Studies. *Visualizing Risk: The Use of Graphical Elements in Risk Analysis and Communications*; Center for Security Studies: Zurich, Switzerland, 2012.
25. Baybutt, P. Guidelines for designing risk matrices. *Process. Saf. Prog.* **2017**, *37*, 49–55. [CrossRef]
26. Thomas, P.; Bratvold, R.B.; Eric, B.J. The Risk of Using Risk Matrices. *SPE Econ. Manag.* **2014**, *6*, 56–66. [CrossRef]
27. Baybutt, P. Designing risk matrices to avoid risk ranking reversal errors. *Process. Saf. Prog.* **2015**, *35*, 41–46. [CrossRef]
28. Oboni, C.; Oboni, F. Is It True that PIGs Fly When Evaluating Risks of Tailings Management Systems? In Proceedings of the 16th International Conference on Tailings and Mine Waste, Keystone, CO, USA, 14–17 October 2012; pp. 95–105.
29. Levine, E. Improving risk matrices: The advantages of logarithmically scaled axes. *J. Risk Res.* **2012**, *15*, 209–222. [CrossRef]
30. Smith, E.D.; Siefert, W.T.; Drain, D. Risk matrix input data biases. *Syst. Eng.* **2008**, *12*, 344–360. [CrossRef]
31. Duijm, N.J. Recommendations on the use and design of risk matrices. *Saf. Sci.* **2015**, *76*, 21–31. [CrossRef]

32. Bao, C.; Li, J.; Wu, D. A fuzzy mapping framework for risk aggregation based on risk matrices. *J. Risk Res.* **2016**, *21*, 539–561. [CrossRef]
33. Oboni, F.; Oboni, C. *Tailings Dam Management for the Twenty-First Century*; Springer Nature Switzerland AG: Cham, Switzerland, 2020; ISBN 978-3-030-19447-5.
34. Canadian Geotechnical Society. *Canadian Foundation Engineering Manual*, 4th ed.; Canadian Geotechnical Society: Richmond, BC, Canada, 2006; ISBN 9780920505281.
35. Global Tailings Review. *Global Industry Standard on Tailings Management*; International Council on Mining and Metals: London, UK; United Nations Environment Programme: Nairobi, Kenya; Principles of Responsible Investment: London, UK, 2020.
36. ICOLD. *Internal Erosion of Existing Dams, Levees and Dikes, and Their Foundations*; International Commission on Large Dams (ICOLD): Paris, France, 2017; Volume Bulletin 164.
37. Task Committee on Low-Level Radioactive Waste Management of the Technical Committee on Nuclear Effects Management of Inactive Uranium Mill Tailings. *J. Environ. Eng.* **1986**, *112*, 490–537. [CrossRef]
38. *DOE Lakeview Oregon Final Completion Report Volume 1*; Department of Energy (DOE) Albuquerque Operations Office: Albuquerque, NM, USA, 1991.
39. Mathes, D.E. Lessons-Learned from the 20-Year Uranium Mill Tailings Remedial Action Surface Project. In Proceedings of the Waste Management, Tuscon, AZ, USA, 28 February–4 March 1999.
40. IAEA. *The Long Term Stabilization of Uranium Mill Tailings*; International Atomic Energy Agency (IAEA): Vienna, Austria, 2004.
41. ICOLD. *Tailings Dam Committee Sustainable Design and Post-Closure Performance of Tailings Dams*; International Committee on Large Dams (ICOLD): Paris, France, 2013; Volume Bulletin 153.
42. CDA. *Application of Dam Safety Guidelines to Mining Dams*; Canadian Dam Association (CDA): Markham, ON, Canada, 2014.
43. OSTDC. De-Licensing of Oil Sands Tailings Dams. In Technical Guidance Document; Oil Sands Tailings Dam Committee (OSTDC). 2014. Available online: https://era.library.ualberta.ca/items/8131d11c-4402-4bed-b0ab-0d3833b95af6/view/7165e9fb-e003-4ce6-8a92-efaf8355ec2b/2014-2003-2028-20--20Report-20--20Oil-20Sands-20Dams-20De-Licensing-20Technical-20Guide.pdf (accessed on 6 November 2021).
44. INAP. Global Cover System Design Technical Guidance Document; International Network for Acid Prevention (INAP). 2017. Available online: https://www.inap.com.au/wp-content/uploads/global-cover-system-design.pdf (accessed on 6 November 2021).
45. Hadjigeorgiou, J. Understanding, Managing and Communicating Geomechanical Mining Risk. In Proceedings of the First International Conference on Mining Geomechanical Risk, Australian Centre for Geomechanics, Perth, WA, Australia, 9–11 April 2019; pp. 3–20.
46. Griffin, B. Risk Assessment. In *Guidelines for Mine Waste Dump and Stockpile Design*; Hawley, M., Cunning, J., Eds.; CSIRO Publishing: Clayton South, VIC, Australia, 2017; pp. 187–195.
47. Reagan, R.T.; Mosteller, F.; Youtz, C. Quantitative meanings of verbal probability expressions. *J. Appl. Psychol.* **1989**, *74*, 433–442. [CrossRef] [PubMed]
48. NOAMI. *The Policy Framework in Canada for Mine Closure and Management of Long-Term Liabilities: A Guidance Document*; National Orphaned/Abandoned Mines Initiative (NOAMI): Ottawa, ON, Canada, 2010.
49. Slovic, P.; Weber, E.U. Perception of Risk Posed by Extreme Events. In Proceedings of the Risk Management Strategies in an Uncertain World, Palisades, New York, NY, USA, 12–13 April 2002.
50. Gibson, G.; O'Faircheallaigh, C. *IBA Community Toolkit: Negotiation and Implementation of Impact and Benefit Agreements*; Walter & Duncan Gordon Foundation: Toronto, ON, Canada, 2015.
51. Nikl, L.; Steele, J.; Robertson, L.; Davidson, S.; Schryburt, R. Environmental Consequence Classification of Mine Dam Failures: Updating the Canadian Classification System. In Proceedings of the ASDSO 2018 Dam Safety Conference, Association of State Dam Safety Officials, Seattle, Washington, DC, USA, 9–13 September 2018.
52. Department of Mines and Petroleum. *Environmental Protection Authority Guidelines for Preparing Mine Closure Plans*; Government of Western Australia: Perth, WA, Australia, 2015.
53. Aven, T. Risk assessment and risk management: Review of recent advances on their foundation. *Eur. J. Oper. Res.* **2016**, *253*, 1–13. [CrossRef]
54. Oboni, F.; Oboni, C. The Long Shadow of Human-Generated Geohazards: Risks and Crises. In *Geohazards Caused by Human Activity*; Faris, A., Ed.; InTech: Rijeka, Croatia, 2016; pp. 107–140. ISBN 978-953-51-2802-1.

minerals MDPI

Article

Applying a Generalized FMEA Framework to an Oil Sands Tailings Dam Closure Plan in Alberta, Canada

Haley L. Schafer [1,*], Nicholas A. Beier [1] and Renato Macciotta [1,2]

[1] Department of Civil and Environmental Engineering, University of Alberta, 9211 116 Street NW, Edmonton, AB T6G1H9, Canada; nabeier@ualberta.ca (N.A.B.); macciott@ualberta.ca (R.M.)
[2] School of Engineering Safety and Risk Management, University of Alberta, 9211 116 Street NW, Edmonton, AB T6G1H9, Canada
* Correspondence: hvandalf@ualberta.ca

Abstract: Historically, tailings facilities have been designed primarily with consideration of the mine's active life. This is problematic, as the lifespan of a tailings dam may far exceed the life of the mine. Over time, it is expected that these structures will transform into a mine waste structure and then eventually a landform. In Alberta, Canada, dam owners can submit a decommissioning, closure, and abandonment (DCA) plan and completion reports to apply for the facility to be de-registered as a dam. If successful, the structure would be considered a solid waste structure and may be reclassified and regulated as a mine waste dump. The Alberta Energy Regulator expects DCAs to be accompanied and supported by risk assessments that consider long-term physical failure modes, including failure modes that may not be applicable during operations, in accordance with Manual 019. To help support the process of de-registering a tailings dam, a risk management tool, referred to as a Generalized Failure Modes Effects (G-FMEA) framework, was developed and presented in the Journal of Minerals in the Special Issue Tailings Dams: Design, Characterization, Monitoring, and Risk Assessment. The G-FMEA was designed to be used for assessing risks of an external tailings facility in closure, with the goal of assessing the long-term risk of geotechnical failure to support the process of de-registration. In Alberta, a number of tailings dams are undergoing closure and reclamation activities. This paper applies the developed G-FMEA framework to an oil sands tailings dam in Alberta to demonstrate the application of the framework. The paper assesses two specific failure modes of two different elements, including clogging of the drains and surface erosion of the berm. The failure modes are assessed over different timescales to demonstrate how the consequence, likelihood, and risk rating may change over time. The results of this process are discussed in the context of the potential for the facility to be de-registered as a dam.

Keywords: tailings dams; closure; Alberta; failure modes effects analysis; geotechnical risk; oil sands dam

1. Introduction

Many tailings storage facilities (TSFs) in Alberta are in the process of being reclaimed and closed. The goal following closure is for the dam owner to submit a decommissioning, closure, and abandonment (DCA) plan and completion reports to apply for the facility to undergo de-registration as a dam [1]. Following this process, the structure would be reclassified and regulated as a mine waste dump under the Oil Sands Conservation Act or the Coal Conservation Act [1]. The Alberta Energy Regulator (AER), who is responsible for the regulation of energy dams, developed Manual 019: Decommissioning, Closure, and Abandonment of Dams at Energy Projects (Manual 019) to accompany the Alberta Dam and Canal Safety Directive (the Directive). Manual 019 specifies that this risk assessment must consider long-term physical failure modes, including failure modes that may not be applicable during operations [1]. It is important to note that Manual 019 indicates that "risk cannot be reduced to zero for oil sands tailings facilities, or any manmade structure for that matter" and that the ideal lowest consequence category would be comparable to natural analogues with an extremely low likelihood [1]. It is expected that active management should continue until the regulator accepts the completion report, at which point the operator is no longer required to actively manage and report on the structure as a dam [1]. In some cases, a tailings dam may require long-term active management if the residual risk is deemed to be too high [1].

To help support the process of de-registering a tailings dam as a dam in Alberta, a risk management tool (the Generalized Failure Modes Effects (G-FMEA) framework) was developed. The tool is summarized by Schafer et al. [2] in the Journal of Minerals within this Special Issue. Deregistration, defined here, is when the regulating governmental body assesses a dam structure to evaluate if it can be de-classified as a dam, so that it is no longer regulated as a dam. The term de-register is not a regulatory or legal term and has been commonly used in industry to describe this process. Throughout the literature, the terms de-license (i.e., Oil Sands Tailings Dam Committee (OSTDC) [3]) and deregulate (i.e., Oberle et al. [4]) have also been used to describe this process.

There are three different configurations of TSFs that may exist at an oil sands mine as defined by AER [1]:

> *"(1) External tailings facilities where perimeter containment is formed by perimeter dams and rising topography in some cases. The tailings are deposited entirely on natural topography above ground.*
>
> *(2) Tailings stored in pit but a full height dam separates the tailings from the receiving environment.*
>
> *(3) Tailings stored in pit but capacity of the pit container is increased by constructing a perimeter dam along a portion or the entire pit crest."*

The goal of the G-FMEA is to assess the long-term geotechnical risks of an external tailings facility (ETF) during the closure phase to support the process of de-registration [2]. However, the risk management tool should be able to be adapted for other configurations to account for additional risks that exist for those scenarios if needed. The G-FMEA should be used at the conceptual design stage, if possible, to assess the closure plan and evaluate the potential for de-registration [2]. It should be updated throughout the project life and can also be used for dams that have already been constructed or are orphaned [2]. This paper demonstrates how the G-FMEA can be used and applied in practice for a case study oil sands external tailings facility in Alberta, Canada. The G-FMEA charts from Schafer et al. [2] are used to evaluate the potential failure modes of an oil sands ETF closure plan using an element approach. This process demonstrates how the framework can be used to go from a generalized list of failure modes and applied directly to a case study. This process requires an evaluation of serviceability failure of each element. The paper then shows how to apply the risk matrix framework to two select failure modes (clogging of drains and surface erosion of the berm) and how the risk of these failure modes changes over time. The site-specific G-FMEA is used to show how the framework can be used to support the closure process at different points of time (conceptual design or de-registration). Overall, the paper serves to demonstrate how risk management can be integrated into the process of closure design to support de-registration. Ultimately, this supports long-term closure goals with a risk management focus.

2. Background

The G-FMEA serves as a screening tool for the closure phase of an ETF, where hazards that are assessed as acceptable will not require additional analyses, but hazards that are assessed to have a higher risk rating (or where there are a number of low-risk hazards) could require more detailed and/or quantitative risk assessments. This may include reliability analyses for critical risks. However, challenges with quantifying uncertainties in climatic conditions in the long-term would likely render uncertainty (particularly epistemic uncertainty) difficult to manage within reliability analyses.

The objective of the G-FMEA is: "*How should geotechnical risks associated with an external tailings facility in Alberta be managed in the long-term to achieve an acceptable closure plan, such that the facility is able to be de-registered as a dam?*" [2]. The G-FMEA includes charts (provided in Schafer et al. [2]) for the drainage system, foundation, dam body, and landform that provide failure modes for different elements present in each system. These charts form the backbone of the G-FMEA framework. Applying the G-FMEA framework to a site involves first defining the system, developing a block diagram, and breaking the system down into key elements. Each element should have the function defined, as well as what constitutes serviceability failure. Once this step has been completed, each element can be screened for applicable failure modes using the G-FMEA charts. A risk assessment is then conducted for every applicable failure mode, which involves determining the likelihood and consequence rating for each temporal scale and determining the risk rating. Appropriate risk mitigation options should be assessed and applied as appropriate.

The G-FMEA should be conducted for different assessment periods, including the immediate term, short term, medium term, and long term [2]. The goal with this is to account for time dependence, depreciation of system elements, and changing risk profiles, which are common critiques of failure modes effects analysis [2].

Schafer et al. [2] presented a likelihood rating table (Table 1), consequence rating table (Table 2), and risk categories (Table 3) to be used in conjunction with the G-FMEA. Schafer et al. [2] provides guidance on how a risk matrix can be color-coded and demonstrates the application through an example risk matrix. The example risk matrix, presented in Figure 1, will be used for the oil sands case study presented in this paper. The example risk matrix has a threshold line where hazards that have a risk rating higher than the threshold line may require additional analysis to adequately assess the risk. It should be noted that risk matrices used in practice should be developed with input from all relevant stakeholders.

Table 1. Likelihood rating (Reprinted from Schafer et al. [2]).

Likelihood Rating	Qualitative Interpretation Guidance [1]	Quantitative Interpretation Guidance	Annualized Probability of Occurrence
Almost certain	Almost certain that an incident will occur given the circumstances. Very high probability of one or more occurrences per year.	Higher than 10% probability in a year	$p \geq 0.1$
Likely	High likelihood. Commonly observed at similar facilities.	Higher than 10% probability in 10 years	$p \geq 0.01$
Possible	Has occurred a number of times within the industry and at least once at the site (or similar facilities in the region).	Higher than 1% probability in 10 years	$p \geq 0.001$
Unlikely	Has occurred before within the industry, but not at the site.	Less than a 1% probability in 10 years	$p < 0.001$

Table 1. *Cont.*

Likelihood Rating	Qualitative Interpretation Guidance [1]	Quantitative Interpretation Guidance	Annualized Probability of Occurrence
Rare	Low likelihood of occurrence, but not impossible. Has not occurred at the site but has occurred in industry.	Less than a 1% probability in 100 years	$p < 0.0001$
Very rare	Very low likelihood of occurrence, but not impossible. Occurrence cannot be deemed non-credible	Less than a 1% probability in 1000 years	$p < 0.00001$
Close to non-credible	Extremely remote likelihood of occurrence. Although the mechanisms are technically plausible for the occurrence, it is seen as near non-credible.	Less than a 1% probability in 10,000 years	$p < 0.000001$

[1] Industry encompasses the mining industry as a whole.

Table 2. Consequence rating (Reprinted from Schafer et al. [2]).

Consequence Rating [1]	Consequence of Failure of Element on the Rest of the System	Degree of Human Intervention Required	Environment	Community
Slight	Failure of element does not have cascading consequences.	Structural integrity maintained. No intervention or maintenance required.	No movement of tailings beyond the structure footprint.	No impact on local community.
Minor	Failure of element may have cascading consequences that do not result in global failure.	Structural integrity maintained. Minor or localized intervention or maintenance required.	Released tailings are not toxic [2], and/or minimal loss of habitat (<5%) of species of special interest [3], and/or acceptable restoration of water bodies and environment feasible in a short time frame (<5 years).	Impact [4] on local community for less than 1 year.
Moderate	Failure of element has cascading consequences that do not result in global failure.	Intervention or maintenance required to limit impact of cascading consequences.	Released tailings are not toxic [2], and/or moderate loss of habitat (5–20%) of species of special interest [3], and/or acceptable restoration of water bodies and environment feasible in a short time frame (<5 years).	Short-term (<5 years) impact [4] on local community.

Table 2. *Cont.*

Consequence Rating [1]	Consequence of Failure of Element on the Rest of the System	Degree of Human Intervention Required	Environment	Community
Major	Global failure of tailings dam with minor release of tailings.	Intervention or maintenance required to maintain function of structure as a whole.	Released tailings are toxic [2], and/or significant loss of habitat (20–50%) of species of special interest [3], and/or acceptable restoration of water bodies and environment feasible in a moderate time frame (5–25 years).	Medium-term (5–25 years) impact [4] on local community.
Severe	Global failure of tailings dam with catastrophic release of tailings.	Structural repair not possible.	Released tailings are toxic [2], and/or very significant loss of habitat (>50%) of species of special interest [3], and/or acceptable restoration of water bodies and environment unlikely within an extended time frame (>25 years).	Long-term (>25 years) impact [4] on local community. Fatalities.

Notes: [1] Assigned consequence should reflect the most likely outcome. If assigning consequence with consideration of the worst case or a combination of discrete outcomes, this must be declared. [2] Toxicity assessment of tailings should consider an assessment of the fluids and solids (leaching potential, acidity, radioactivity). [3] Species of special interest is defined as a species that lives in the inundation area that would be greatly impacted by habitat loss (preferable to select a species that is provincially or federally listed). [4] Community impacts must be determined through meaningful engagement with stakeholders and may include a consideration of health, loss of access/destruction of traditional lands, housing, destruction/damage of farmland, harm to livestock, damage to water or soil resources, impacts to trapping and fishing, loss of animals, overall cultural impact, and employment. Reputation, legal aspects, and economics are not considered in this consequence table as they are considered site- and corporation-specific. It may be necessary to assess these aspects on a site-specific basis.

Table 3. Risk category (Reprinted from Schafer et al. [2]).

Risk Category	Description of Risk Category
Low	Risk minimal. Monitor risks. Acceptable closure plan.
Moderate	Risk tolerable with controls. Assess risk mitigation options and monitor these risks. Minor re-design of closure plan may be required to accommodate risk mitigation.
High	Risk undesirable. Risk mitigation should be employed to ALARP to reduce risk category. Closure plan may require alteration to accommodate risk mitigation.
Extreme	Risk intolerable. Risk mitigation required immediately to reduce risk category. Requires more detailed risk analysis. Closure plan requires alteration.

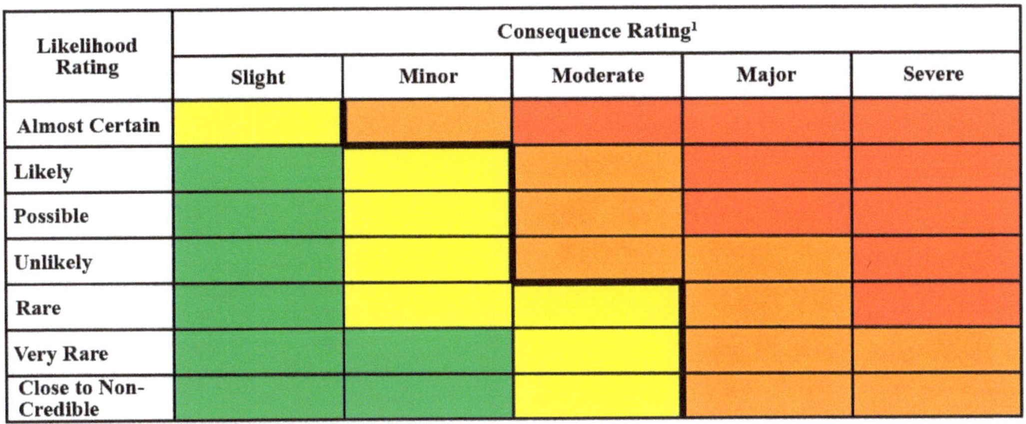

[1]Assign risk rating for each consequence category.
[2]This matrix is provided for illustration purposes. Site specific risk matrices should be developed using Steps 1-7 and include meaningful engagement with stakeholders to ensure that risk tolerance of these stakeholders is represented.

Figure 1. Example risk matrix considering major hazard aversion and threshold for quantitative analysis (Reprinted from Schafer et al. [2]).

3. Application of G-FMEA to a Case Study

The G-FMEA was used to evaluate an oil sands case study in Alberta, Canada. The purpose of this was to demonstrate the utility of the G-FMEA process.

3.1. Site Description (System Definition)

The first step of conducting the G-FMEA involves defining the system as described by Schafer et al. [2]. This section provides a base-level description of the site for information purposes. The ETF is a sand dam with a starter dyke constructed of overburden. The facility was constructed with a mixture of upstream and centerline construction. In upstream construction, tailings are discharged upstream from the crest of the dam (beginning first with the starter dyke) to form a beach, which becomes the foundation for subsequent raises [5]. In centerline construction, fill is placed on the downstream slope of the dam and tailings are discharged in the upstream direction to form a beach [5]. Consequently, the centerline of the dam remains at the same position relative to the starter dyke [5]. Analysis of the facility for this paper was conducted at a section of the ETF that had been constructed using upstream construction with hydraulic fill deposition and cell construction. Cell construction involves discharging tailings to a cell where the solids settle, and water and fines are decanted [6]. Dozers are used to maintain containment dykes and spread and compact the sand in the cell by using vibration [6,7]. The water and fines that do not settle flow to the end of the cell and through a weir structure to the beach [8].

The dam is composed of coarse sand tailings (CST) and beach deposits, including beach above water (BAW) and beach below water (BBW). CST is made up of the coarse fraction of whole tailings and is formed by cycloning the tailings [9]. The BAW and BBW deposits are formed when tailings are discharged into the pond. The coarser tailings will settle first on the beach, leading to the formation of these two different deposits. The BAW and BBW can have variable gradation, solids content, hydraulic conductivity, and density [6]. At the location of the analyzed section, the dam contains fluid fine tailings (FFT). The dam has a toe berm constructed of lean oil sands (fines and a modest amount of bitumen) [10], with benched slopes to provide stability to the facility. The foundation of the ETF at the analyzed section is typical for the Fort McMurray area and consists of the McMurray formation, Pleistocene sands, and McMurray formation tidal flat mud and mud flat. It should be noted that there is a weak tailings layer at the base of the tailings deposit and the upstream constructed lifts of the dam. The weak tailings layer was formed due to initial deposition of tailings into standing water, has a high bitumen and fines content, and is considered liquefiable. This layer creates stability issues, which led to the design and construction of the toe berm that is currently in place. The section used for analysis is extensively instrumented and monitored.

Drainage at the facility consists of two levels of 200 mm diameter perforated collector pipes surrounded by a woven geotextile sock drain that run parallel to the dam centreline. Smaller outtake pipes connect to the collector pipes at approximately 150 m intervals and discharge water to a perimeter ditch.

As the facility operations proceed towards closure, it is expected that the facility will be infilled with CST, partially displacing the FFT. It is expected that this process will result in a zone of mixed CST and FFT. This zone will be covered by approximately 4 m of CST that will be placed and compacted using methods similar to cell construction to create a trafficable surface. The compacted CST will then be capped with mine waste to enable construction of the landform. The vegetative cover for the reclamation surface is not yet defined. The reclamation surface is expected to have hummocks and a system of drainage channels that direct flow towards the outlet of the facility. The outlet is designed to have a base width of 100 m and side slopes of 15 horizontal:1 vertical.

The case study cross-section analyzed is provided in Figure 2.

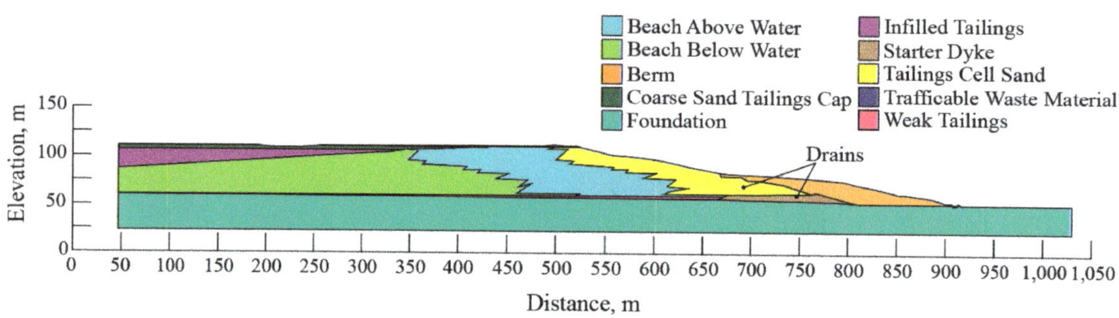

Figure 2. Case study cross-section.

3.2. Block Diagram, Function Definition, and Serviceability Failure Description

The next three steps of conducting the G-FMEA involve developing a block diagram, breaking the subsystem down into key elements and functions, and defining serviceability failure for each element. The block diagram is used to break the system down into functional subsystems that can be analyzed and serves to show the relationships between those subsystems. The block diagram for the case study is shown in Figure 3. Table 4 provides descriptions of the elements, the function before and after closure, and serviceability failure for each element based on the defined function.

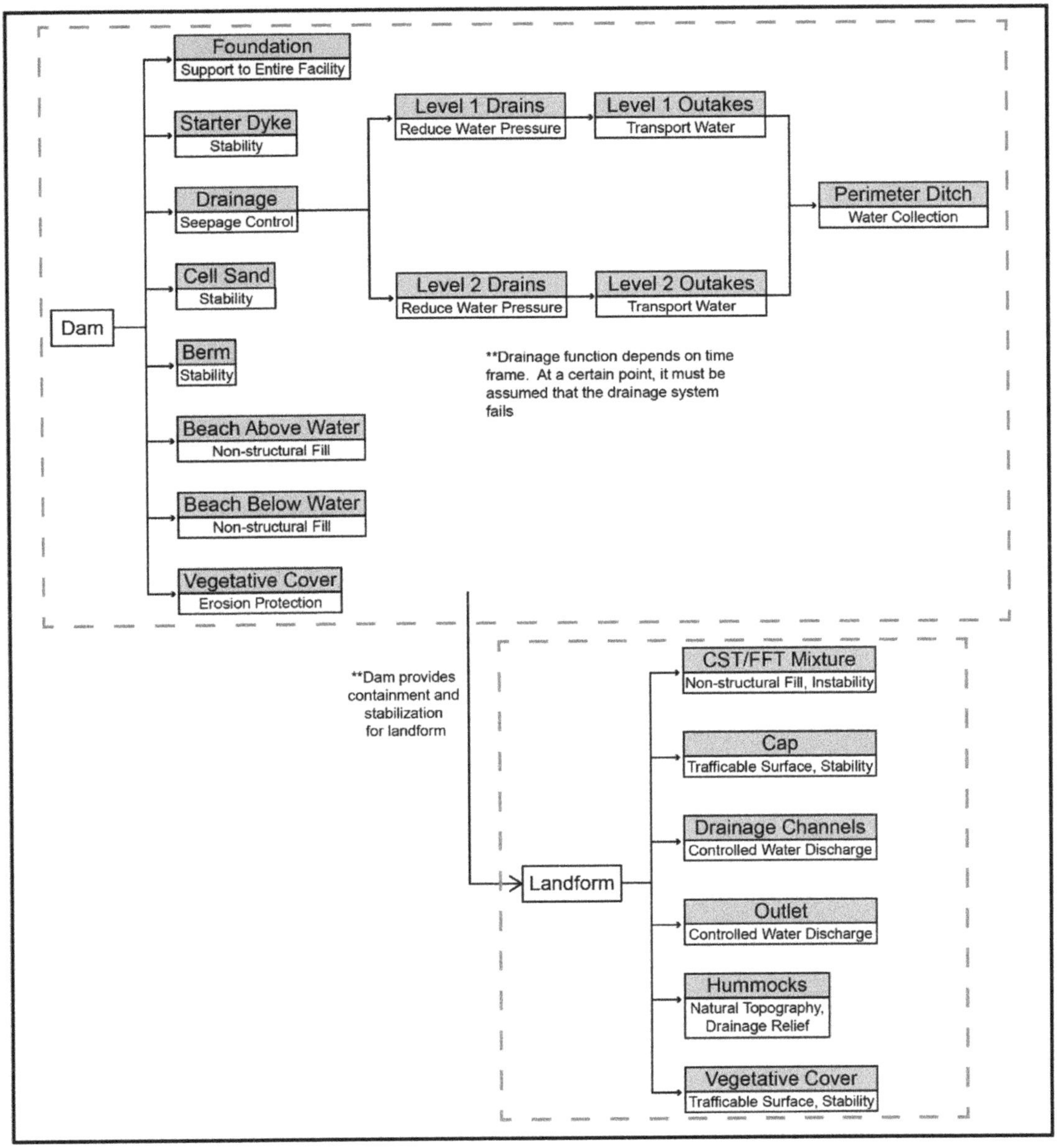

Figure 3. Case study block diagram.

Table 4. Case study element functions and failure description.

System	Elements	Function (before Closure)	Function (after Closure)	Failure Description
Dam	Foundation	Supports capacity of the dam.	Supports capacity of the dam.	Instability associated with movements of the soil mass.
	Starter dyke	Allowed for initial containment of tailings. Provides stability.	Provides stability.	Instability.
	Drainage system		See drainage system.	
	Cell sand	Containment, dyke stability, provides stability.	Provides stability.	Instability.
	Berm	Provides stabilization against liquefaction with weak tailings layer present at base of facility.	Provides stabilization against liquefaction, with weak tailings layer present at base of facility.	Instability.
	Beach above water	Containment, dyke stability.	Non-structural fill.	Instability.
	Beach below water	Containment of tailings and dam raises.	Non-structural fill. Potential source of instability.	Instability.
	Vegetative cover	Erosion stabilization and protection of slope.	Erosion stabilization and protection of slope.	Fails to provide erosion protection.
Drainage System	Level 1 drains	Control the phreatic surface in the dam.	Controls phreatic surface in the dam up to a certain point of time (must be determined with modelling). After this point, the drains are assumed to fail and will no longer serve a function in closure.	Drains fail to control phreatic surface.
	Level 2 drains	Control the phreatic surface in the dam.	Controls phreatic surface in the dam up to a certain point of time (must be determined with modelling). After this point, the drains are assumed to fail and will no longer serve a function in closure.	Drains fail to control phreatic surface.
	Outtakes—level 1	Transmit water from drains to perimeter ditch.	Controls phreatic surface in the dam up to a certain point of time (must be determined with modelling). After this point, the drains are assumed to fail and will no longer serve a function in closure.	Drains fail to control phreatic surface.

Table 4. *Cont.*

System	Elements	Function (before Closure)	Function (after Closure)	Failure Description
	Outtakes—level 2	Transmit water from drains to perimeter ditch.	Controls phreatic surface in the dam up to a certain point of time (must be determined with modelling). After this point, the drains are assumed to fail and will no longer serve a function in closure.	Drains fail to control phreatic surface.
	Perimeter ditch	Collect water from drains and intercept seepage and sediment.	Water and sediment collection.	Fails to collect, intercept, and transport seepage.
Landform	Cap (overburden and coarse sand tailings)	N/A	Trafficable surface, stability to landform, limits infiltration.	Landform instability.
	Infilled coarse sand tailings mixed with FFT	N/A	Non-structural fill. Potential source of instability.	Landform instability.
	Hummocks	N/A	Provide topographic variation and diversity. Designed to keep the water table low in upland areas.	Fails to provide topographic variation and diversity. Fails to help control the water table.
	Drainage channels	N/A	Directs flow to the outlet to be carried away from the facility.	Fails to direct flow to the outlet.
	Outlet	N/A	To ensure controlled discharge under exceptional inflow conditions.	Fails to control discharge.
	Vegetative cover	N/A	Stabilization of landform. Erosion protection. Ecosystem creation. Wildlife habitat.	Fails to provide erosion protection.

3.3. Site-Specific FMEA

Each element discussed in Table 4 was assessed using the appropriate chart from the G-FMEA to assign failure modes to the elements. Appendix A shows the 128 failure modes that were identified for the dam elements. The next steps in the FMEA are to assess the applicability of the failure modes, determine the failure effects and consequences, determine the risk rating for each temporal scale, and determine corresponding risk mitigation as required. This step has not been completed in full but has been completed for demonstration purposes for failure mode 34 and 87 in Table A2, which are summarized in Table 5. These failure modes were selected due to the way that they may change and develop over time. A detailed description of the clogging of the level 2 drains due to particulate clogging (failure mode 34) is provided in Section 3.3.1, and a description of the surface erosion failure of the berm is provided in Section 3.3.2. It should be noted that this work is not accompanied by a level of confidence assessment for each risk rating (as recommended in the G-FMEA steps), as this work is currently underway by a researcher at the University of Alberta.

Table 5. Demonstration failure modes.

Element	Failure Mode Identification	Failure Mode Description	Potential Trigger/Cause	Failure Effects
Level 2 Drains	34	Clogging	Particulate clogging (sediment from perimeter ditch or other)	Lack of control of phreatic surface (potential rise in phreatic surface), increase in seepage, pond on reclamation surface, internal erosion, global instability, release of water into downstream shell, erosion on downstream slope
Berm	87	Surface erosion from wind and overland flow resulting in rills, gullies, or sheet erosion	Destruction of vegetation (FM 112), rainfall, melting of snow, wind, increased seepage on downstream slope from failure of drainage system, increased seepage from internal erosion in embankment	Slope failures (shallow surficial movement, slumps), change in downstream slope angle, blockage in perimeter channel with sediment, development of negative drainage, development of large erosion scarps

3.3.1. Clogging of the Level 2 Drains

The level 2 drains consist of a 200 mm perforated collector pipe surrounded by a woven geotextile sock placed on a bed of sand and gravel, and they are located in the cell sand. The function of the level 2 drains before closure is to control the phreatic surface of the dam. Following closure, the function of the drains changes over time, and they are required to contribute to the control of the phreatic surface for a period of time. At this mine site, it is assumed that the drains will fail in the long term, and the facility has been designed to support this assumption. However, the question of how long the drains will need to be operational to support the lowering of the phreatic surface is critical. Seepage modelling to support this question is currently underway.

Schafer et al. [11] interviewed industry professionals on tailings dam closure. Notably, interview participants had conflicting comments regarding the progression of the phreatic surface over time [11]. Some participants indicated that it is expected that the phreatic surface will go down in the long term, while others indicated that there are a number of conditions under which this may not be true, such as the drains clogging [11]. This shows the importance of assessing drain performance over time, as it may have a critical role in how the phreatic surface changes over time, which ultimately may impact the stability of the structure.

Drains may fail in a number of ways, as shown in Table A2, which may include a breakage of the pipe due to buckling or physical degradation, clogging (chemical, biological, particulate, failure from the outtakes), clogging of the surround (biological, chemical, particulate), or a breakage of the connection between the drain and outlet pipe. This section focuses on the clogging of the level 2 drains due to particulate clogging.

As per the G-FMEA framework, the analysis should take place over four temporal scales: immediate-term, short-term, medium-term, and long-term. For this tailings facility, it is expected that the immediate-term and short-term period will fall within the adaptive management period, as the mine site will still be mining following closure of the facility. The medium-term period will fall within the reactive management period, and the long-term occurs at about 1000 years. For the purposes of this analysis, it was assumed that the drains were required to control the phreatic surface for the immediate-term, short-term, and medium-term periods. As such, failure occurs when the drain is no longer able to control the phreatic surface. For the long-term period, the drains were assumed to fail and are no longer considered to be a controlling feature of stability. Consequently, long-term risk assessment was not conducted for this failure mode.

The completed risk assessments for the immediate-term, short-term, and medium-term temporal scales for particulate clogging of the level 2 drains are provided in Table 6. As the temporal scale increases risk assessment, there is an increase in likelihood of the drains failing due to particulate clogging, from very rare in the immediate term to possible in the medium term.

Table 6. Immediate-term, short-term, and medium-term risk assessments for particulate clogging of level 2 drains.

Temporal Scale	Likelihood	Consequences				Risk Rating				Controls
		Element Failure Consequence	Human Intervention	Environment	Community	Element Failure Consequence	Human Intervention	Environment	Community	
Immediate term	Very rare	Minor	Minor	Slight	Slight	L	L	L	L	No control needed
Short term	Rare	Minor	Minor	Slight	Slight	M	M	L	L	No control needed
Medium term	Possible	Moderate	Moderate	Slight	Minor	H	H	L	M	Modelling to predict drain performance and location of phreatic surface over time, monitoring of pore water pressures

The consequences also change as the temporal scale increases. The phreatic surface has continued to drop in the facility. For the purposes of this work, it is assumed that the phreatic surface may be below the drain location at some spots throughout the facility at the time of closure. This hypothesis should be supported by modelling and onsite monitoring, which are beyond the scope of this work. Using these assumptions, it is expected that the failure of the level 2 drains may result in a minor rise in the phreatic surface, but this is not expected to result in global failure. In the immediate term and short term, it is expected that failure of the element may have cascading consequences. However, these cascading consequences are not expected to result in global failure due to other controls in place during the adaptive management phase, resulting in a minor consequence rating. This is coupled with the potential for minor or localized intervention, which is possible due to the presence of staff on site. Due to the expected consequences, a rating of slight was assigned in the environment and community categories.

In the medium term, if the facility requires the level 2 drains to function for stability, it is expected that failure could result in a rise in the phreatic surface, an increase in seepage, and formation of ponds on the reclamation surface, but these consequences are not expected to result in global failure, resulting in a moderate consequence rating. However, this failure may require human intervention or maintenance to limit the extent of the cascading consequences (i.e., if extensive ponds formed on the reclamation surface), resulting in a moderate consequence rating. A slight consequence rating was assigned to the environment category as no tailings are expected to move beyond the footprint from this failure. It is expected that there may be a short-term impact on the local community if ponds form on the reclamation surface, depending on the end land use, resulting in a minor consequence rating. Modelling must be conducted to determine the influence of drain failure on the phreatic surface and if drain function is required to maintain stability in the medium term (i.e., to keep the phreatic surface low such that the potentially liquefiable units de-saturate). Modelling is an effort to reduce the uncertainty associated with the risk assessment, not necessarily reducing the likelihood or consequences. However, understanding the failure mode may result in the likelihood or consequence rating being reduced. Following the completion of the modelling, more targeted controls to reduce the likelihood and consequence can be developed, if needed.

As shown in Table 6, this results in risk ratings that increase as the temporal scale increases for the failure consequences, human intervention, community, and end land use consequence categories. The risk rating for the environment consequence category remains low for the different temporal scales as tailings are not expected to go beyond the facility footprint due to a failure of the level 2 drains.

3.3.2. Surface Erosion of the Berm

The berm consists of lean oil sands with benched slopes to provide stability. The function of the berm before and after closure is to provide stabilization against liquefaction due to a weak tailings layer present at the base of the facility. Failure of the berm (instability) may occur due to a number of different failure modes, as shown in Table A2, including vertical deformation, seismic liquefaction, static liquefaction, shear failure, internal erosion, surface erosion, and toe erosion. This section focuses on surface erosion from wind and overland flow.

Unlike the particulate clogging of the level 2 drains, the risk assessment for the surface erosion of the berm was completed over all four temporal scales. The completed risk assessments for all four temporal scales for surface erosion of the berm are provided in Table 7. It is common practice during operations for erosion gullies to be repaired on an ongoing basis as they develop in the downstream, bare sand slopes. Similarly, erosion events may be less critical in the immediate term and short term due to the opportunity for maintenance.

Slingerland et al. [12,13] conducted a study to investigate erosion of a sand dam in the Alberta oil sands using a Landscape Evolution Model (LEM) to aid in assessing the long-term geomorphic stability. The results of the study for a 200-year simulation period indicated that the morphology of the structure was well established within 60 years, with gullies most frequently forming in areas of concentrated flow (i.e., horizontally concave dam sections) and retrogressively travelling inward towards the pond [12,13]. The development of erosional features on a dam may be impacted by the dam height (including the length of the slope), the grain size distribution of the material, and the climate [13]. The results showed that some potential areas of concern in the long term may be substantial sediment discharge off-site, the formation of deep gullies (greater than 15 m deep), and the filling of drainage channels [13]. The results also suggested that it would be advantageous to grade dams such that surface water is collected and directed toward armored swales and transported to the perimeter channel, as opposed to having a uniform slope that allows sheet flow [13].

Table 7. Immediate-term, short-term, medium-term, and long-term risk assessments for surface erosion of the berm.

Temporal Scale	Likelihood	Consequences				Risk Rating				Controls
		Element Failure Consequence	Human Intervention	Environment	Community	Element Failure Consequence	Human Intervention	Environment	Community	
Immediate term	Possible	Minor	Minor	Minor	Slight	M	M	M	L	N/A
Short term	Possible	Minor	Minor	Minor	Slight	M	M	M	L	N/A
Medium term	Likely	Moderate	Minor	Minor	Minor	H	M	M	M	LEM to evaluate geomorphic evolution, monitoring, maintenance thresholds
Long term	Likely	Moderate	Moderate	Moderate	Moderate	H	H	H	H	LEM to evaluate geomorphic evolution, monitoring, maintenance thresholds

Slingerland et al. [12,13] also showed that vegetation played a huge role in reducing erosion. Overall, major gullies started to form before the vegetation matured, with no major gullies forming following vegetation maturity [12]. Erosion gullies began to form near the bottom of the slope as the soil typically has a higher water content, resulting in the flow transitioning from dispersive flow to centralized, efficient flow paths [12]. This centralized flow allows the water to gain speed and increases its erosive power [12]. Erosion gullies also formed from surface water being directed and concentrated in one location and then finding the easiest route to travel [12]. Overall, the results of the study indicated that erosion in the long term may result in the structure not functioning as designed, resulting in failure of the structure to retain tailings. This indicates a potential need for ongoing maintenance of the structure.

The section being analyzed (and taken to be representative of the facility) has mature vegetation on the downstream slope. Consequently, it is expected that erosion events will occur with a 'possible' likelihood in the immediate term and the short term. As time progresses, it is expected that the likelihood will change from 'possible' to 'likely' as climate change occurs, leading to more extreme events and a transition of vegetation. Assessing the likelihood of surface erosion in the long term is further complicated by the potential for anthropogenic activities, which may alter the projected landscape and amplify the likelihood of surface erosion (i.e., clearcutting all trees on the downstream slope).

Table 7 shows how the risk rating increases as the temporal scale increases in the different consequence categories. In all cases, it is not expected that erosion of the berm would result in a global failure of the facility. For the element failure consequence, there is an increase from a moderate risk rating to a high risk rating due to a change in the consequence category rating. Reviewing Table 2, it can be seen that a consequence rating of minor and moderate have the same definition for the element failure consequence category. The moderate consequence category was assigned for the medium term and long term due to the unknowns associated with these time frames. For the human intervention category, there is a change from a moderate risk rating in the immediate term, short term, and medium term to a high risk rating in the long-term. This is due to a moderate consequence rating assigned in the human intervention category for the long-term temporal scale, due to the potential need to employ intervention or maintenance to limit cascading consequences and prevent global failure. In the environment category, the risk rating changes from moderate in the immediate. term, short term, and medium term to a high risk rating in the long term. This is due to a change in the consequence rating from minor to moderate. It should be noted that this assumed that the tailings are not toxic. For the community, the risk rating increases from low in the immediate term and short term to moderate in the medium term to high in the long term. It is assumed that no fatalities will occur due to a failure of the element, but that there may be short-term impacts to the community as a failure may impact the post-mining land use, require intervention and maintenance, and result in tailings being carried away from the footprint, which has the potential to impact fishing and so on.

For controls, this failure mode should be analyzed with an LEM to evaluate geomorphic evolution, monitoring to evaluate erosion progression in the vegetated slope and confirm if it aligns with the LEM, and defined maintenance thresholds, if necessary.

4. Discussion

Risk management consists of two main components, including risk assessment and risk control, and is a systematic process where risks are recognized and assessed, and appropriate control measures are implemented [14]. The G-FMEA framework outlines the steps for conducting both risk assessment and risk control. Step 8 of the process involves assessing appropriate risk mitigation options based on the risk rating for each temporal scale and then repeating the risk assessment process as needed. Controls are briefly discussed for the two different failure modes that had risk assessments completed using the risk matrix. These controls focus on two key elements, including detailed modelling and monitoring and maintenance. The monitoring and maintenance requirements must be clearly outlined and must be in line with the long-term closure plan. This means that if the long-term closure goals include walk away closure with no maintenance, then the use of monitoring and maintenance as a control is not appropriate, in contrast to a closure plan that includes targeted monitoring and maintenance. Once the risk mitigation/controls are outlined and completed (in the case of modelling), the risk assessment can be repeated to re-assess the risk. This process may result in a need to alter the closure plan to adequately reduce the risk. The success of risk mitigation methods may depend on where the ETF is at in its life cycle. The earlier risk management practices are employed in the closure planning process, the easier it is to make meaningful changes that will positively impact the long-term risk. For example, Slingerland [15] investigated the impact of different dam slope shapes on surface erosion using LEM and found that the shape of the downstream slope had a substantial impact on the erodibility of the slope, where uniform slopes contribute to the development of focused drainage patterns and thus an increase in erosion. Slingerland [15] noted that implementing these dam shapes into the initial dyke design is an important element to success. Once the dam is farther into its life cycle, these types of design changes to aid in mitigating the risks associated with erosion would be more difficult to employ, and maintenance controls may need to be relied on more heavily.

While the G-FMEA was not completed in full for the case study dam, the application of the framework was demonstrated. Completion of the G-FMEA may indicate that some failure modes require additional risk assessment to adequately quantify the risk. The framework and example risk matrix outlined by Schafer et al. [2] states that a threshold can be used with the risk matrix where hazards that fall above the threshold may require further risk assessment methods. This is due to the known pitfalls of risk matrices. If a threshold value is adopted, careful consideration should be given to hazards that fall above the threshold, particularly those with a high risk rating, where as low as reasonably practical (ALARP) principles may be employed. Assessment of risk using risk matrices should always be conducted with awareness of the shortfalls of risk matrices.

Once the G-FMEA has been completed for all of the failure modes, this can be used to assess the success of the closure plan to support de-registration. If this is in the earlier stages of design, this process can be highly effective at making meaningful changes to the design of the facility and the closure plan to reduce long-term risks. This process would be highly effective if integrated into the process of landform design, which is defined as "the integrated, multidisciplinary design and construction of mining landforms and landscapes, directed by a dedicated team working with different mine operations groups and others over the life of the mine and beyond" [16]. Landform design has a focus on integration—including integration across time scales and disciplines [16]. The process of integrating risk management of the closure plan into the overall design of the facility and designing with the end in mind is the best tool for successful de-registration. The G-FMEA can then be re-visited and updated, as necessary, over the life of the facility. In all cases, an essential element is that the risk mitigation options are assessed and implemented with consideration of the long-term custodial transfer plans.

Once a facility has reached the stage of reclamation and closure, as with the oil sands dam presented in this case study, the completed G-FMEA can be used to help support the de-registration process. Manual 019 released by the AER in 2020 requires dam owners to complete risk assessment to support DCA plans and completion reports [1]. A requirement of the DCA plan is to assess risk using FMEA or a similar method, and this should include identification of hazards and potential failure modes, analyzing the hazards, assessment of the consequences and likelihoods, characterizing the overall acceptability of the residual risk, and redesigning to reduce the residual risk if needed [1]. The G-FMEA can be used for this purpose to aid in assessing and managing the risks associated with closure.

5. Summary and Conclusions

Globally, mining companies and regulators are working towards how to close tailings facilities in a responsible and sustainable way to achieve the goal of physical, chemical, ecological, and social stability, with no further risks to life or the environment, as outlined by ICOLD [17]. Getting to this point requires collaboration between industry, regulators, and academia. The Alberta Energy Regulator (AER) has developed new guidance for closure of tailings dams, referred to as Manual 019, which requires risk assessments that consider long-term physical failure modes [1]. To help support the effort of de-registering a tailings dam in Alberta, Schafer et al. [2] developed the G-FMEA Framework to evaluate the geotechnical risks associated with closure of external tailings facilities and assess if the closure plan is sufficient to support de-registration. The framework was developed through collaboration between industry, the regulator, and academia.

This paper shows how the G-FMEA framework developed by Schafer et al. [2] can be applied to a case study site. The case study site used for this work is an external tailings facility located at an oil sands site in northern Alberta. For the oil sands facility, the system definition and block diagram were developed and used to develop failure modes for the individual dam elements using the generalized charts from the G-FMEA. From there, select failure modes (clogging of drains and surface erosion of the berm) were used to demonstrate how the risk matrix can be used to assign risk ratings. The failure modes were assessed over different temporal scales to demonstrate how the risk may change over time. In both cases, the risk rating increases as the temporal scale increases (from immediate-term to long-term). In part, uncertainty contributes to the increasing likelihood, and modelling may be a useful exercise to aid in informing risk decisions. Full completion of the G-FMEA can be used to assess a closure plan based on the residual risk and if this closure plan will support the de-registration, which is a vital step to fulfilling long-term closure goals and satisfying the social license to operate. A level of confidence rating was not applied to the individual risk ratings as this work is currently being developed at the University of Alberta. The G-FMEA is intended to be used and applied throughout the life of a structure—from the planning phase to assessing risks associated with closure.

Limitations of the G-FMEA process lie in the tremendous amount of information required to do a comprehensive evaluation of a tailings facility for de-registration, especially at the conceptual planning stage when data is more limited. The G-FMEA also uses a risk matrix for evaluation of risk. Risk matrices have a number of pitfalls that risk matrix developers and users must be aware of that are outlined comprehensively in Schafer et al. [2].

The G-FMEA framework takes an element approach to evaluating the risk of a tailings facility and encourages risk assessors to evaluate the interactions between individual components. This is different from many FMEAs being conducted in industry that look at the tailings facility as an entire structure and evaluate global stability (ultimate failure), rather than breaking the structure down into individual elements and evaluating function failure of each element. While the overall process may be time intensive, it is necessary to evaluate the long-term risk of a tailings facility. Specifically, incorporating the G-FMEA into the early processes of design may allow for closure-based decision making to be used in the overall design process (i.e., if erosion is identified as being critical to long-term stability, the downstream slopes could be modified at the onset to minimize long-term erosion).

The overall G-FMEA framework may be adapted for other structures and jurisdictions, as needed. This would require an adaptation of the G-FMEA generalized charts, the likelihood table, consequence table, and risk rating presented in Schafer et al. [2]. Regardless, the overall methodology is transferable across the mining industry.

Author Contributions: Application of G-FMEA to case study, H.L.S.; writing—original draft preparation, H.L.S.; writing—review and editing, N.A.B. and R.M.; project administration, N.A.B.; supervision, N.A.B. and R.M. All authors have read and agreed to the published version of the manuscript.

Funding: This research was funded by the Vanier Banting Secretariat and the Alberta Energy Regulator (Service Agreement: 17SA-OP010).

Data Availability Statement: Not applicable.

Acknowledgments: The authors would like to thank Tim Eaton (Alberta Energy Regulator), Scott Martens (Canadian Natural Resources Ltd.), and Gord McKenna (McKenna Geotechnical Inc.) for their ongoing support throughout this research.

Conflicts of Interest: The authors declare no conflict of interest. The funders had no role in the design of the study; in the collection, analyses, or interpretation of data; in the writing of the manuscript, or in the decision to publish the results.

Appendix A

Table A1. Recommended G-FMEA worksheet (Reprinted from Schafer et al. [2]).

Element	Failure mode identification	Failure mode description	Potential trigger/cause	Screening assessment of failure mode	Is this failure mode applicable?	If yes, is there sufficient data to evaluate the risk? List any resource gaps.	Failure effects	Immediate-term/Short-term/Medium-term/Long-term Assessment *													Controls	Remarks
									Consequences				Risk Rating				Level of Confidence					
								Likelihood	Element failure consequence	Human intervention	Environment	Community	Element failure consequence	Human intervention	Environment	Community	Element failure consequence	Human intervention	Environment	Community		

* Likelihood rating, consequence rating, risk rating, level of confidence, and controls must be determined for each failure mode for the short-term assessment, medium-term assessment, and long-term assessment.

Table A2. Application of G-FMEA to a case study site.

Element	Failure Mode Identification	Failure Mode Description	Potential Trigger/Cause	Screening Assessment of Failure Mode	Failure Effects
Perimeter ditch	1	Blockage	Sedimentation	Is there erosion protection in place? What is the slope of the ditch? Is it sufficient to keep particles suspended? What is the strength of material? Have there been failures in this material before?	Flooding, discharge of process-affected water to the environment, blockage of level 1/level 2 drain outlet, rise in phreatic surface, increase in seepage, pond on reclamation surface, internal erosion, global instability, toe erosion
	2	Blockage	Sloughing/slope failure of walls	What is the slope of the side slopes? What is the strength of material? Have there been failures in this material before?	Flooding, discharge of process-affected water to the environment, blockage of level 1/level 2 drain outlet, rise in phreatic surface, increase in seepage, pond on reclamation surface, internal erosion, global instability, toe erosion

Table A2. *Cont.*

Element	Failure Mode Identification	Failure Mode Description	Potential Trigger/Cause	Screening Assessment of Failure Mode	Failure Effects
	3	Blockage	Beaver dams	Are there beavers in the area?	Flooding, discharge of process-affected water to the environment, blockage of level 1/level 2 drain outlet, rise in phreatic surface, increase in seepage, pond on reclamation surface, internal erosion, global instability, toe erosion
	4	Blockage	Icing	Is the mine located in an area that could experience icing?	Flooding, discharge of process-affected water to the environment, blockage of level 1/level 2 drain outlet, rise in phreatic surface, increase in seepage, pond on reclamation surface, internal erosion, global instability, toe erosion
	5	Reduction in cross-sectional area	Sloughing/slope failure of walls	What is the slope of the side slopes? What is the strength of material? Have there been failures in this material before?	Reduced capacity, erosion, potential flooding, discharge of process affected water to the environment, blockage of level 1/level 2 drain outlet, rise in phreatic surface, increase in seepage, pond on reclamation surface, internal erosion, global instability, toe erosion
	6	Reduction in cross-sectional area	Excessive vegetation	Will the ditch regularly have water running through it or will it stay dry for a portion of the year? Are there deterrents in place to prevent the growth of vegetation?	Reduced capacity, erosion, potential flooding, discharge of process affected water to the environment, blockage of level 1/level 2 drain outlet, rise in phreatic surface, increase in seepage, pond on reclamation surface, internal erosion, global instability, toe erosion
	7	Change in slope	Erosion	Is there erosion protection in place?	Water discharge velocity change, local instability of dam
	8	Change in slope	Differential settlement	Does the material have the potential to consolidate or settle over time? Is it a cut into natural ground or is the material placed?	Creation of secondary channels, localized areas of erosion, local instability of dam
	9	Sand channel buoyancy	Freezing conditions in channels composed of sand	Are the drainage channels constructed of sand? Could the channel experience freezing?	Flooding, discharge of process affected water to the environment, blockage of level 1/level 2 drain outlet, rise in phreatic surface, increase in seepage, pond on reclamation surface, internal erosion, global instability, toe erosion

Table A2. *Cont.*

Element	Failure Mode Identification	Failure Mode Description	Potential Trigger/Cause	Screening Assessment of Failure Mode	Failure Effects
	10	Breakage of pipe	Break in pipe	Is the pipe capable of breaking?	Release of water into downstream shell, lack of control of phreatic surface (potential rise in phreatic surface), increase in seepage, pond on reclamation surface, internal erosion, global instability, erosion on downstream slope
	11	Breakage of pipe	Buckling	Is the pipe capable of buckling?	Release of water into downstream shell, lack of control of phreatic surface (potential rise in phreatic surface), increase in seepage, pond on reclamation surface, internal erosion, global instability, erosion on downstream slope
	12	Breakage of pipe	Physical degradation	Is the pipe capable of physically degrading over time?	Release of water into downstream shell, lack of control of phreatic surface (potential rise in phreatic surface), increase in seepage, pond on reclamation surface, internal erosion, global instability, erosion on downstream slope
Outtakes-Level 1	13	Clogging	Chemical or biological	Is chemical or biological clogging possible?	Failure to transmit water to perimeter ditch, clogging of level 1 drains, lack of control of phreatic surface (potential rise in phreatic surface), increase in seepage, pond on reclamation surface, internal erosion, global instability, release of water into downstream shell, erosion on downstream slope
	14	Clogging	Particulate clogging (sediment from perimeter ditch or other)	Is sediment capable of clogging the pipe?	Failure to transmit water to perimeter ditch, clogging of level 1 drains, lack of control of phreatic surface (potential rise in phreatic surface), increase in seepage, pond on reclamation surface, internal erosion, global instability, release of water into downstream shell, erosion on downstream slope
	15	Blockage	Perimeter channel (FM 1–6) or other (snow, debris, etc.)	Is the pipe outlet close enough to the base of the perimeter channel that it could become blocked?	Failure to transmit water to perimeter ditch, clogging of level 1 drains, lack of control of phreatic surface (potential rise in phreatic surface), increase in seepage, pond on reclamation surface, internal erosion, global instability

Table A2. *Cont.*

Element	Failure Mode Identification	Failure Mode Description	Potential Trigger/Cause	Screening Assessment of Failure Mode	Failure Effects
Level 1 Drains	16	Breakage of pipe	Break in pipe	Is the pipe capable of breaking?	Lack of control of phreatic surface (potential rise in phreatic surface), increase in seepage, pond on reclamation surface, internal erosion, global instability, release of water into downstream shell, erosion on downstream slope
	17	Breakage of pipe	Buckling	Is the pipe capable of buckling?	Lack of control of phreatic surface (potential rise in phreatic surface), increase in seepage, pond on reclamation surface, internal erosion, global instability, release of water into downstream shell, erosion on downstream slope
	18	Breakage of pipe	Physical degradation	Is the pipe capable of physically degrading over time?	Lack of control of phreatic surface (potential rise in phreatic surface), increase in seepage, pond on reclamation surface, internal erosion, global instability, release of water into downstream shell, erosion on downstream slope
	19	Clogging	Chemical or biological	Is chemical or biological clogging possible?	Lack of control of phreatic surface (potential rise in phreatic surface), increase in seepage, pond on reclamation surface, internal erosion, global instability, release of water into downstream shell, erosion on downstream slope
	20	Clogging	Particulate clogging (sediment from perimeter ditch or other)	Is sediment capable of clogging the pipe?	Lack of control of phreatic surface (potential rise in phreatic surface), increase in seepage, pond on reclamation surface, internal erosion, global instability, release of water into downstream shell, erosion on downstream slope
	21	Clogging	Failure in level 1 outtakes (FM 13–15)	Is it possible for level 1 outtakes to fail?	Lack of control of phreatic surface (potential rise in phreatic surface), increase in seepage, pond on reclamation surface, internal erosion, global instability, release of water into downstream shell, erosion on downstream slope
	22	Clogging surround (woven sock or sand and gravel bed)	Biological, chemical, or particulate clogging	Is it possible for the material surrounding the pipe to become clogged?	Lack of control of phreatic surface (potential rise in phreatic surface), increase in seepage, pond on reclamation surface, internal erosion, global instability, release of water into downstream shell, erosion on downstream slope

Table A2. *Cont.*

Element	Failure Mode Identification	Failure Mode Description	Potential Trigger/Cause	Screening Assessment of Failure Mode	Failure Effects
	23	Breakage of connection between a drain and outlet pipe	Overloading, degradation of connection, poor installation	How are the drain and outtake connected?	Lack of control of phreatic surface (potential rise in phreatic surface), increase in seepage, pond on reclamation surface, internal erosion, global instability, release of water into downstream shell, erosion on downstream slope
Outtakes-Level 2	24	Breakage of pipe	Break in pipe	Is the pipe capable of breaking?	Release of water into downstream shell, lack of control of phreatic surface (potential rise in phreatic surface), increase in seepage, pond on reclamation surface, internal erosion, global instability, erosion on downstream slope
	25	Breakage of pipe	Buckling	Is the pipe capable of buckling?	Release of water into downstream shell, lack of control of phreatic surface (potential rise in phreatic surface), increase in seepage, pond on reclamation surface, internal erosion, global instability, erosion on downstream slope
	26	Breakage of pipe	Physical degradation	Is the pipe capable of physically degrading over time?	Release of water into downstream shell, lack of control of phreatic surface (potential rise in phreatic surface), increase in seepage, pond on reclamation surface, internal erosion, global instability, erosion on downstream slope
	27	Clogging	Chemical or biological	Is chemical or biological clogging possible?	Failure to transmit water to perimeter ditch, clogging of level 2 drains, lack of control of phreatic surface (potential rise in phreatic surface), increase in seepage, pond on reclamation surface, internal erosion, global instability, release of water into downstream shell, erosion on downstream slope
	28	Clogging	Particulate clogging (sediment from perimeter ditch or other)	Is sediment capable of clogging the pipe?	Failure to transmit water to perimeter ditch, clogging of level 2 drains, lack of control of phreatic surface (potential rise in phreatic surface), increase in seepage, pond on reclamation surface, internal erosion, global instability, release of water into downstream shell, erosion on downstream slope

Table A2. *Cont.*

Element	Failure Mode Identification	Failure Mode Description	Potential Trigger/Cause	Screening Assessment of Failure Mode	Failure Effects
	29	Blockage	Perimeter channel (FM 1–6) or other (snow, debris, etc.)	Is the pipe outlet close enough to the base of the perimeter channel that it could become blocked?	Failure to transmit water to perimeter ditch, clogging of level 2 drains, lack of control of phreatic surface (potential rise in phreatic surface), increase in seepage, pond on reclamation surface, internal erosion, global instability
Level 2 Drains	30	Breakage of pipe	Break in pipe	Is the pipe capable of breaking?	Lack of control of phreatic surface (potential rise in phreatic surface), increase in seepage, pond on reclamation surface, internal erosion, global instability, release of water into downstream shell, erosion on downstream slope
	31	Breakage of pipe	Buckling	Is the pipe capable of buckling?	Lack of control of phreatic surface (potential rise in phreatic surface), increase in seepage, pond on reclamation surface, internal erosion, global instability, release of water into downstream shell, erosion on downstream slope
	32	Breakage of pipe	Physical degradation	Is the pipe capable of physically degrading over time?	Lack of control of phreatic surface (potential rise in phreatic surface), increase in seepage, pond on reclamation surface, internal erosion, global instability, release of water into downstream shell, erosion on downstream slope
	33	Clogging	Chemical or biological	Is chemical or biological clogging possible?	Lack of control of phreatic surface (potential rise in phreatic surface), increase in seepage, pond on reclamation surface, internal erosion, global instability, release of water into downstream shell, erosion on downstream slope
	34	Clogging	Particulate clogging (sediment from perimeter ditch or other)	Is sediment capable of clogging the pipe?	Lack of control of phreatic surface (potential rise in phreatic surface), increase in seepage, pond on reclamation surface, internal erosion, global instability, release of water into downstream shell, erosion on downstream slope

Table A2. *Cont.*

Element	Failure Mode Identification	Failure Mode Description	Potential Trigger/Cause	Screening Assessment of Failure Mode	Failure Effects
	35	Clogging	Failure in level 2 outtakes (FM 27–29)	Is it possible for the level 2 outtakes to fail?	Lack of control of phreatic surface (potential rise in phreatic surface), increase in seepage, pond on reclamation surface, internal erosion, global instability, release of water into downstream shell, erosion on downstream slope
	36	Clogging surround (woven sock or sand and gravel bed)	Biological, chemical, or particulate clogging	Is it possible for the material surrounding the pipe to become clogged?	Lack of control of phreatic surface (potential rise in phreatic surface), increase in seepage, pond on reclamation surface, internal erosion, global instability, release of water into downstream shell, erosion on downstream slope
	37	Breakage of connection between a drain and outlet pipe	Overloading, degradation of connection, poor installation	How are the drain and outtake connected?	Lack of control of phreatic surface (potential rise in phreatic surface), increase in seepage, pond on reclamation surface, internal erosion, global instability, release of water into downstream shell, erosion on downstream slope
Foundation	38	Heave (seepage forces create zero effective stress condition)	Embankment loading, excessive rainfall, embankment seepage	What are the current hydraulic gradients and maximum possible due to geometry? What are the materials present? Are there cohesionless soils confined by an overlying lower permeability layer?	Global instability
	39	Vertical deformation from collapse of karst formation	Collapse of karst formation	Is there karst present in the foundation?	Cracking (transverse cracks—perpendicular to dam crest are larger problems than longitudinal cracks) in dam, internal erosion in dam, crest subsidence
	40	Vertical deformation caused by settlement of material	Consolidation	Will the materials in the foundation consolidate over time? How much consolidation has already occurred? Does the material have the potential to collapse?	Cracking (transverse cracks—perpendicular to dam crest are larger problems than longitudinal cracks) in dam, internal erosion in dam, crest subsidence

Table A2. *Cont.*

Element	Failure Mode Identification	Failure Mode Description	Potential Trigger/Cause	Screening Assessment of Failure Mode	Failure Effects
	41	Excessive/uncontrolled seepage through foundation or foundation/dam contact	Excessive rainfall	Is there potential for seepage through the foundation? What is the permeability of the materials?	Erosion of downstream toe, increase in porewater pressure in dam, global instability
	42	Shear failure along pre-existing shear plane from changing shear stress	Loading/unloading of foundation, earthquake, subsurface stress changes (geothermal development, in situ oil or gas production, wastewater injection, etc.)	Are there pre-existing shear planes? Is there the potential for anthropogenic loading or unloading events? Is the material erodible?	Slumping of downstream slope, translational slide, rotational slide, static liquefaction
	43	Shear failure along new shear plane from changing shear stress	Loading/unloading of foundation, earthquake, subsurface stress changes (geothermal development, in situ oil or gas production, wastewater injection, etc.)	Is there the potential for anthropogenic loading or unloading events? Is the material erodible?	Slumping of downstream slope, translational slide, rotational slide, static liquefaction
	44	Shear failure along pre-existing shear plane from changing shear strength	Degradation/weathering, porewater pressure change, progressive failure of strain-softening materials, brittle failure of contractive materials	Are there pre-existing shear planes? Is there the potential for degradation or weathering of the material? Is the material strain-softening or brittle?	Slumping of downstream slope, translational slide, rotational slide, static liquefaction

Table A2. *Cont.*

Element	Failure Mode Identification	Failure Mode Description	Potential Trigger/Cause	Screening Assessment of Failure Mode	Failure Effects
	45	Shear failure along new shear plane from changing shear strength	Degradation/weathering, porewater pressure change, progressive failure of strain-softening materials, brittle failure of contractive materials	Is there the potential for degradation or weathering of the material? Is the material strain-softening or brittle?	Slumping of downstream slope, translational slide, rotational slide, static liquefaction
	46	Internal erosion in foundation or dam/foundation contact from global backward erosion	Failure of soil above or around a backward erosion pipe to hold a roof, heave, high hydraulic gradients, design/construction defect, presence of non-plastic soils in the foundation	Are there non-plastic soils in the foundation?	Static liquefaction, global instability, unravelling/sloughing of downstream face, sub vertical cavities
	47	Internal erosion in foundation or dam/foundation contact from backward erosion piping	Heave, high hydraulic gradients, design/construction defect, presence of non-plastic soils that are capable of holding a roof	Are there non-plastic soils in the foundation and soils capable of 'holding a roof'?	Enlargement of pipe, global instability, static liquefaction
	48	Internal erosion in foundation or dam/foundation contact from contact erosion	Parallel flow in coarser layer to the interface between the coarse-grained and fine-grained soil, high hydraulic gradients, design/construction defect	Is there a contact between a coarse-grained and a fine-grained soil? Is the geometrical and hydraulic condition for contact erosion met?	Global instability, static liquefaction, settlement of the crest, loss of stability or unravelling, eroded material can clog the permeable layer and increase the porewater pressure (could result in hydraulic fracture and uplift of the downstream toe or a rise in the phreatic surface), development of a pipe

Table A2. *Cont.*

Element	Failure Mode Identification	Failure Mode Description	Potential Trigger/Cause	Screening Assessment of Failure Mode	Failure Effects
	49	Internal erosion in foundation or dam/foundation contact from suffusion	High hydraulic gradients, design/construction defect, presence of widely gap-graded or non-plastic gap-graded soils	Is the material widely gap-graded or gap-graded non plastic?	Global instability, seepage on the downstream slope, settlement of the crest, permeability may increase as erosion progresses or decrease if clogging occurs
	50	Internal erosion in foundation or dam/foundation contact from concentrated leak	Fracture in foundation soil, hydraulic fracture, high hydraulic gradient, cracks at dam/foundation contact from vertical deformation in foundation or poor construction practices or differential settlement, design/construction defects	Is there a crack or gap that could allow for a concentrated leak to develop?	Global instability, development of a pipe
	51	Thawing of foundation permafrost	Climate change	Is there permafrost in the foundation?	Cracking (transverse cracks—perpendicular to dam crest are larger problems than longitudinal cracks) in dam, piping in dam, crest subsidence
Starter Dyke	52	Vertical deformation (differential or otherwise) from consolidation	Consolidation/settlement	Does the material have the potential to consolidate? How much consolidation has occurred already? How much is expected to occur?	Release of pore water and loss of height (potential for pond to develop on reclamation surface), development of cracks above starter dyke, internal erosion, overtopping
	53	Seismic liquefaction	Earthquakes, induced seismicity, construction traffic, blasting	Seismic events in area? Induced seismicity? Density of material (contractive or dilative)? Saturated or unsaturated? Hydraulically placed or compacted?	Global instability

Table A2. *Cont.*

Element	Failure Mode Identification	Failure Mode Description	Potential Trigger/Cause	Screening Assessment of Failure Mode	Failure Effects
	54	Static liquefaction from changing shear stress	Loading/unloading; overloading, including increasing the load, construction activities at the crest, fill placement at toe; over steepening of downstream slope or toe, including erosion or excavation (slumping of downstream slope from shear failure); foundation shear (FM 42–45), surface erosion of cell sand or berm on downstream slope (FM 74–75, 87–89); excessive and uncontrolled seepage through foundation resulting in erosion of toe (FM 41); slumping of downstream slope from shear failure (FM 42–45, 58, 59, 69, 70, 82, 83, 96, 97, 107, 108)	Density of material (contractive or dilative)? Hydraulically placed or compacted? Saturated or unsaturated? Is there a likelihood for anthropogenic contributions in the future (i.e., unexpected construction)? Is the site remote? What is the material of the downstream slope? Is there a nearby river?	Global instability
	55	Static liquefaction from changing mean effective stress	Change in pore pressures caused by a phreatic surface change (failure of level 1 and level 2 drains (FM 16–23 and 30–37))	Density of material (contractive or dilative)? Saturated or unsaturated? Hydraulically placed or compacted? Does control of the phreatic surface rely on drain function? Could drains become clogged or fail in the future?	Global instability

Table A2. *Cont.*

Element	Failure Mode Identification	Failure Mode Description	Potential Trigger/Cause	Screening Assessment of Failure Mode	Failure Effects
	56	Static liquefaction from long-term change in material properties resulting in changing shear strength	Changing shear strength caused by degradation/weathering, progressive failure, porewater pressure change, failure of level 1 and level 2 drains (FM 16–23 and 30–37)	Density of material (contractive or dilative)? Hydraulically placed or compacted? Saturated or unsaturated? To what extent could weathering or degradation of the material occur? Will it result in an increase or decrease in strength? Will a change in the phreatic surface impact the strength of the material?	Global instability
	57	Static liquefaction from changing shear stress and mean effective stress	Lateral extrusion	Density of material (contractive or dilative)? Hydraulically placed or compacted? Saturated or unsaturated? Weak layers interbedded in tailings? Could the tailings 'squish' out like toothpaste during loading?	Global instability
	58	Shear failure from changing shear stress	Loading/unloading crest, toe, upstream, downstream; surface erosion of cell sand or berm on downstream slope (FM 74–75, 87–89); excessive and uncontrolled seepage through foundation resulting in erosion of toe (FM 41)	Is there potential for anthropogenic contributions (i.e., excavations or construction)? Erodibility of material?	Slumping of downstream slope, translational slide, rotational slide

Table A2. *Cont.*

Element	Failure Mode Identification	Failure Mode Description	Potential Trigger/Cause	Screening Assessment of Failure Mode	Failure Effects
	59	Shear failure from changing shear strength	Degradation/weathering, porewater pressure change, change in permeability over time, failure of drains, progressive failure of strain-softening materials, brittle failure of contractive materials, failure of level 1 and level 2 drains (FM 16–23 and 30–37)	Is there potential for weathering or degradation of materials? Does the porewater pressure rely on drain performance? Could drains fail over time? Are the materials strain-softening or brittle?	Slumping of downstream slope, translational slide, rotational slide
	60	Internal erosion in dam from contact erosion	Parallel flow in coarser layer to the interface between the coarse-grained and fine-grained soil, high hydraulic gradients, design/construction defects	Is there a contact between a coarse and fine-grained soil? Is there a filter in place?	Global instability, static liquefaction, settlement of the crest, loss of stability or unravelling, eroded material can clog the permeable layer and increase the porewater pressure (could result in hydraulic fracture and uplift of the downstream toe or a rise in the phreatic surface), development of a pipe
	61	Internal erosion in dam from suffusion	High hydraulic gradients, design/construction defect, presence of widely gap-graded or non-plastic gap-graded soils	Is the material widely gap-graded or gap-graded non plastic?	Global instability, seepage on the downstream slope, settlement of the crest, permeability may increase as erosion progresses or decrease if clogging occurs
	62	Internal erosion in dam—concentrated leak	Cracks from foundation (FM 39–40, 51), tunnels created by burrowing animals, hydraulic fracture, high hydraulic gradient, design/construction defects	Is there a crack or gap that could allow for a concentrated leak to develop?	Global instability, development of a pipe

Table A2. *Cont.*

Element	Failure Mode Identification	Failure Mode Description	Potential Trigger/Cause	Screening Assessment of Failure Mode	Failure Effects
Cell Sand	63	Vertical deformation (differential or otherwise) from consolidation	Consolidation/settlement	Does the material have the potential to consolidate? How much consolidation has occurred already? How much is expected to occur?	Release of porewater and loss of height (potential for pond to develop on reclamation surface), development of cracks, internal erosion
	64	Seismic liquefaction	Earthquakes, induced seismicity, construction traffic, blasting	Seismic events in area? Induced seismicity? Density of material (contractive or dilative)? Saturated or unsaturated? Hydraulically placed or compacted?	Global instability
	65	Static liquefaction from changing shear stress	Loading/unloading; overloading, including increasing the load, construction activities at the crest, fill placement at toe; over steepening of downstream slope or toe, including erosion or excavation (slumping of downstream slope from shear failure); foundation shear (FM 42–45); shear in the starter dyke (FM 58–59); surface erosion of cell sand or berm on downstream slope (FM 74–75. 87–89); excessive and uncontrolled seepage through foundation resulting in erosion of toe (FM 41); slumping of downstream slope from shear failure (FM 42–45, 58, 59, 69, 70, 82, 83, 96, 97, 107, 108)	Density of material (contractive or dilative)? Hydraulically placed or compacted? Saturated or unsaturated? Is there a likelihood for anthropogenic contributions in the future (i.e., unexpected construction)? Is the site remote? What is the material of the downstream slope? Is there a nearby river?	Global instability

Table A2. *Cont.*

Element	Failure Mode Identification	Failure Mode Description	Potential Trigger/Cause	Screening Assessment of Failure Mode	Failure Effects
	66	Static liquefaction from changing mean effective stress	Change in pore pressures caused phreatic surface change (i.e., failure of level 1 and level 2 drains (FM 16–23 and 30–37))	Density of material (contractive or dilative)? Saturated or unsaturated? Hydraulically placed or compacted? Does control of the phreatic surface rely on drain function? Could drains become clogged or fail in the future?	Global instability
	67	Static liquefaction from long-term change in material properties resulting in changing shear strength	Change in shear strength caused by degradation/weathering, progressive failure, porewater pressure change, failure of level 1 and level 2 drains (FM 16–23 and 30–37)	Density of material (contractive or dilative)? Hydraulically placed or compacted? Saturated or unsaturated? To what extent could weathering or degradation of the material occur? Will it result in an increase or decrease in strength? Will a change in the phreatic surface impact the strength of the material?	Global instability
	68	Static liquefaction from changing shear stress and mean effective stress	Lateral extrusion	Density of material (contractive or dilative)? Hydraulically placed or compacted? Saturated or unsaturated? Weak layers interbedded in tailings? Could the tailings 'squish' out like toothpaste during loading?	Global instability
	69	Shear failure from changing shear stress	Loading/unloading crest, toe, upstream, downstream; surface erosion of berm on downstream slope (FM 87–89); excessive and uncontrolled seepage through foundation resulting in erosion of toe (FM 41)	Is there potential for anthropogenic contributions (i.e., excavations or construction)? Erodibility of material?	Slumping of downstream slope, translational slide, rotational slide

Table A2. *Cont.*

Element	Failure Mode Identification	Failure Mode Description	Potential Trigger/Cause	Screening Assessment of Failure Mode	Failure Effects
	70	Shear failure from changing shear strength	Degradation/weathering, porewater pressure change, change in permeability over time, failure of drains, progressive failure of strain-softening materials, brittle failure of contractive materials, failure of level 1 and level 2 drains (FM 16–23 and 30–37)	Is there potential for weathering or degradation of materials? Does the porewater pressure rely on drain performance? Could drains fail over time? Are the materials strain-softening or brittle?	Slumping of downstream slope, translational slide, rotational slide
	71	Internal erosion in dam from contact erosion	Parallel flow in coarser layer to the interface between the coarse-grained and fine-grained soil, high hydraulic gradients, design/construction defects	Is there a contact between a coarse and fine-grained soil? Is there a filter in place?	Global instability, static liquefaction, settlement of the crest, loss of stability or unravelling, eroded material can clog the permeable layer and increase the porewater pressure (could result in hydraulic fracture and uplift of the downstream toe or a rise in the phreatic surface), development of a pipe
	72	Internal erosion in dam from suffusion	High hydraulic gradients, design/construction defect, presence of widely gap-graded or non-plastic gap-graded soils	Is the material widely gap-graded or gap-graded non plastic?	Global instability, seepage on the downstream slope, settlement of the crest, permeability may increase as erosion progresses or decrease if clogging occurs

Table A2. *Cont.*

Element	Failure Mode Identification	Failure Mode Description	Potential Trigger/Cause	Screening Assessment of Failure Mode	Failure Effects
	73	Internal erosion in dam from concentrated leak	Cracks from vertical deformation in foundation (FM 39–40, 51), cracks from vertical deformation in starter dyke (FM 52), cracks from vertical deformation in BAW (FM 90), cracks from vertical deformation in BBW (FM 101), tunnels created by burrowing animals, hydraulic fracture, high hydraulic gradient, design/construction defects	Is there a crack or gap that could allow for a concentrated leak to develop?	Global instability, development of a pipe
	74	Surface erosion from wind and overland flow resulting in rills, gullies, or sheet erosion	Destruction of vegetation (FM 112), rainfall, melting of snow, wind, increased seepage on downstream slope from failure of drainage system, increased seepage from internal erosion in embankment	Is the material susceptible to erosion? Is there a vegetative cover or erosion protection?	Slope failures (shallow surficial movement, slumps), change in downstream slope angle, blockage in perimeter channel with sediment, development of negative drainage, development of large erosion scarps

Table A2. *Cont.*

Element	Failure Mode Identification	Failure Mode Description	Potential Trigger/Cause	Screening Assessment of Failure Mode	Failure Effects
	75	Surface erosion—spring sapping (headward erosion of gullies due to concentration of seepage forces at the locus of the gully which accentuates erosion)	Destruction of vegetation (FM 112), increased seepage on downstream slope from failure of level 1 or 2 drains (FM 10–37), increased seepage from internal erosion of starter dyke, cell sand, BAW, BBW (FM 60–61, 71–72, 98–99, 109–110)	Is the material susceptible to erosion? Is there a vegetative cover or erosion protection? Will seepage daylight on the downstream slope?	Slope failures (shallow surficial movement, slumps), change in downstream slope angle, blockage in perimeter channel with sediment, development of negative drainage, development of large erosion scarps
Berm	76	Vertical deformation (differential or otherwise) from consolidation	Consolidation/settlement	Does the material have the potential to consolidate? How much consolidation has occurred already? How much is expected to occur?	Release of pore water and loss of height (potential for pond to develop on reclamation surface), development of cracks, internal erosion
	77	Seismic liquefaction	Earthquakes, induced seismicity, construction traffic, blasting	Seismic events in area? Induced seismicity? Density of material (contractive or dilative)? Saturated or unsaturated? Hydraulically placed or compacted?	Global instability

Table A2. *Cont.*

Element	Failure Mode Identification	Failure Mode Description	Potential Trigger/Cause	Screening Assessment of Failure Mode	Failure Effects
	78	Static liquefaction from changing shear stress	Loading/unloading; overloading, including increasing the load, construction activities at the crest, fill placement at toe; over steepening of downstream slope or toe, including erosion or excavation (slumping of downstream slope from shear failure); foundation shear (FM 42–45); shear in the starter dyke (FM 58–59); shear in cell sand (FM 69–70); surface erosion of cell sand or berm on downstream slope (FM 74–75, 87–89); excessive and uncontrolled seepage through foundation resulting in erosion of toe (FM 41); slumping of downstream slope from shear failure (FM 42–45, 58, 59, 69, 70, 82, 83, 96, 97, 107, 108)	Density of material (contractive or dilative)? Hydraulically placed or compacted? Saturated or unsaturated? Is there a likelihood for anthropogenic contributions in the future (i.e., unexpected construction)? Is the site remote? What is the material of the downstream slope? Is there a nearby river?	Global instability
	79	Static liquefaction from changing mean effective stress	Change in pore pressures caused by a phreatic surface change (failure of level 1 and level 2 drains (FM 16–23 and 30–37))	Density of material (contractive or dilative)? Saturated or unsaturated? Hydraulically placed or compacted? Does control of the phreatic surface rely on drain function? Could drains become clogged or fail in the future?	Global instability

Table A2. *Cont.*

Element	Failure Mode Identification	Failure Mode Description	Potential Trigger/Cause	Screening Assessment of Failure Mode	Failure Effects
	80	Static liquefaction from long-term change in material properties resulting in changing shear strength	Changing shear strength caused by degradation/weathering, progressive failure, porewater pressure change, failure of level 1 and level 2 drains (FM 16–23 and 30–37)	Density of material (contractive or dilative)? Hydraulically placed or compacted? Saturated or unsaturated? To what extent could weathering or degradation of the material occur? Will it result in an increase or decrease in strength? Will a change in the phreatic surface impact the strength of the material?	Global instability
	81	Static liquefaction from changing shear stress and mean effective stress	Lateral extrusion	Density of material (contractive or dilative)? Hydraulically placed or compacted? Saturated or unsaturated? Weak layers interbedded in tailings? Could the tailings 'squish' out like toothpaste during loading?	Global instability
	82	Shear failure from changing shear stress	Loading/unloading crest, toe, upstream, downstream; surface erosion of cell sand or berm on downstream slope (FM 74–75, 87–89); excessive and uncontrolled seepage through foundation resulting in erosion of toe (FM 41)	Is there potential for anthropogenic contributions (i.e., excavations or construction)? Erodibility of material?	Slumping of downstream slope, translational slide, rotational slide

Table A2. *Cont.*

Element	Failure Mode Identification	Failure Mode Description	Potential Trigger/Cause	Screening Assessment of Failure Mode	Failure Effects
	83	Shear failure from changing shear strength	Degradation/weathering; porewater pressure change, change in permeability over time, failure of drains, progressive failure of strain-softening materials, brittle failure of contractive materials, failure of level 1 and level 2 drains (FM 16–23 and 30–37)	Is there potential for weathering or degradation of materials? Does the porewater pressure rely on drain performance? Could drains fail over time? Are the materials strain-softening or brittle?	Slumping of downstream slope, translational slide, rotational slide
	84	Internal erosion in dam from contact erosion	Parallel flow in coarser layer to the interface between the coarse-grained and fine-grained soil, high hydraulic gradients, design/construction defects	Is there a contact between a coarse and fine-grained soil? Is there a filter in place?	Global instability, static liquefaction, settlement of the crest, loss of stability or unravelling, eroded material can clog the permeable layer and increase the porewater pressure (could result in hydraulic fracture and uplift of the downstream toe or a rise in the phreatic surface), development of a pipe
	85	Internal erosion in dam from suffusion	High hydraulic gradients, design/construction defect, presence of widely gap-graded or non-plastic gap-graded soils	Is the material widely gap-graded or gap-graded non plastic?	Global instability, seepage on the downstream slope, settlement of the crest, permeability may increase as erosion progresses or decrease if clogging occurs

Table A2. *Cont.*

Element	Failure Mode Identification	Failure Mode Description	Potential Trigger/Cause	Screening Assessment of Failure Mode	Failure Effects
	86	Internal erosion in dam from concentrated leak	Cracks from vertical deformation in foundation (FM 39–40, 51), cracks from vertical deformation in starter dyke (FM 52), cracks from vertical deformation in cell sand (FM 63)), tunnels created by burrowing animals, hydraulic fracture, high hydraulic gradient, design/construction defects	Is there a crack or gap that could allow for a concentrated leak to develop?	Global instability, development of a pipe
	87	Surface erosion from wind and overland flow resulting in rills, gullies, or sheet erosion	Destruction of vegetation (FM 112), rainfall, melting of snow, wind, increased seepage on downstream slope from failure of drainage system, increased seepage from internal erosion in embankment	Is the material susceptible to erosion? Is there a vegetative cover or erosion protection?	Slope failures (shallow surficial movement, slumps), change in downstream slope angle, blockage in perimeter channel with sediment, development of negative drainage, development of large erosion scarps
	88	Surface erosion from spring sapping (headward erosion of gullies due to concentration of seepage forces at the locus of the gully which accentuates erosion)	Destruction of vegetation (FM 112); increased seepage on downstream slope from failure of level 1 or 2 drain system (FM 10–37); excessive erosion from internal erosion of berm, starter dyke, cell sand, BAW, BBW (FM 60–61, 71–72, 98–99, 109–110)	Is the material susceptible to erosion? Is there a vegetative cover or erosion protection? Will seepage daylight on the downstream slope?	Slope failures (shallow surficial movement, slumps), change in downstream slope angle, blockage in perimeter channel with sediment, development of negative drainage, development of large erosion scarps

Table A2. *Cont.*

Element	Failure Mode Identification	Failure Mode Description	Potential Trigger/Cause	Screening Assessment of Failure Mode	Failure Effects
	89	Toe erosion	Flow action from perimeter ditch or nearby river, release of a dam from a beaver, flood event, river changing course over time, destruction of vegetation (FM 110), excessive seepage and toe erosion (FM 41), excessive erosion from internal erosion (FM 84–85)	Is the material susceptible to erosion? Is there a vegetative cover or erosion protection? Is there a nearby perimeter ditch or river? Is there known animal activity in the area?	Slope failures (shallow surficial movement, slumps), change in downstream slope angle, blockage in perimeter channel with sediment, development of negative drainage, development of large erosion scarps, beaver bafflers
Beach above water	90	Vertical deformation (differential or otherwise) from consolidation	Consolidation/settlement	Does the material have the potential to consolidate? How much consolidation has occurred already? How much is expected to occur?	Release of pore water and loss of height (potential for pond to develop on reclamation surface), development of cracks, internal erosion
	91	Seismic liquefaction	Earthquakes, induced seismicity, construction traffic, blasting	Seismic events in area? Induced seismicity? Density of material (contractive or dilative)? Saturated or unsaturated? Hydraulically placed or compacted?	Global instability

Table A2. *Cont.*

Element	Failure Mode Identification	Failure Mode Description	Potential Trigger/Cause	Screening Assessment of Failure Mode	Failure Effects
	92	Static liquefaction from changing shear stress	Loading/unloading; overloading, including increasing the load, construction activities at the crest, fill placement at toe; over steepening of downstream slope or toe, including erosion or excavation (slumping of downstream slope from shear failure); foundation shear (FM 42–45); shear in the BBW (FM 107–108); surface erosion of cell sand or berm on downstream slope (FM 74–75, 87–89); excessive and uncontrolled seepage through foundation resulting in erosion of toe (FM 41); slumping of downstream slope from shear failure (FM 42–45, 58, 59, 69, 70, 82, 83, 96, 97, 107, 108)	Density of material (contractive or dilative)? Hydraulically placed or compacted? Saturated or unsaturated? Is there a likelihood for anthropogenic contributions in the future (i.e., unexpected construction)? Is the site remote? What is the material of the downstream slope? Is there a nearby river?	Global instability
	93	Static liquefaction from changing mean effective stress	Change in pore pressures caused by a phreatic surface change (failure of level 1 and level 2 drains (FM 16–23 and 30–37))	Density of material (contractive or dilative)? Saturated or unsaturated? Hydraulically placed or compacted? Does control of the phreatic surface rely on drain function? Could drains become clogged or fail in the future?	Global instability

Table A2. *Cont.*

Element	Failure Mode Identification	Failure Mode Description	Potential Trigger/Cause	Screening Assessment of Failure Mode	Failure Effects
	94	Static liquefaction from long-term change in material properties resulting in changing shear strength	Changing shear strength caused by degradation/weathering, progressive failure, porewater pressure change, failure of level 1 and level 2 drains (FM 16–23 and 30–37)	Density of material (contractive or dilative)? Hydraulically placed or compacted? Saturated or unsaturated? To what extent could weathering or degradation of the material occur? Will it result in an increase or decrease in strength? Will a change in the phreatic surface impact the strength of the material?	Global instability
	95	Static liquefaction from changing shear stress and mean effective stress	Lateral extrusion	Density of material (contractive or dilative)? Hydraulically placed or compacted? Saturated or unsaturated? Weak layers interbedded in tailings? Could the tailings 'squish' out like toothpaste during loading?	Global instability
	96	Shear failure from changing shear stress	Loading/unloading crest, toe, upstream, downstream; surface erosion of cell sand on downstream slope (FM 74–75), surface erosion of berm on downstream slope (FM 87–89); excessive and uncontrolled seepage through foundation resulting in erosion of toe (FM 41)	Is there potential for anthropogenic contributions (i.e., excavations or construction)? Erodibility of material?	Slumping of downstream slope, translational slide, rotational slide

Table A2. *Cont.*

Element	Failure Mode Identification	Failure Mode Description	Potential Trigger/Cause	Screening Assessment of Failure Mode	Failure Effects
	97	Shear failure from changing shear strength	Degradation/weathering, porewater pressure change, change in permeability over time, failure of drains, progressive failure of strain-softening materials, brittle failure of contractive materials, failure of level 1 and level 2 drains (FM 16–23 and 30–37)	Is there potential for weathering or degradation of materials? Does the porewater pressure rely on drain performance? Could drains fail over time? Are the materials strain-softening or brittle?	Slumping of downstream slope, translational slide, rotational slide
	98	Internal erosion in dam from contact erosion	Parallel flow in coarser layer to the interface between the coarse-grained and fine-grained soil, high hydraulic gradients, design/construction defects	Is there a contact between a coarse and fine-grained soil? Is there a filter in place?	Global instability, static liquefaction, settlement of the crest, loss of stability or unravelling, eroded material can clog the permeable layer and increase the porewater pressure (could result in hydraulic fracture and uplift of the downstream toe or a rise in the phreatic surface), development of a pipe
	99	Internal erosion in dam from suffusion	High hydraulic gradients, design/construction defect, presence of widely gap-graded or non-plastic gap-graded soils	Is the material widely gap-graded or gap-graded non plastic?	Global instability, seepage on the downstream slope, settlement of the crest, permeability may increase as erosion progresses or decrease if clogging occurs

Table A2. *Cont.*

Element	Failure Mode Identification	Failure Mode Description	Potential Trigger/Cause	Screening Assessment of Failure Mode	Failure Effects
	100	Internal erosion in dam—concentrated leak	Cracks from vertical deformation in BBW (FM 101), cracks from vertical cracks in foundation (FM 39–40, 50), tunnels created by burrowing animals, hydraulic fracture, high hydraulic gradient, design/construction defects	Is there a crack or gap that could allow for a concentrated leak to develop?	Global instability, development of a pipe
	101	Vertical deformation (differential or otherwise) from consolidation	Consolidation/settlement	Does the material have the potential to consolidate? How much consolidation has occurred already? How much is expected to occur?	Release of pore water and loss of height (potential for pond to develop on reclamation surface), development of cracks, internal erosion
Beach below water	102	Seismic liquefaction	Earthquakes, induced seismicity, construction traffic, blasting	Seismic events in area? Induced seismicity? Density of material (contractive or dilative)? Saturated or unsaturated? Hydraulically placed or compacted?	Global instability

Table A2. *Cont.*

Element	Failure Mode Identification	Failure Mode Description	Potential Trigger/Cause	Screening Assessment of Failure Mode	Failure Effects
	103	Static liquefaction from changing shear stress	Loading/unloading; overloading, including increasing the load, construction activities at the crest, fill placement at toe; over-steepening of downstream slope or toe, including erosion or excavation (slumping of downstream slope from shear failure); foundation shear (FM 42–45); surface erosion of cell sand or berm on downstream slope (FM 74–75, 87–89); excessive and uncontrolled seepage through foundation resulting in erosion of toe (FM 41); slumping of downstream slope from shear failure (FM 42–45, 58, 59, 69, 70, 82, 83, 96, 97, 107, 108)	Density of material (contractive or dilative)? Hydraulically placed or compacted? Saturated or unsaturated? Is there a likelihood for anthropogenic contributions in the future (i.e., unexpected construction)? Is the site remote? What is the material of the downstream slope? Is there a nearby river?	Global instability

Table A2. *Cont.*

Element	Failure Mode Identification	Failure Mode Description	Potential Trigger/Cause	Screening Assessment of Failure Mode	Failure Effects
	104	Static liquefaction from changing mean effective stress	Change in pore pressures caused by a phreatic surface change (failure of level 1 and level 2 drains (FM 16–23 and 30–37))	Density of material (contractive or dilative)? Saturated or unsaturated? Hydraulically placed or compacted? Does control of the phreatic surface rely on drain function? Could drains become clogged or fail in the future?	Global instability
	105	Static liquefaction from long-term change in material properties resulting in changing shear strength	Changing shear strength caused by degradation/weathering, progressive failure, porewater pressure change, failure of level 1 and level 2 drains (FM 16–23 and 30–37)	Density of material (contractive or dilative)? Hydraulically placed or compacted? Saturated or unsaturated? To what extent could weathering or degradation of the material occur? Will it result in an increase or decrease in strength? Will a change in the phreatic surface impact the strength of the material?	Global instability
	106	Static liquefaction from changing shear stress and mean effective stress	Lateral extrusion	Density of material (contractive or dilative)? Hydraulically placed or compacted? Saturated or unsaturated? Weak layers interbedded in tailings? Could the tailings 'squish' out like toothpaste during loading?	Global instability

Table A2. *Cont.*

Element	Failure Mode Identification	Failure Mode Description	Potential Trigger/Cause	Screening Assessment of Failure Mode	Failure Effects
	107	Shear failure from changing shear stress	Loading/unloading crest, toe, upstream, downstream; surface erosion of cell sand on downstream slope (FM 74–75); surface erosion of berm on downstream slope (FM 87–89); excessive and uncontrolled seepage through foundation resulting in erosion of toe (FM 41)	Is there potential for anthropogenic contributions (i.e., excavations or construction)? Erodibility of material?	Slumping of downstream slope, translational slide, rotational slide
	108	Shear failure from changing shear strength	Degradation/weathering, porewater pressure change, change in permeability over time, failure of drains, progressive failure of strain-softening materials, brittle failure of contractive materials, failure of level 1 and level 2 drains (FM 16–23 and 30–37)	Is there potential for weathering or degradation of materials? Does the porewater pressure rely on drain performance? Could drains fail over time? Are the materials strain-softening or brittle?	Slumping of downstream slope, translational slide, rotational slide
	109	Internal erosion in dam from contact erosion	Parallel flow in coarser layer to the interface between the coarse-grained and fine-grained soil, high hydraulic gradients, design/construction defects	Is there a contact between a coarse and fine-grained soil? Is there a filter in place?	Global instability, static liquefaction, settlement of the crest, loss of stability or unravelling, eroded material can clog the permeable layer and increase the porewater pressure (could result in hydraulic fracture and uplift of the downstream toe or a rise in the phreatic surface), development of a pipe

Table A2. *Cont.*

Element	Failure Mode Identification	Failure Mode Description	Potential Trigger/Cause	Screening Assessment of Failure Mode	Failure Effects
	110	Internal erosion in dam from suffusion	High hydraulic gradients, design/construction defect, presence of widely gap-graded or non-plastic gap-graded soils	Is the material widely gap-graded or gap-graded non plastic?	Global instability, seepage on the downstream slope, settlement of the crest, permeability may increase as erosion progresses or decrease if clogging occurs
	111	Internal erosion in dam—concentrated leak	Cracks from vertical deformation in foundation (FM 39–40, 51), tunnels created by burrowing animals, hydraulic fracture, high hydraulic gradient, design/construction defects	Is there a crack or gap that could allow for a concentrated leak to develop?	Global instability, development of a pipe
Vegetative cover	112	Destruction of vegetation	Suffocation by eroded material, forest fires, pests and disease, climate change, large storm event, anthropogenic contributions, surface erosion in cell sand or berm (FM 74–75, FM 87–89), evolution of vegetation over time due to climate change	Does the resistance to erosion rely on the vegetation? Is the area susceptible to forest fires? Are there pests/disease that could lead to vegetation destruction? What is the downstream slope? Is the area remote? Are there surrounding communities that could lead to destruction of vegetation (i.e., recreational vehicles)?	Increase in surface erosion, instability of downstream slope, global instability, change in overall evapotranspiration and water balance impacts (infiltration versus runoff)
Cap (overburden and coarse sand tailings)	113	Excessive settlement	Consolidation	Does the material have the potential to settle over time?	Formation of ponds on reclamation surface, overtopping, piping (increase in seepage forces and gradients)
	114	Differential settlement	Consolidation, poor construction practices	Does the material have the potential to settle over time? Are there areas that have the potential to settle more than others?	Formation of ponds on reclamation surface, failure to direct surface water runoff towards drainage channels, development of cracks, formation of preferential flow paths, localized depressions, increased infiltration, increased phreatic surface

Table A2. *Cont.*

Element	Failure Mode Identification	Failure Mode Description	Potential Trigger/Cause	Screening Assessment of Failure Mode	Failure Effects
Infilled coarse sand tailings mixed with FFT	115	Excessive settlement	Consolidation	Does the material have the potential to settle over time?	Formation of ponds on reclamation surface, overtopping, piping (increase in seepage forces and gradients)
	116	Differential settlement	Consolidation, poor construction practices	Are there areas that have the potential to settle more than others?	Formation of ponds on reclamation surface, overtopping, piping (increase in seepage forces and gradients), failure of drainage channels to behave as intended, localized depressions
Hummocks	117	Shear failure from changing shear stress	Loading/unloading crest, toe, surface erosion on slopes, failure of underlying material to support hummock	Is there potential for anthropogenic contributions (i.e., excavations or construction)? Erodibility of material?	Slumping, translational slide, rotational slide, blockage of drainage channels
	118	Shear failure from changing shear strength	Degradation/weathering, porewater pressure change, change in permeability over time, failure of drains, progressive failure of strain-softening materials, brittle failure of contractive materials	Is there potential for weathering or degradation of materials? Are the materials strain-softening or brittle?	Slumping of slopes, translational slide, rotational slide, blockage of drainage channels
	119	Surface erosion from wind and overland flow resulting in rills, gullies, or sheet erosion	Destruction of vegetation, rainfall, melting of snow, wind	Is the material susceptible to erosion? Is there a vegetative cover or erosion protection?	Slope failures (shallow surficial movement, slumps), change in downstream slope angle, blockage in drainage channel with sediment, development of negative drainage, development of large erosion scarps

Table A2. *Cont.*

Element	Failure Mode Identification	Failure Mode Description	Potential Trigger/Cause	Screening Assessment of Failure Mode	Failure Effects
Drainage channels	120	Washout of erosion protection (riprap)	Precipitation event larger than design events (including extreme or repeat events)	What precipitation events are the channels designed for? What is the chance of exceedance over 1000 years? How susceptible are the underlying materials to erosion?	Excessive erosion (erosion gullies, etc.), change in slope of drainage channels, erosion and release of materials underlying drainage channels
	121	Blockage (complete or partial)	Debris, beaver dam, icing, sedimentation, slumping from slope failure, ingress of vegetation, slope failure/excessive erosion from nearby hummock	Are there beavers in the area? Is there a chance for a slope failure? Could debris be carried downstream and deposited in the channels, resulting in a complete or partial blockage? Is the mine located in an area that could experience icing?	Formation of a pond upstream of the drainage channel, blockage breakthrough resulting in flooding, overtopping from pond formation, reversion back to a pond, piping through dam (increase in seepage forces and gradient)
	122	Sand channel buoyancy	Freezing conditions in channels composed of sand	Are the drainage channels constructed of sand? Could the channel experience freezing?	Flooding
	123	Erosion control failure	Improper design/construction, differential settlement	Do the drainage channels rely on erosion control for stability? Is there the chance for differential settlement of the channel?	Excessive erosion (erosion gullies, etc.), change in slope of drainage channel, erosion and release of materials underlying drainage channel, formation of secondary channel
Outlet	124	Washout of erosion protection (riprap)	Precipitation event larger than design event (including extreme or repeat events), flood following sand channel buoyancy event in drainage channel	What precipitation event is the outlet designed for? What is the chance of exceedance over 1000 years? How susceptible are the underlying materials to erosion?	Excessive erosion (erosion gullies, etc.), change in slope of outlet, erosion and release of materials underlying outlet

Table A2. *Cont.*

Element	Failure Mode Identification	Failure Mode Description	Potential Trigger/Cause	Screening Assessment of Failure Mode	Failure Effects
	125	Blockage (complete or partial)	Debris, beaver dam, icing, sedimentation, slumping from slope failure, ingress of vegetation, increase in depositional material due to failure of erosion protection in drainage channels upstream	Are there beavers in the area? Is there a chance for a slope failure? Could debris be carried downstream and deposited in the outlet, resulting in a complete or partial blockage? Is the mine located in an area that could experience icing?	Formation of a pond upstream of the outlet, blockage breakthrough resulting in flooding, overtopping from pond formation, reversion back to a pond, piping through dam (increase in seepage forces and gradient)
	126	Sand channel buoyancy	Freezing conditions in channels composed of sand	Is the outlet constructed of sand? Could the channel experience freezing?	Flooding
	127	Erosion control failure	Improper design/construction, differential settlement	Does the outlet rely on erosion control for stability? Is there the chance for differential settlement of the channel?	Excessive erosion (erosion gullies, etc.), change in slope of outlet, erosion and release of materials underlying outlet, formation of secondary channel
Vegetative cover	128	Destruction of vegetation	Suffocation by eroded material, forest fires, pests and disease, climate change, large storm event, anthropogenic contributions	Does the resistance to erosion rely on the vegetation? Is the area susceptible to forest fires? Are there pests/disease that could lead to vegetation destruction? Is the area remote? Are there surrounding communities that could lead to destruction of vegetation (i.e., recreational vehicles)?	Increase in surface erosion, deposition of material in drainage channels that could lead to a blockage via sedimentation, development of negative drainage on reclamation surface, ponding of water near dam crest, internal erosion (increase in seepage forces and hydraulic gradient)

References

1. AER. *Manual 019 Decommissioning, Closure, and Abandonment of Dams at Energy Projects*; Alberta Energy Regulator (AER): Calgary, AB, Canada, 2020.
2. Schafer, H.L.; Beier, N.A.; Macciotta, R. A Failure Modes and Effects Analysis Framework for Assessing Geotechnical Risks of Tailings Dam Closure. *Minerals* **2021**, *11*, 1234. [CrossRef]
3. OSTDC De-Licensing of Oil Sands Tailings Dams. Technical Guidance Document. Oil Sands Tailings Dam Committee (OSTDC). Available online: https://era.library.ualberta.ca/items/8131d11c-4402-4bed-b0ab-0d3833b95af6/view/7165e9fb-e003-4%0Ace6-8a92-efaf8355ec2b/2014-2003-2028-20--20Report-20--20Oil-20Sands-20Dams-20De-Licensing-20Technical-20Guide.pdf (accessed on 30 November 2021).
4. Oberle, B.; Brereton, D.; Mihaylova, A. *Towards Zero Harm: A Compendium of Papers Prepared for the Global Tailings Review*; Global Tailings Review: St. Fallen, Switzerland, 2020.
5. Vick, S.G. *Planning, Design, and Analysis of Tailings Dams*; BiTech: Vancouver, BC, Canada, 1990.
6. Lighthall, P.C.; Watts, B.D.; Rice, S. Deposition Methods for Construction of Hydraulic Fill Tailings Dams. In Proceedings of the 4th Vancouver Geotechnical Society Symposium, Vancouver, BC, Canada, 26 May 1989.
7. Martens, S.; Lappin, T. Compaction of Upstream Construction Tailings Dam Beaches Using Dozers. In Proceedings of the 61st Canadian Geotechnical Conference, Edmonton, AB, Canada, 21–24 September 2008; pp. 746–751.
8. Yano, L.T.; Fair, A.E. Overview of Hydraulic Construction Techniques Utilized at the Syncrude Canada Limited Tailings Pond. In *Proceedings of the Hydraulic Fill Structures Conference*; American Society of Civil Engineers: Fort Collins, CO, USA, 1988.
9. CTMC Oil Sands Tailings Technology Deployment Roadmaps—Volume 1—Project Summary Report. Consortium of Tailings Management Consultants (CTMC). Available online: https://www.cosia.ca/sites/default/files/attachments/Tailings%20Roadmap%20Volume%201%20June%202012.pdf (accessed on 30 November 2021).
10. Pollock, G.W.; Mettananda, D.C.A.; MacGowan, T. Suncor's New Approach to Tailings Containment—Sand Dump 8. In Proceedings of the 2014 CDA Annual Conference, Banff, AB, Canada, 5–7 June 2014.
11. Schafer, H.L.; Beier, N.A.; Macciotta, R. Closure and the Long-Term Behaviour of Tailings Dams: Using Industry Experience to Fill in the Gaps. In Proceedings of the 72nd Canadian Geotechnical Conference, St. John's, NL, Canada, 29 September–2 October 2019.
12. Slingerland, N.; Beier, N.A.; Wilson, G.W. Landscape Evolution Modelling of Large Sand Tailings Dams. In Proceedings of the 2018 International Mine Closure Conference, Leipzig, Germany, 3–7 September 2018; pp. 341–348.
13. Slingerland, N.; Isidoro, A.; Fernandez, S.; Beier, N.A. Geomorphic Analysis for Tailings Dam Design in Consideration of a 1000-Year Closure Design Life. In Proceedings of the 2nd International Congress on Planning for Closure of Mining Operations, Santiago, Chile, 7–9 November 2018.
14. *ISO 31000:2009(E)*; Risk Management—Principles and Guidelines. International Organization for Standardization (ISO): Geneva, Switzerland, 2009.
15. Slingerland, N. Geomorphic Landform Design and Long-Term Assessment of Tailings Storage Facilities in the Athabasca Oil Sands. Ph.D. Thesis, University of Alberta, Edmonton, AB, Canada, 2019.
16. LDI. *Mining with the End in Mind: Landform Design for Sustainable Mining. Position Paper 2021-01*; Landform Design Institute (LDI): Delta, BC, Canada, 2021.
17. ICOLD Tailings Dam Committee. *Sustainable Design and Post-Closure Performance of Tailings Dams. Bulletin 153*; International Committee on Large Dams (ICOLD): Paris, France, 2013.

minerals

MDPI

Article

Assessment of the Self-Compaction Effect in Filtered Tailings Disposal under Unsaturated Condition

Ricardo Gallardo Sepúlveda [1,2,*], Esteban Sáez Robert [1,3] and Javier Camacho-Tauta [4]

1 Department of Structural and Geotechnical Engineering, Pontificia Universidad Católica de Chile, Macul 8331150, Chile; esaez@ing.puc.cl
2 Faculty of Engineering, Escuela de Ingeniería Civil, Pontificia Universidad Católica de Valparaíso, Valparaíso 2340000, Chile
3 National Research Center of Integrated Natural Disaster Management (CIGIDEN), Vicuna Mackenna 4860, Santiago 8940000, Chile
4 Faculty of Engineering, Universidad Militar Nueva Granada, Bogotá 110111, Colombia; javier.camacho@unimilitar.edu.co
* Correspondence: ricardo.gallardo@pucv.cl

Abstract: Filtered tailing is a relatively-environmentally friendly technique due to the recovery of water during the process of dehydration. This technique was recently developed and therefore there are few studies available compared to research on conventional tailings. This work focuses on the behaviour of a filtered copper tailing under an unsaturated condition, studying the effect of self-compaction by self-weight which is triggered by the deposition process through 1D consolidation tests. In addition, the impact of confinement on the stiffness and shear strength was studied by means of unsaturated drained triaxial tests. The initial matric suction was determined based on soil-water characteristic curves for two initially loose dry densities and the water content at which the material leaves the dehydration process. The unsaturated consolidation tests showed that for vertical stresses above 100 kPa and with ratios between the applied matric suction and air entry value lower than twelve, the material approaches saturated condition, reaching up to 95% degree of saturation. On the other hand, for the unsaturated drained triaxial tests, the effect of the matric suction generates increases in the shear strength and the stiffness of the material up to 200% and 600%, respectively. However, the increase tends to disappear as confinement increases.

Keywords: filtered tailings; unsaturated soils; matric suction; monotonic unsaturated tests; air entry value

1. Introduction

The high consumption of water in the mining industry is a constant and relevant problem. Therefore, various mining waste disposal strategies, such as filtered tailings, are being developed to optimize and reduce water consumption. This type of tailing is considered to be high-density tailing [1] since they are subjected to a dehydration process and reach a solid concentration higher than 77% [2,3]. This dehydration process can be carried out by mechanical or vacuum filtration, and the material usually achieves a water content below its saturation condition. Therefore, the material cannot be pumped, and it must be transported to the site of collection or final disposal by means of conveyor belts or trucks [4].

High-density tailings, such as filtered tailings, are naturally subjected to a certain degree of desiccation in arid climates [1,5]. Therefore, these tailings should be characterized using the soil-water characteristic curve (SWCC) associated with the drying path used. According to Oldecop and Rodari [6], lower water content in mining waste deposits means structurally more stable deposits which are less vulnerable to seismic events and a lower probability of generating acid drainage. However, according to Ulrich and Coffin [7], eventual saturation and the consequent development of pore pressures can also occur in this type of deposit. In fact, Wilson and Robertson [8] warn that this new technique is not immune to potential failures, as two high-density tailing facilities have had problems in both design and operating conditions. Therefore, there is still a great need to advance on the understanding of the hydromechanical behaviour of these deposits, particularly about the rates and heights of deposition that ensure adequate performance [9].

Table 1 and Figure 1 show the existing filtered tailing facilities in Chile and their geographical locations according to the update carried out in 2020 by the National Geology and Mining Service (https://www.sernageomin.cl/datos-publicos-deposito-de-relaves, accessed on 25 February 2022). According to the information available in East and Fernandez [10] and Ulrich and Coffin [7], the second-largest project associated with this type of deposit worldwide is located in Chile; there are only 15 of these operations in the country, of which only seven are currently active. Therefore, this technology is relatively new compared to conventional deposition techniques, and it is necessary to study and evaluate the behaviour of filtered tailing deposits under unsaturated conditions during the construction, operation, and closure phases [11].

Table 1. Facilities with filtered tailings in Chile.

* Tons (Mt)	Facility	Region	Mineral	Conditions	Approval Date
14.8	El Peñón	II	Gold	Inactive	01-2003
129.6	La Coipa	III	Gold, silver	Inactive	02-2005
49.8	Ampliación El Peñón	II	Gold, silver	Active	12-2010
138.2	Mantos Blancos	II	Copper	Active	07-2018
1.2	Tambo de Oro	IV	Gold	Active	03-2018
2.8	Filtrados 2	III	Copper, molybdenum	In construction	04-2019
3.3	El Toqui	XI	Zinc, gold	Inactive	06-2011
0.73	Guanaco	II	Copper	Active	08-2016
1.9	Convertidor Teniente	III	Copper, molybdenum	Active	12-2016
39.4	Planta San José	III	Copper	In construction	03-2016
21	Manto Verde	II	Copper	Active	** 02-2021
6.8	Faena Tambillos	IV	Copper	Active	07-2018
3.7	Escondida	II	Copper	Inactive	11-2016
1.1	Diego de Almeyda	III	Copper	Inactive	10-2017
24.5	Kozan	III	Copper	In construction	10-2019
22.2	Salares Norte	III	Gold, silver	In construction	04-2020

* Quantity authorized to exploit; ** Updated by the Superintendency of the Environment of the Government of Chile.

The existing studies of filtered tailings have mainly focused on the analysis of hydromechanical properties and flow modelling [12,13], considerations and practical experiences for design, operation, and field evaluation [7,10,14–17] and studies on the hydrological behaviour and undrained resistance of mixed waste rock and filtered tailings [18–20]. However, there are very few studies on the eventual increase of the saturation by self-compaction or about shear strength decrease at long term, due to the increase of confinement that the material could reach.

This research is focused on the characterization of the hydromechanical behaviour of filtered copper tailings without mechanical compaction and the effects of the unsaturated condition in both long-term compressibility and shear strength. The analysis was carried out by studying stiffness and strength, through one-dimensional consolidation and drained triaxial tests in conditions of total and partial saturation. One-dimensional consolidation tests are performed to evaluate the increase in the saturation degree as a result of self-compaction, simulating accelerated deposition rates, and to study the evolution of the compressibility of the material as a function of the increase in matric suction due to the drying these materials present in the field [21]. On the other hand, triaxial tests quantify the impact on the shear strength and stiffness at long term under unsaturated conditions as a function of the confinement. The consolidation analysis was performed adopting the matric suction corresponding to the water content of the material when deposited in the disposal, and then the matric suction is increased to simulate in-situ drying. For triaxial path, the initial matric suction was selected to estimate the minimum impact that an eventual saturation would have on the shear strength of the material. The samples were extracted from fresh tailings from a copper sulphide mill plant concentrator located in northern Chile. This operation has a concentration by flotation plant which generates 18,000 tons of tailings per day.

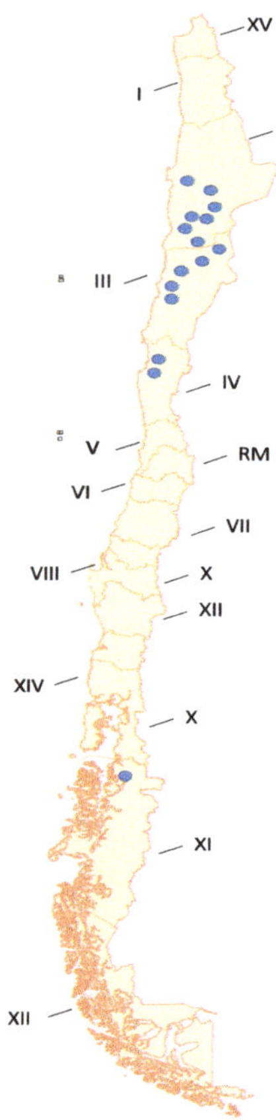

Figure 1. Geographical locations of filtered tailing facilities in Chile.

2. Materials and Methods

2.1. Geotechnical Characterization of the Material

The main physical characteristics of the studied material are presented in Table 2, while the particle size distribution and the preparation densities are shown in Figure 2. The tested samples were made by the moist tamping method at densities corresponding to 86% and 81% of those obtained in the Standard Proctor test. These densities agree with the values reported from on-site tests carried out by the company that supplied the material and represents the loose condition of the material as it is piled up in the field. Furthermore, the water content of the samples was approximately 16% as indicated by the company.

Figure 3 shows the mineralogical composition of the material obtained by X-ray diffraction (XRD) using a Bruker D8 Advance diffractometer and a scan range between 10–80°. Qualitative analysis was performed through EVA and TOPAS software packages. This test shows the presence of hematite and albite, enhancers of acid drainage [27,28], and neutralizing minerals such as calcite, which consume the acid generated in mining deposits (acid mine drainage) through the formation of bicarbonate and carbonic acid [29].

Table 2. Physical characteristics of filtered tailing.

Property	Value	Testing Procedure
Gs	2.97	ASTM D854-14 [22]
D10, D30, D50, D60 (μm)	3, 15, 65, 112	ASTM D2487-17e1 [23]
Cu, Cc	37.3, 1	ASTM D2487-17e1 [23]
LL, PL, PI	18, 15, 3	ASTM D4318-17e1 [24]
γd,max (kN/m^3)S.P *, M.P **	20.3, 21.95	ASTM 2012 D698-12; D1557-12 [25,26]
Clasification (USCS)	ML	ASTM D2487-17e1 [23]

* Standard Proctor test, ** Modified Proctor test.

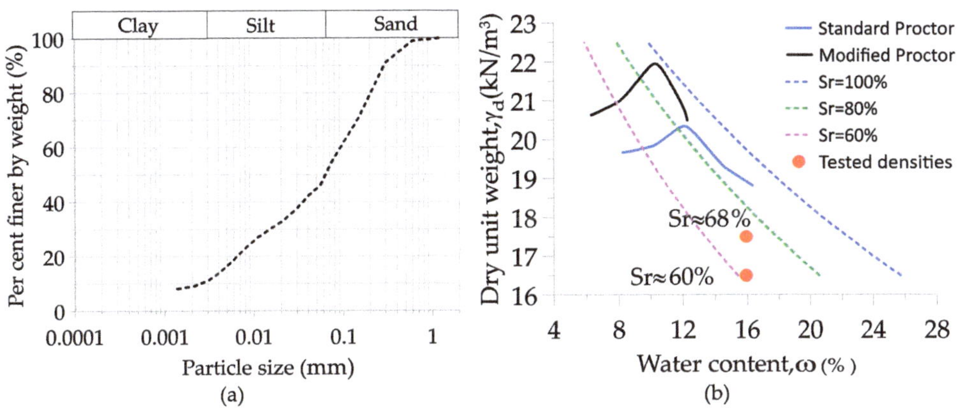

Figure 2. (**a**) Particle size distribution of the material tested. (**b**) Water content and initial mounting dry unit weight.

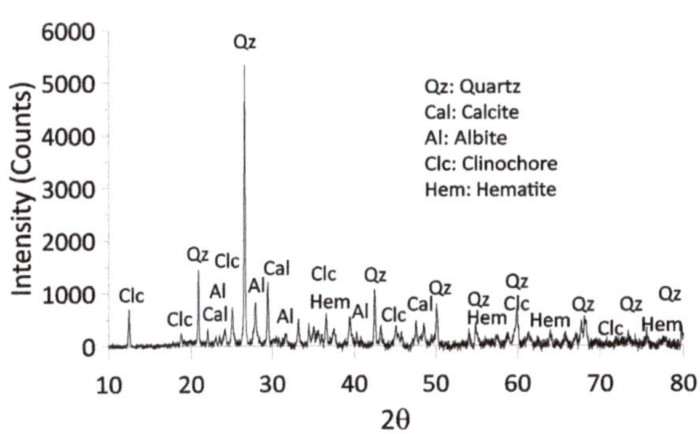

Figure 3. Mineralogical composition of the studied material.

2.2. Acid Drainage Potential

Acid drainage is generated by the drying of material over a long time in an environment with abundant oxygen [30–32]; that is precisely the condition of the soil in all of the tests carried out, mainly due to the incorporation of air pressure necessary to impose a constant matric suction during the tests. The acid drainage phenomenon was experienced strongly in the first consolidation tests. To avoid this phenomenon, which leads to a reduction in the useful life of the equipment used, a leaching test was carried out on the material using water with an acidic pH (2) and a basic pH (10) (Table 3). These tests were conducted according to the following methodology: (i) 1 g of tailing material was prepared in 50 mL of solution (acidic or neutral water), (ii) the preparation was shaken on an orbital shaker for 1 h at 150 RPM and then filled up to 250 mL with the same solution; and (iii) the material was filtered on 8- to 12-μ cellulose filter papers and (iv) atomic absorption analysis was carried out. In this investigation, a Perkin Elmer AAnalyst 100® was used. The process is summarized as a flowchart in the Figure 4. As shown in Table 3, the use of water with an acidic pH (<7) triggers the generation of acid drainage and can be mitigated by using water with a basic pH. For this reason, the tests were then carried out with industrial water (with a pH of 7.3) provided by the same mining company. This water controlled the generation of acid drainage. However, as a precaution, at the end of each test, the water was circulated through high air entry porous discs (HAEPD) to clean any residual acid drainage generated during the execution of the test.

Table 3. Leaching test results.

pH	Leached Iron (%)	Leached Copper (%)
2	0.012	0.08
10	<0.00004	<0.00005

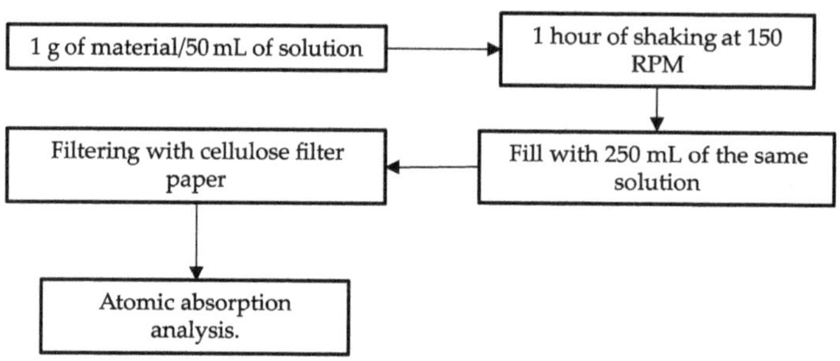

Figure 4. Leaching test flowchart.

2.3. Experimental Program

The effect of the matric suction on the stiffness and compressibility of the material, and the potential increase in saturation under monotonic loading, were evaluated to replicate the stress path considering a hypothetical storage pile. The SWCC was constructed for the two target densities up to 1.6 MPa and then the matric suction was found as a function of the moisture content at which the material is deposited in field (16%). One-dimensional consolidation and drained triaxial tests were performed using the axis translation technique [33]. One-dimensional saturated and unsaturated consolidation tests were carried out with matric suctions from 50 kPa to 400 kPa, while unsaturated triaxial tests were subjected to matric suction of 50 kPa. The purpose of studying the material under the consolidation path with increasing suctions is to simulate the environmental drying effect of the material and assess the impact on the compressibility and the evolution of its saturation degree by the increase of vertical stress. Regarding the drained triaxial tests, the effect of the initial matric suction on the stiffness and the shear strength was evaluated at several confinement stresses, since the initial condition is the closest to the saturation condition which should correspond to the minimum shear strength of the material. Of course, the hydro-mechanical path in the field is more complex, since the environmental drying effect is added to the water flow inside the disposal due to self-compaction, increasing the humidity in the lower zones and drying the upper layers. Moreover, none of these effects are 1D. The simplified path explored in this work are intended to isolate these combined effects to improve the understanding the behavior of the material in the disposal. Figure 5 summarizes the experimental program.

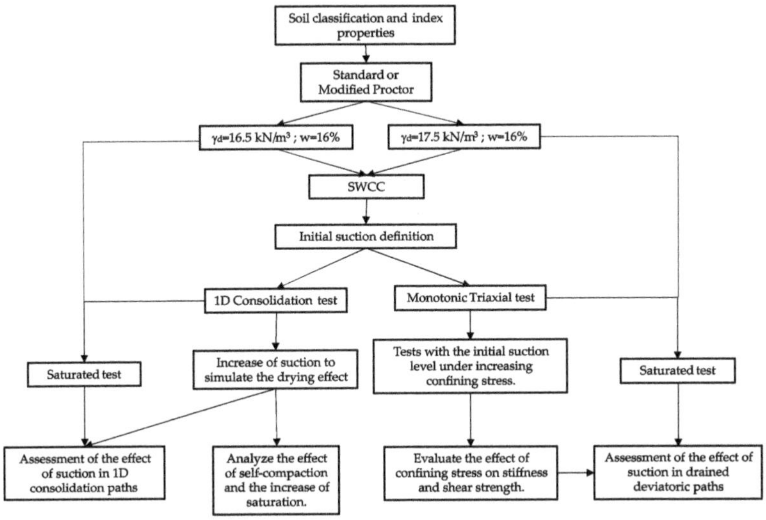

Figure 5. Flowchart of the experimental program.

The construction of the SWCC was carried out with the following techniques: (i) a tensiometer was used from 0 to 90 kPa in a device that records weight and matric suction variations (commercial equipment manufactured by UGT) on samples of 72 mm of diameter and 61.4 mm of height; (ii) an automatic oedometer was used for the axis translation technique from 50 kPa to 400 kPa (commercial equipment manufactured GDS Instruments) on samples of 63 mm of diameter and 22 mm of height; (iii) the contact filter paper test [34] was used from 50 kPa to 1.6 MPa (using Whatman 42 filter paper); and (iv) the equilibrium vapor testing method was used to characterize the residual zone with values from 4 MPa to 110 MPa. For techniques (iii) and (iv), we used samples of 50 mm of diameter and 18 mm of height. In technique (iv), a hermetic desiccator instrument with high-precision load cells was developed to continuously monitor the variation in weight. The exchange of salts must be rapid to avoid a significant change of the temperature and relative humidity (RH) inside of the desiccator, which could modify the measurement of the weight. Figure 6 shows the combination of methods used to construct the SWCC.

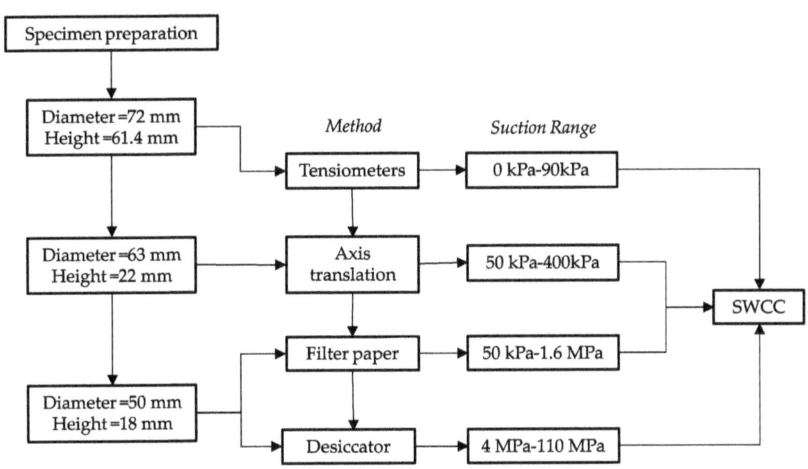

Figure 6. Combination of methods to obtain SWCC.

In the present investigation, the samples inside of desiccator were dried in an oven for 24 h and then equilibrated with the salt reaching the highest moisture content, with the purpose to start from the wettest possible condition. The salt exchange time was less than two minutes, and the salts were exchanged in decreasing order with relative humidity to simulate the drying effect. Inside the desiccator, there were three samples for each test density (i.e., six samples) to ensure repeatability of the test. The temperature and relative humidity within the closed circuit were continuously monitored, and Kelvin's law was used to determine the total suction [35] according to the following equation:

$$\psi = -\rho_w RT/\omega_v \ln(RH). \quad (1)$$

where ρ_w is the density of water, R is the universal gas constant (8.31432 J/mol K°), T is the temperature inside of the desiccator, ω_v is the molecular mass of water vapor (18.016 g/mol), and RH is the relative humidity inside the desiccator. Figure 7 shows the equipment used to obtain the SWCCs through all of the methodologies mentioned above.

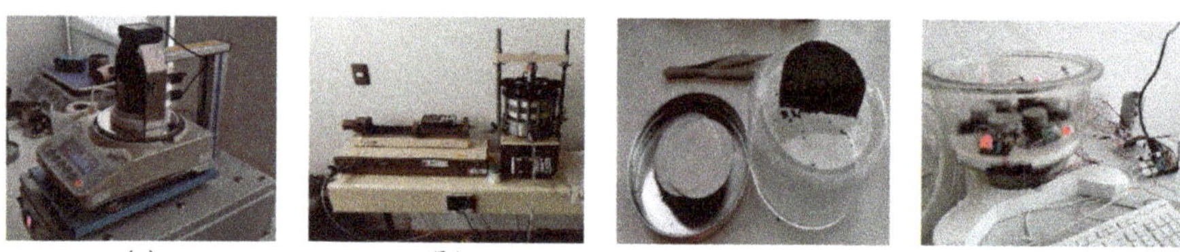

(a) (b) (c) (d)

Figure 7. Equipment used to obtain the SWCC. (**a**) Ku-pf; (**b**) GDS oedometer; (**c**) filter paper; (**d**) equilibrium vapor.

The residual zone is one of the most complex to characterize due to the high level of suction that must be achieved and quantified. To reach these levels of suction, the equilibrium vapor technique was used. Table 4 and Figure 8 shows the salts used for this technique and the moisture balance process for both densities. Figure 8 shows the average weight variation evolution of the samples that were placed in the desiccator (3 samples for each density). The salts were exchanged from the higher RH to the lower RH as shown in Table 4. Vertical lines indicate each salt replacement in the system. The results of the SWCC show that the combination of the four techniques is suitable for obtaining a continuous curve and consistent with the proposed models.

Table 4. Salts used to determine the total suction by the equilibrium vapor technique.

Salt	*RH* (%) Reached	*T* (°C) Equilibrium	Total Suction (MPa)
K_2SO_4	96.2%	20.2	4.1
KCl	92.3%	19.1	23.5
K_2CO_3	51.2%	17.6	110.1

The tested specimens were 63 mm in diameter and 22 mm heigh for the oedometer and 50 mm in diameter and 100 mm heigh for the triaxial tests. Table 5 summarizes the tests carried out, where the matric suction ψ corresponds to the difference between the air pressure u_a and the backpressure u_w applied to the sample. The axis translation technique was applied using a porous stone with a high air entry value (HAEPD) of 1500 kPa until a variation of less than 0.0004 mm^3/s was obtained [36]. This condition was considered equilibrium for this investigation. The time to reach this equilibrium depended on the permeability of the HAEPD, was well as the matric suction applied. This balance must be achieved at each stage of load increase.

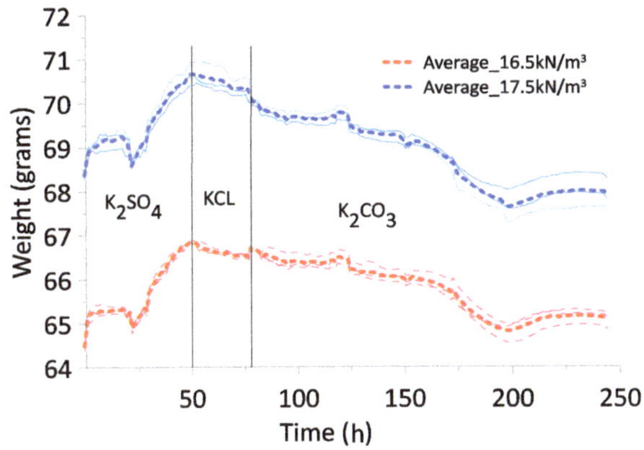

Figure 8. Variations in weight relative to test duration and salts used for each density.

Table 5. Tests carried out under saturated and unsaturated conditions.

Test	Sample ID	Cell Pressure (kPa)	Back Pressure (kPa)	Air Pressure (kPa)	γ_d (kN/m^3)
1D Consolidation	-	0	0	0	16.5
	-	150	100	150	16.5
	-	200	100	200	16.5
	-	300	100	300	16.5
	-	500	100	500	16.5
	-	0	0	0	17.5
	-	150	100	150	17.5
	-	200	100	200	17.5
	-	300	100	300	17.5
	-	500	100	500	17.5
CID	Sat_1	450	400	0	16.5
	Sat_2	500	400	0	16.5
	Sat_3	600	400	0	16.5
	Unsat_1	110	10	60	16.5
	Unsat_2	160	10	60	16.5
	Unsat_3	260	10	60	16.5
	Sat_4	1040	990	0	17.5
	Sat_5	1040	940	0	17.5
	Sat_6	1040	840	0	17.5
	Unsat_4	110	10	60	17.5
	Unsat_5	160	10	60	17.5
	Unsat_6	260	10	60	17.5

2.4. Conventions Considered for the Presentation of the Results

The measurement of the variation in the water within the sample (ΔV_w) and its vertical deformation ΔH allow for calculating the degree of saturation (S_r), the void ratio (e), and the volumetric water content (θ) throughout the consolidation and drained triaxial tests. The initial matric suction corresponds to the water content of the material deposited in the ground (16%) and according to the SWCC corresponding to 50 kPa (see Figure 10). This initial matric suction is, in principle, the most saturated condition that the material will have in the field since the evaporative drying process should prevail over the potential external water inputs. Indeed, most of this kind of deposits are located in the desert area of northern Chile where annual precipitation are in the range of 30 to 50 mm/year in average [37].

To determine the saturation in the consolidation tests, Equation (1) was used:

$$S_r(\%) = \frac{V_w}{V_v} = \frac{V_{wo} + \Delta V_{wBP}}{V_{vo} - \Delta H \cdot A_S} \quad (2)$$

where V_{wo} is the initial volumetric water content for each test density, ΔV_{wBP} is the is the change in the volume of water within the specimen and is measured by the volumetric backpressure controller at each load increment, V_{vo} is the initial void volume, ΔH is the variation of the height of the specimen and A_s is the cross-sectional area of specimen to test. The volumetric back pressure controller allows checking for any issues with the air diffusion through the high air entry porous stone and accurately measuring any change in the actual volume of water in the sample.

In triaxial tests and after an equilibrated state of consolidation, the shear loading phase was executed at a rate of 0.0015 mm/min, given the low permeability of the HAEPD, and up to 20% of axial deformation. To analyze the data the approach by Fredlund et al. [38] and the extended in mean net stress term by Wang et al. [39] was used. The equations are the following:

$$q = M \cdot p_{net} + q_s \quad (3)$$

$$\sigma_3 = \sigma_c - u_a \quad (4)$$

$$\sigma_1 = \sigma_c + q - u_a \quad (5)$$

$$q = \sigma_1 - \sigma_3 \quad (6)$$

$$p_{net} = \frac{\sigma_1 + 2 \cdot \sigma_3}{3} = \frac{q}{3} + \sigma_c - u_a \quad (7)$$

where q is the deviatoric stress, M is the projection of the critical state line (CSL) in this plane, p_{net} is the mean net stress, q_s is the final intercept of this projection with the q axis, σ_3 is the net confinement stress, σ_c is the cell pressure, u_a is the air pressure applied to sample, and σ_1 is the net principal stress. The change volume in the sample were obtained according to the following expressions:

$$V = V_a + V_w + V_s \quad (8)$$

$$V_w = V_{wo} + \Delta V_{wBP} \quad (9)$$

where V_s is the volume of solid for each test density.

The air volume (V_a) in the sample was obtained according to the gas law, thanks to the digital pressure/air volume controller used [40], according to the following expressions:

$$V_a = V_{a,system} - (V_{p1} - V_{p2}) - V_C - \Delta V_C \quad (10)$$

$$V_{a,system} = \frac{k\,R\,T}{u_a + u_{atm}} \quad (11)$$

$$kRT = (u_a + u_{atm})(V_a + V_C + \Delta V_C + V_{p1} + V_{p2}) \quad (12)$$

where T is the system temperature, u_{atm} is the atmospheric pressure measured directly by the equipment, V_c is the initial volume of air in the controller (after equilibrium was reached in the consolidation phase), ΔV_C is the air volume variation recorded by the controller, V_{p1} is the volume of air in the line connecting the upper cap with the outlet valve of the triaxial cell, and V_{p2} is the air volume in the line connecting the cell valve to the digital air pressure/volume controller. Figure 9 illustrates the designations of the different volumes in the previous expression.

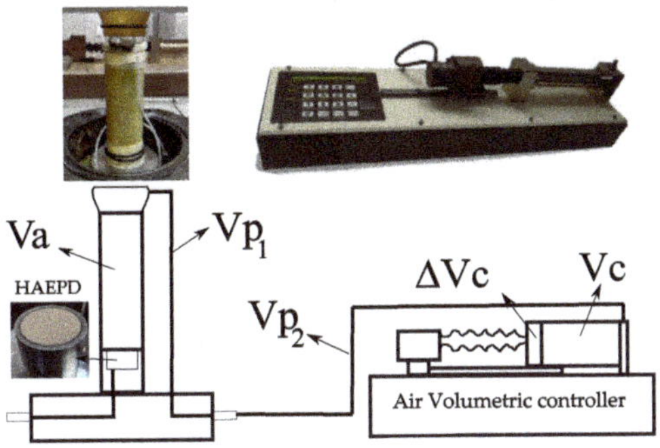

Figure 9. Designations of different air volumes.

3. Results and Discussions

3.1. SWCC of Studies Densities

Figure 10 shows the SWCC as a function of the volumetric water content θ (the quotient between the volume of water V_w and the total volume of V_T of the sample), for the two dry densities proposed in this investigation and three commonly used adjustment models: Gardner [41], Van Genuchten [42], and Fredlund and Xing [43]. From Figure 10, comparing the denser material (17.5 kN/m^3) with the looser material (16.5 kN/m^3), the air entry value (AEV) decreased from 8 kPa to 4 kPa. The value of parameters of each curve are shown in Table 6. According to Vanapalli et al. [44], this point separates the material's behavior from a saturated state to an unsaturated state, indicating that the denser material (17.5 kN/m^3) remains in saturated behaviour in a wider range of suction than the looser material. However, over 700 kPa of suction, the behaviour of the materials with both densities were similar, resulting in very similar residual volumetric water content ($\theta_r \approx 0.07$).

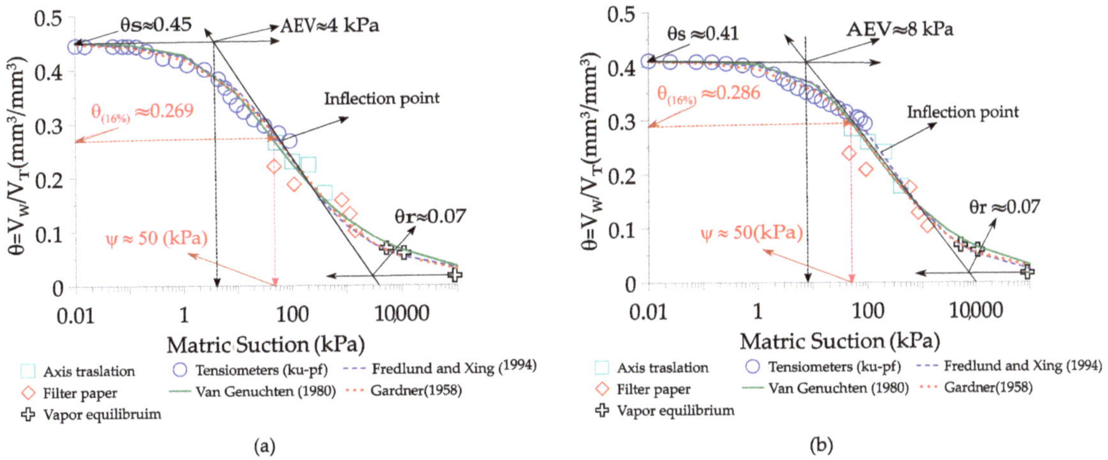

Figure 10. Soil-water characteristic curves [41–43]. (**a**) Dry unit weight 16.5 kN/m^3. (**b**) Dry unit weight 17.5 kN/m^3.

The impact of dry density on the properties of the water retention curve is related to the relative size of the pores. In the sample with the highest density, the pore size is relatively small, inducing a higher air entry value. Therefore, higher suction levels must be achieved for the same saturation degree regarding lower density [45,46]. On the other hand, Gao and Sun [47] and Cai et al. [48] indicated that samples with different dry densities showed very similar behaviours at significant levels of matric suction, a phenomenon that was observed above 700 kPa in this research.

Table 6. SWCC adjustment parameters for each model. θ_s and θ_r are the volumetric contents of the saturated and residual water, respectively.

Model	Equation	Variable	γ_d = 16.5 kN/m^3	γ_d = 17.5 kN/m^3
Gardner (1958) [41]	$\theta = \theta_r + \frac{\theta_s - \theta_r}{1+\left(\frac{\psi}{\alpha}\right)^n}$	θ_s	0.45	0.41
		θ_r	0.01	0.01
		α	100	200
		n	0.55	0.6
Van Genuchten (1980) [42]	$\theta = \theta_r + \frac{\theta_s - \theta_r}{\left[1+\frac{\psi}{\alpha}^n\right]^m}$	θ_s	0.45	0.41
		θ_r	0.01	0.01
		α	8	18
		n	1.28	1.34
		m	0.24	0.24
Fredlund and Xing (1994) [43]	$\theta = \theta_r + \frac{\theta_s - \theta_r}{\left\{ln\left(e+\left(\frac{\psi}{\alpha}\right)^n\right)\right\}^m}$	θ_s	0.45	0.41
		θ_r	0.01	0.01
		α	60	200
		n	0.66	0.66
		m	2	2.4

ψ (kPa) is the suction, according to authors of models α (kPa) is a soil parameter that defines the inflection point that is greater than the AEV, and n and m are curve fitting parameters.

3.2. Hydromechanical Behaviour of the Material at a Constant Matric Suction in the 1D Consolidation Test

Details of the 1D consolidations carried out are shown in Figure 11. Figure 11a,c show the evolution of the void ratio as a function of vertical stress, while Figure 11b,d show the evolution of saturation degree as the applied vertical stress increases. The compression (Cc) and expansion indexes (Cs) for both densities are displayed in Table 7. These indexes are considered apparent to differentiate them from those corresponding to a saturated condition.

The oedometric curves in saturated condition shows an approximately normally consolidated behavior (NC) in both dry densities. However the looser density has a compressibility index 26% larger than the highest dry density tested.

The effect of matric suction shows that when a matric suction exceeds the level of the air entry value, the soil becomes stiffer and the compressibility index is slightly reduced compared to the value at 50 kPa (up to 5% for an initial dry density of 16.5 kN/m^3 and 16% for the dry density of 17.5 kN/m^3, see Table 7). However, the effect on saturation is much more significant, reaching up to 95% for suctions in the range of 50 to 100 kPa and initial dry density of 17.5 kN/m^3 since the lower void ratio. Also, it can be seen the contribution of the matric suction which shows a more clearly pre-consolidated "type" behavior.

Regarding the unloading, it was observed that the matric suction did not have a significant effect, and the expansion indexes were equivalent from a practical point of view to the one corresponding to the saturated condition. This may have occurred since the unloading was carried out at high stress (1600 kPa), and as the evolution of compressibility suggests, the material at that stress level has reached a particle arrangement corresponding to a pseudo-elastic behaviour under unloading. Regarding the effect of the density in the looser sample, the reduction in the void ratio reached approximately 50% of the preparation density if the matric suction did not exceed 50 kPa. In the denser sample, this value was closer to 40% but with matric suction up to 100 kPa.

In terms of the evolution of saturation, it was observed that the value remained stable up to approximately 100 to 200 kPa of vertical stress. However, for higher vertical stresses, the saturation increased at both densities. For a density of 17.5 kN/m^3 and a vertical stress higher than 1600 kPa with matric suctions between 50 and 100 kPa, the material reached a saturation condition of up to 95%. For a density of 16.5 kN/m^3, the saturation achieved was of about 90%, and only for a matric suction of 50 kPa. Hence, for a ratio between the matric suction and the air entry value of less than twelve (12), the material increases its saturation degree, reaching up to 95%. This behaviour has already been observed by other authors [49,50]. However, in those studies, the water content in the sample was correlated and not directly measured, in which differs from this investigation.

Compared to the lower density sample, the denser sample has less porosity, and the reorganization of the solid skeleton reduces the space available for water and therefore it is easier to increase its saturation. The comparison of the results at low suctions against the saturated condition indicated that when the saturation exceeded approximately 80%, the changes in the soil were such as those observed in the saturated condition with vertical stress greater than 800 kPa. This pattern was observed in the looser material at a matric suction of only 50 kPa, while for the denser material, this behaviour takes place at suctions from 50 kPa and 100 kPa.

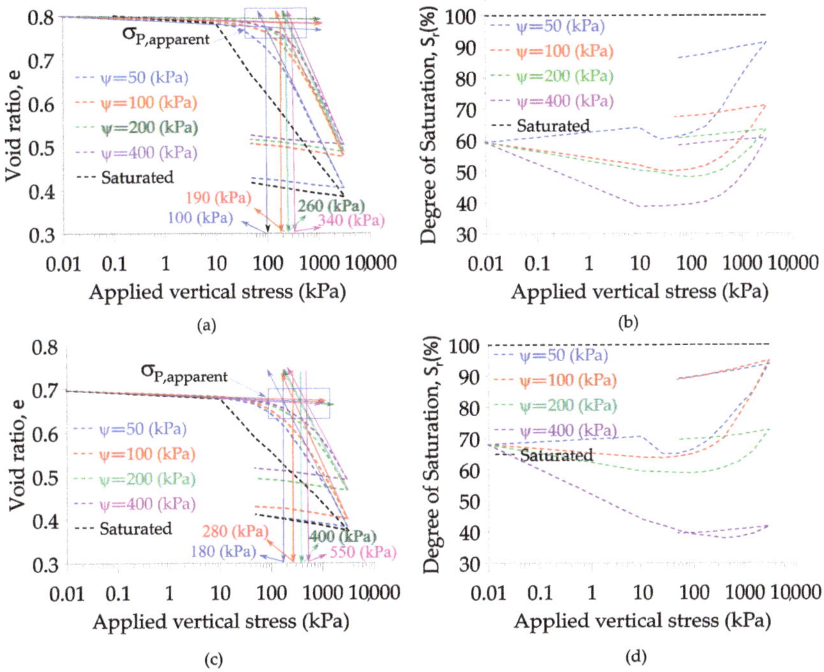

Figure 11. 1D consolidation tests and the degree of saturation reached. (**a**) Variation in the void ratio vs. vertical stress at a dry unit weight of 16.5 kN/m^3. (**b**) Variation in the void ratio vs. vertical stress at a dry unit weight of 16.5 kN/m^3. (**c**) Degree of saturation in each increment of vertical stress at a dry unit weight of 17.5 kN/m^3. (**d**) Degree of saturation in each increment of vertical stress at a dry unit weight of 17.5 kN/m^3.

Another way to quantify the effect of matric suction on the compressibility of the material is through the calculation of an apparent pre-consolidation stress using the Casagrande method for each curve. Since the material is remoulded and due to the matric suction, this pre-consolidation is considered apparent ($\sigma_{P,apparent}$). The results are presented in Figure 12, were the values of apparent pre-consolidation stresses correspond to the fit shown in Figure 11a,c. This curve shows an increase in apparent pre-consolidation stresses as matric suction increases, for both densities under study. The slope of this curve decreases with the increase suction, suggesting an asymptotic tendency at very high suction. According to our results, it is suggested to consider suction above 400 kPa to determine the maximum achievable apparent pre-consolidation stress. These results follow the same trend as those presented by Zou et al. [51] and Burton et al. [52]. However, they were obtained for natural soils at a constant moisture content.

Table 7. Compression and expansion index as a function of matric suction.

γ_d (kN/m^3)	ψ (kPa)	C_C	C_S
16.5	0	0.152	0.018
	50	0.235	0.016
	100	0.230	0.017
	200	0.226	0.016
	400	0.224	0.016
17.5	0	0.120	0.021
	50	0.231	0.015
	100	0.226	0.016
	200	0.209	0.016
	400	0.202	0.015

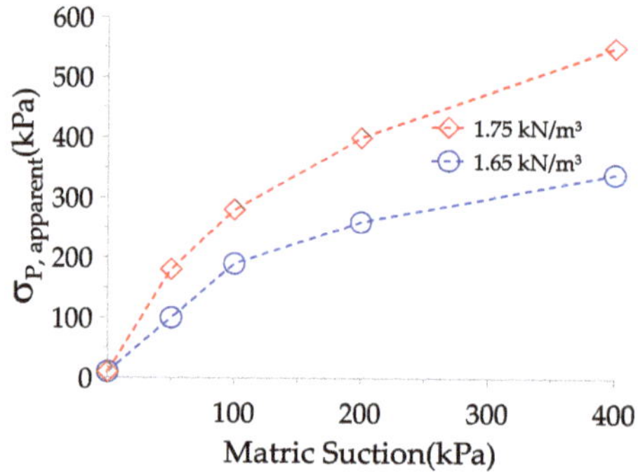

Figure 12. Effect of matric suction on the apparent pre-consolidation stress.

3.3. *Hydromechanical Behaviour of Material at a Constant Matric Suction in Drained Triaxial Tests*

A series of drained triaxial tests of material at a constant matric suction were carried out to assess the influence on the stiffness and the shear strength. Figures 13 and 14 present the results for the two initial densities studied and the results of the saturated condition as a reference. The tests were carried out at a constant matric suction of 50 kPa in accordance with the SWCC presented above and the humidity expected in the field. First, despite the saturation degree, a contractive behaviour was observed for all of the confinement imposed on the looser preparation (16.5 kN/m^3). For the denser case, a slight dilative behaviour was observed at an axial deformation of approximately 6%. From the point of view of shear behaviour, both densities showed an increase in shear strength and stiffness, as suggested by other authors [31,53].

Regarding the isotropic consolidation line (ICL) in Figures 13d and 14d, the effect of suction on its slope is only evident for the lower density case (16.5 kN/m^3). For the denser case, the slopes are very similar suggesting that the effect of suction on compressibility becomes less relevant as the material has a stiffer soil skeleton. In the case of the large deformation states from the triaxial tests at failure, the trends are more erratic, and it is not obvious to distinguish a clear trend from these results. In general, changes in slopes with respect to the saturated case are observed for both initial densities.

Figures 13b and 14b shows the shear failure envelopes and their parameters for both preparation densities. A slight reduction in the friction angle was observed for both densities when a constant matric suction of 50 kPa is applied. In addition, an apparent cohesion of about 25–30 kPa appears due to applied matric suction. Figure 15 shows the initial and the evolution of the degree of saturation during the shearing showing that there is no significant variation in this stage. Nevertheless, the saturation increases with the confinement.

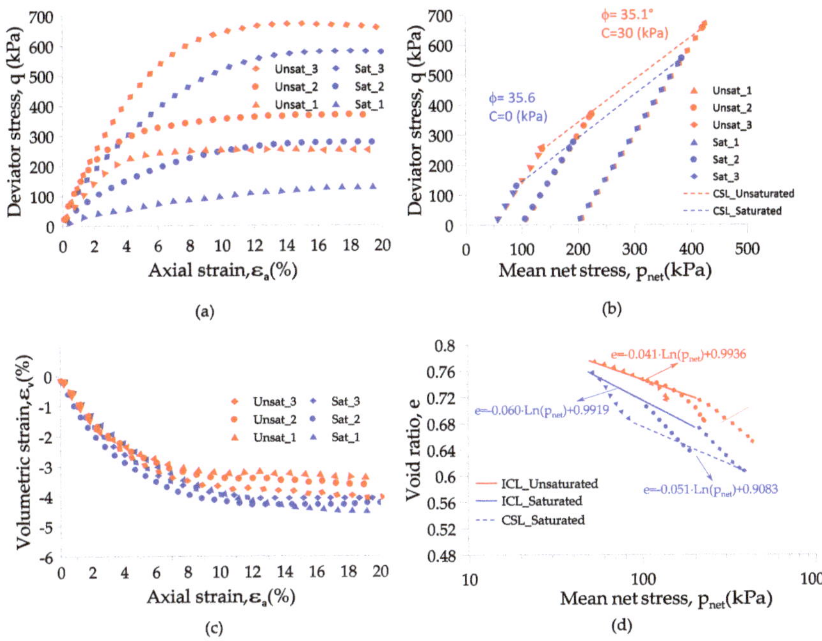

Figure 13. Behavior of the soil studied at a dry unit weight of 16.5 kN/m^3. (**a**) Diagram of stress relative to axial strain (q vs. ε_a). (**b**) Diagram of p_{net} vs. q. (**c**) Volumetric strain versus axial strain. (**d**) Diagram of p_{net} vs. void ratio during the shear phase.

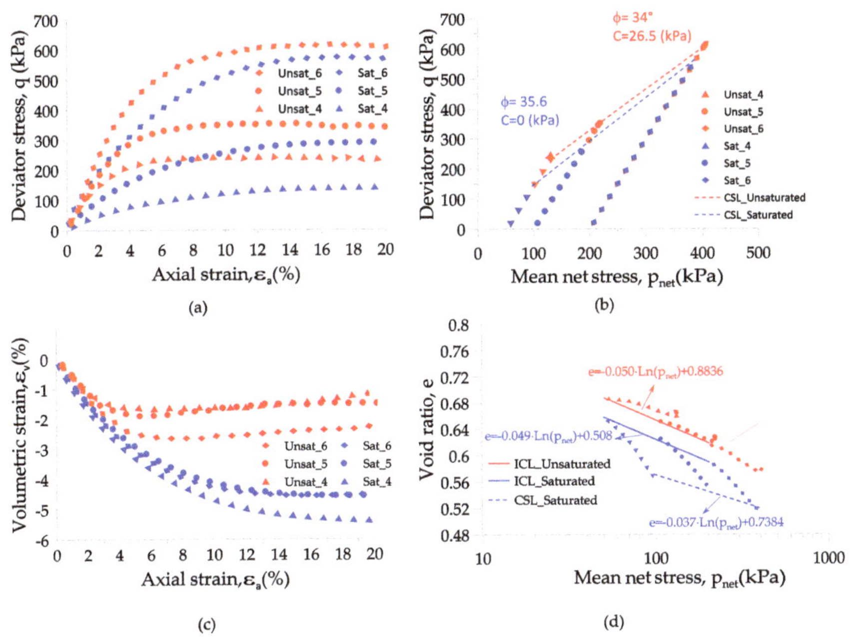

Figure 14. Behavior of the soil studied at a dry unit weight of 17.5 kN/m^3. (**a**) Diagram of stress versus axial strain (q vs. ε_a). (**b**) Diagram of p_{net} vs. q. (**c**) Volumetric strain versus axial strain. (**d**) Diagram of p_{net} vs. void ratio during the shear phase.

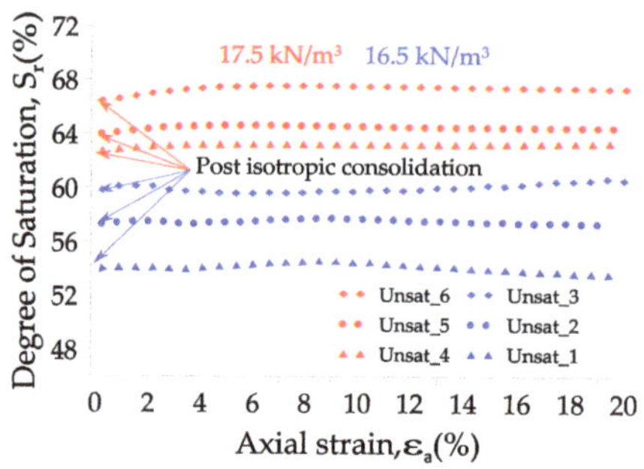

Figure 15. Behavior of the saturation at both densities in the shear phase.

Difference in stiffness and strength compared to the saturated condition tends to decrease as the confining pressure increases. To quantify this difference, Figure 16a,b shows the ratio between the secant modulus at 50% of the peak load (E50) and the large strain deviatoric stress under saturated and unsaturated conditions, respectively. These figures show that at confinements of about 50 kPa, the increase of stiffness and deviatoric stress could reach up to 600% and 200%, respectively, while at confinements of 200 kPa, the unsaturated condition is only approximately 180% more rigid with an increase of 20% the deviatoric stress for the dry density of 16.5 kN/m^3. This behaviour was more evident for the preparation at a dry unit weight of 17.5 kN/m^3 due to the increase of saturation in comparison with the lower density (see Figure 15). Then, by applying a matric suction closer to the air entry value, the material behavior approaches the saturated case. These effects were observed in the consolidation tests under the same level of matric suction (see Figure 11a,c).

In this work the matric suctions were maintained as a constant to facilitate the interpretation of the results. However, in the field the material could receive external supplies of water due to climatic conditions or from flows within deposited dry stacks or also experience a more significant reduction in humidity in arid regions. Furthermore, in seismic countries such as Chile, there are also questions about the cyclic behaviour of this type of uncompacted tailing material. These and other factors are expected to be addressed in future phases of this research.

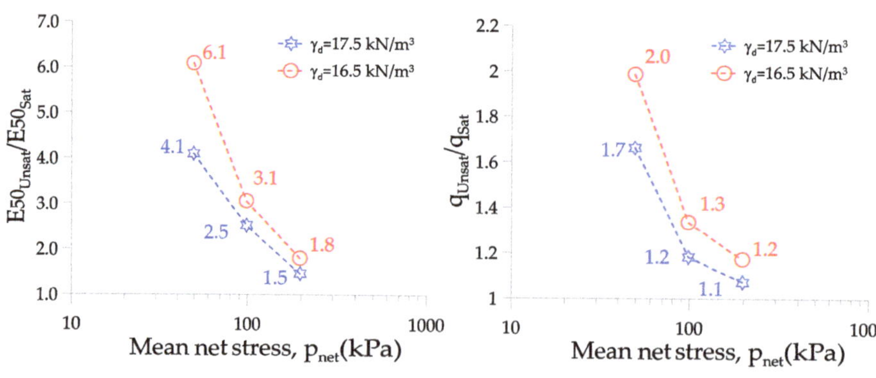

Figure 16. Stiffness and large strain deviatoric stress ratios for the unsaturated vs. saturated conditions of both dry densities studied. (**a**) Modulus of stiffness at 50% maximum load. (**b**) Large strain deviatoric stress.

4. Conclusions

The response of the soil to the consolidation tests at a constant matric suction showed an increase in the apparent pre-consolidation stress as the matric suction increased. However, the soil became saturated, reaching values of up to 95% for matric suctions lower than twelve times the air entry value. These results allow a partial answer to some doubts raised by authors such as Ulrich and Coffin [7], Wilson and Robertson [8] and Fredlund and Houston [11] regarding a possible saturation at the base of this type of tailing deposit.

The contractive behaviour of the material in the unsaturated and initially loose conditions was not affected by a matric suction of 50 kPa, maintaining a predominantly contractive behaviour under drained triaxial paths. In addition, the stiffness and large strain deviatoric stress of the material increases up to 600% and 200%, respectively, for confining pressures below 100 kPa. However, above 200 kPa of confining, these differences were less significant, and the behavior approaches saturated condition.

From the point of view of the deposition of the material in the field and assuming that the expected matric suction is around 50 kPa, it can be inferred that to avoid a behaviour close to saturation, the deposition height should not exceed 40 m. This means that a confinement below 800 kPa must be achieved to guarantee a saturation degree of the material below 70%. Naturally, this analysis does not consider the external or internal water contributions to the disposal by the flow from upper layers, so this is only a first rough approximation.

Finally, it is possible to conclude that the hydromechanical behaviour of an unsaturated filtered tailing under loose conditions and monotonic drained loading does not prevent the increase of its degree of saturation due to increase of confinement stress. Therefore, the maximum deposition height should be limited and constantly monitored.

Author Contributions: Conceptualization, E.S.R. and R.G.S.; methodology, E.S.R., R.G.S. and J.C.-T.; formal analysis, R.G.S.; investigation, R.G.S.; resources, E.S.R., R.G.S.; data curation, R.G.S.; writing—original draft preparation, R.G.S.; writing—review and editing, E.S.R. and J.C.-T.; supervision, E.S.R. and J.C.-T.; project administration, E.S.R.; funding acquisition, E.S.R. and R.G.S. All authors have read and agreed to the published version of the manuscript.

Funding: This research has been supported by CODELCO, through the Piensa Minería scholarship year 2019, the Pontificia Universidad Católica de Chile, through its doctoral scholarship, the National Research Center for Integrated Natural Disaster Management (CONICYT/FONDAP/15110017) and the FONDEQUIP EQM150024 project. The third author acknowledges the funding given by the Research Vice-presidency of the Universidad Militar Nueva Granada, Colombia, through the Project IMP-ING-2932.

Data Availability Statement: The data presented in this study are available on request from the corresponding author.

Acknowledgments: To Sergio Barrera for supplying the filtered tailing.

Conflicts of Interest: The authors declare no conflict of interest.

References

1. Simms, P. 2013 colloquium of the Canadian geotechnical society: Geotechnical and geoenvironmental behaviour of high-density tailings. *Can. Geotech. J.* **2016**, *54*, 455–468. [CrossRef]
2. Lara Montani, J.; Pornillos, E.; Munoz, H. Geotechnical-Geochemical and operational considerations for the application ofdry stacking tailings deposits—State ofthe art. In Proceedings of the 16th International Seminar on Paste and Thickened Tailings, Belo Horizonte, Brazil, 17–19 June 2013; pp. 249–260.
3. Weatherwx, T.W.; Kipara, A. Paste operations training at Barrick's Bulyanhulu Gold Mine. In Proceedings of the 13th International Conference on Paste and Thickened Tailings, Toronto, ON, Canada, 3–6 May 2010; pp. 41–50.
4. Davies, M.P.; Rice, S. An alternative to conventional tailing management—"Dry stack" filtered tailings. In Proceedings of the 8th International Conference on Tailings and Mine Waste'01, Fort Collins, CO, USA, 15–18 January 2001; pp. 411–420.
5. Daliri, F.; Simms, P.; Sivathayalan, S. Shear and dewatering behaviour of densified gold tailings in a laboratory simulation of multi-layer deposition. *Can. Geotech. J.* **2016**, *53*, 1246–1257. [CrossRef]

6. Oldecop, L.; Rodari, G. Unsaturated Soil Mechanics in Mining. In Proceedings of the Pan Am Unsaturated Soils, Dallas, TX, USA, 12–15 November 2017; pp. 257–280.
7. Ulrich, B.; Coffin, J. Considerations for Tailings Facility Design and Operation Using Filtered Tailings Summary of Significant Operating Filtered Tailings Operations. In Proceedings of the 16th International Seminar on Paste and Thickened Tailings, Belo Horizonte, Brazil, 17–20 June 2013; pp. 201–210.
8. Wilson, G.W.; Robertson, A.M. The Value of Failure. *Geotech. News* **2015**, *33*, 24–28.
9. Ulrich, B. Practical thoughts regarding filtered tailings. In Proceedings of the 22th International Conference on Tailings and Mine Waste, Cape Town, South Africa, 8–10 May 2019; pp. 71–79.
10. East, D.; Fernandez, R. Managing Water to Minimize Risk in Tailings Storage Facility Design, Construction, and Operation. *Mine Water Environ.* **2021**, *40*, 36–41. [CrossRef]
11. Fredlund, D.G.; Houston, S.L. Interpretation of soil-water characteristic curves when volume change occurs as soil suction is changed. In Proceedings of the 1st Pan-American Conference on Unsaturated Soils, Cartagena de Indias, Colombia, 20–22 February 2013; pp. 15–31.
12. Bussière, B. Colloquium 2004: Hydrogeotechnical properties of hard rock tailings from metal mines and emerging geoenvironmental disposal approaches. *Can. Geotech. J.* **2007**, *44*, 1019–1052. [CrossRef]
13. Sánchez, J.A.; Beltra, L.N.; Trujillo-Vela, M.G.; Larrahondo, J.M. Flows of Liquefied Filtered Tailings: Laboratory-Scale Physical and Numerical Modeling. *Int. J. Civ. Eng.* **2020**, *18*, 393–404. [CrossRef]
14. Crystal, C.; Hore, C.; Ezama, I. Filter-Pressed Dry Stacking: Design Considerations Based on Practical Experience. In Proceedings of the 22th International Conference on Tailings and Mine Waste, Keystone, CO, USA, 30 September–2 October 2018; pp. 209–219.
15. Hore, C.; Luppnow, D. Karara Iron Ore TSF—Design considerations for a unique large scale dry stack facility. In Proceedings of the 18th International Conference on Tailings and Mine Waste, Keystone, CO, USA, 5–8 October 2014; pp. 293–304.
16. Robertson, P.K.; Da Fonseca, A.V.; Ulrich, B.; Coffin, J. Characterization of unsaturated mine waste: A case history. *Can. Geotech. J.* **2017**, *54*, 1752–1761. [CrossRef]
17. Whatnall, O.; Barber, K.; Robinson, P. Tailings Filtration Using Viper Filtration Technology—A Case Study. *Min. Metall. Explor.* **2021**, *38*, 1297–1303. [CrossRef]
18. Borja, R.N.; Bareither, C.A. Shear Behavior of Waste Rock and Filtered Tailings Mixtures. In Proceedings of the Geo-Congress 2020: Geotechnical Earthquake Engineering and Special Topics, Minneapolis, MI, USA, 25–28 February 2020; pp. 872–880.
19. Gorakhki, M.R.H.; Bareither, C.A.; Scalia, J.; Aparicio, M. Hydrologic and Environmental Behavior of Geo Waste and Waste Rock in Field Experimental Piles. *Min. Metall. Explor.* **2021**, *38*, 1339–1354. [CrossRef]
20. Ulrich, B.; Coffin, J. Combined tailings and mine waste. In Proceedings of the Tailings and Mine Waste, Vancouver, BC, Canada, 26–28 October 2015; pp. 83–92.
21. Lee, K.; Sills, G. The consolidation of a soil stratum, including self-weight effects and large strains. *Int. J. Numer. Anal. Methods Geomech.* **1981**, *5*, 405–428. [CrossRef]
22. *ASTM Standard D854-14*; ASTM Standard Test Methods for Specific Gravity of Soil Solids by Water Pycnometer. ASTM Int.: West Conshohocken, PA, USA, 2014.
23. *ASTM Standard D2487-17e1*; ASTM Standard Practice for Classification of Coils for Engineering Purposes (Unified Soil Classification System). ASTM Int.: West Conshohocken, PA, USA, 2017.
24. *ASTM Standard D4318-17e1*; ASTM Standard Test Methods for Liquid Limit, Plastic Limit, and Plasticity Index of Soils. ASTM Int.: West Conshohocken, PA, USA, 2017.
25. *ASTM Standard D1557-12*; ASTM Standard Test Methods for Laboratory Compaction Characteristics of Soil Using Modified Effort. ASTM Int.: West Conshohocken, PA, USA, 2012.
26. *ASTM Standard D698-12*; ASTM Standard Test Methods for Laboratory Compaction Characteristics of Soil Using Standard Effort. ASTM Int.: West Conshohocken, PA, USA, 2012.
27. Dold, B. Evolution of acid mine drainage formation in sulphidic mine tailings. *Minerals* **2014**, *4*, 621–641. [CrossRef]
28. Murray, J.; Nordstrom, D.K.; Dold, B.; Kirschbaum, A. Seasonal fluctuations and geochemical modeling of acid mine drainage in the semi-arid Puna region: The Pan de Azúcar Pb–Ag–Zn mine, Argentina. *J. South. Am. Earth Sci.* **2021**, *109*, 103197. [CrossRef]
29. Akcil, A.; Koldas, S. Acid Mine Drainage (AMD): Causes, treatment and case studies. *J. Clean. Prod.* **2006**, *14*, 1139–1145. [CrossRef]
30. Daliri, F.; Paul, S.; Siva, S. A Comparison of Different Laboratory Techniques to Simulate Stress and Moisture History of Hard Rock Mine Tailings. In Proceedings of the Tailings and Mine Waste 2011 Conference, Vancouver, BC, Canada, 6–9 November 2011; pp. 163–175.
31. Daliri, F.; Kim, H.; Simms, P.; Sivathayalan, S. Influence of desiccation and overconsolidation on monotonic and cyclic shear response of thickened gold tailings. In Proceedings of the 14th Pan-American Conference on Soil Mechanics and Geotechnical Engineering, Toronto, ON, Canada, 2–6 October 2011; pp. 1–8.
32. Simms, P.; Grabinsky, M.; Zhan, G. Modelling evaporation of paste tailings from the Bulyanhulu mine. *Can. Geotech. J.* **2007**, *44*, 1417–1432. [CrossRef]
33. Hilf, J.W. An Investigation of Pore-Water in Compacted Cohesive Soils. Technical Memo. No. 654. Ph.D. Thesis, Bureau of Reclamation, Design and Construction Division, U.S. Department of the Interior, Denver, CO, USA, 1956.

34. *ASTM Standard D5298-16*; ASTM Standard Test Method for Measurement of Soil Potential (Suction) Using Filter Paper. ASTM Int.: West Conshohocken, PA, USA, 2016. [CrossRef]
35. Edlefsen, N.E.; Alfred, B.C.A. Thermodynamics of soil moisture. *Hilgardia* **1943**, *15*, 31–298. [CrossRef]
36. Hoyos, L.R.; Pérez-Ruiz, D.D.; Puppala, A.J. Modeling Unsaturated Soil Response under Suction-Controlled True Triaxial Stress Paths. *Int. J. Geomech.* **2012**, *12*, 292–308. [CrossRef]
37. Schumacher, V.; Comin, A.; Sarricolea, P.; Meseguer-ruiz, O. Comparison between observations and gridded data sets over complex terrain in the Chilean Andes: Precipitation and temperature. *Int. J. Climatol.* **2020**, *40*, 5266–5288. [CrossRef]
38. Fredlund, D.; Morgenstern, N.; Widger, R. The shear strength of unsaturated soils. *Can. Geotech. J.* **1978**, *15*, 313–321. [CrossRef]
39. Wang, Q.; Pufahl, D.; Fredlund, D. A study of critical state on an unsaturated silty soil. *Can. Geotech. J.* **2002**, *39*, 213–218. [CrossRef]
40. Adams, B.A.; Wulfsohn, D.; Fredlund, D.G. Air volume change measurement in unsaturated soil testing using a digital pressure-volume controller. *Geotech. Test. J.* **1996**, *19*, 12–21. [CrossRef]
41. Gardner, W. Some steady-state solutions of the unsaturated moisture flow equation with application to evaporation from a water table. *Soil Sci.* **1958**, *85*, 228–232. [CrossRef]
42. Van Genuchten, M.T. A closed-form equation for predicting the hydraulic conductivity of unsaturated soils. *Soil Sci. Soc. Am. J.* **1980**, *44*, 892–898. [CrossRef]
43. Fredlund, D.G.; Xing, A. Equations for the soil-water characteristic curve. *Can. Geotech. J.* **1994**, *31*, 521–532. [CrossRef]
44. Vanapalli, S.K.; Fredlund, D.G.; Pufahl, D.E.; Clifton, A.W. Model for the prediction of shear strength with respect to soil suction. *Can. Geotech. J.* **1996**, *33*, 379–392. [CrossRef]
45. Fredlund, D.G.; Rajardo, H.; Frdlund, M.D. *Unsaturated Soil Mechanics in Engineering Practice*; Wiley Online Library: Hoboken, NJ, USA, 2012; ISBN 978111813359.
46. Heshmati, A.A.; Motahari, M.R. Modeling the Dependency of Suction Stress Characteristic Curve on Void Ratio in Unsaturated Soils. *KSCE J. Civ. Eng.* **2015**, *19*, 91–97. [CrossRef]
47. Gao, Y.; Sun, D. Soil-water retention behavior of compacted soil with different densities over a wide suction range and its prediction. *Comput. Geotech.* **2017**, *91*, 17–26. [CrossRef]
48. Cai, G.; Zhou, A.; Sheng, D. Permeability function for unsaturated soils with different initial densities. *Can. Geotech. J.* **2014**, *51*, 1456–1467. [CrossRef]
49. Roy, S.; Rajesh, S. Test apparatus for rapid determination of soil-water retention curve under isotropic loading condition. *Geotech. Test. J.* **2021**, *44*, 255–273. [CrossRef]
50. Islam, T.; Kodikara, J. Interpretation of the loading—Wetting behaviour of compacted. *Can. Geotech. J.* **2016**, *805*, 783–805. [CrossRef]
51. Zou, W.; Han, Z.; Vanapalli, S.K.; Zhang, J.; Zhao, G. Predicting volumetric behavior of compacted clays during compression. *Appl. Clay Sci.* **2018**, *156*, 116–125. [CrossRef]
52. Burton, G.; Pineda, J.; Sheng, D.; Airey, D.; Zhang, F. Exploring one-dimensional compression of compacted clay under constant degree of saturation paths. *Geotechnique* **2016**, *66*, 435–440. [CrossRef]
53. Kimoto, S.; Oka, F.; Fukutani, J.; Yabuki, T.; Nakashima, K. Monotonic and cyclic behavior of unsaturated sandy soil under drained and fully undrained conditions. *Soils Found.* **2011**, *51*, 663–681. [CrossRef]

minerals

MDPI

Article

Investigation of the Attenuation and Release of Cu^{2+} Ions by Polymer-Treated Tailings

Mohammad Boshrouyeh Ghandashtani [1,*], Mansour Edraki [1], Thomas Baumgartl [2], Allan Costine [3] and Samar Amari [4]

1 Sustainable Minerals Institute, The University of Queensland, St Lucia, QLD 4072, Australia; m.edraki@cmlr.uq.edu.au
2 Geotechnical and Hydrogeological Engineering, Federation University, Churchill, VIC 3842, Australia; t.baumgartl@federation.edu.au
3 CSIRO Mineral Resources, Waterford, WA 6152, Australia; allan.costine@csiro.au
4 Directional Studies and Innovation, Newcrest Mining Ltd., Melbourne, VIC 3004, Australia; samar.amari@newcrest.com.au
* Correspondence: m.boshrouye@uq.edu.au

Abstract: This study investigated the attenuation and release behaviour of copper ions using a standard kaolin-silt slurry as the synthetic tailings in a high solids/high salinity application before and after inline flocculation. A homogenous, synthetic tailings slurry was prepared in a 0.6 M NaCl solution and treated in a low-shear mixer by adding Magnafloc® 336 flocculant. Following the evaluation of morphological properties of both the untreated (UT) and polymer-treated tailings (PT), identical equilibrium tests were performed via the bottle-point method constant concentration technique. The maximum copper ions uptake capacity of polymer-treated tailings was 25% more than the untreated slurry at the equilibrium state in a chemisorption process in which the ions had the capability of binding onto one location on the sorbent, which could be influencing other binding sites on the same sorbent. Polymer treatment resulted in a highly porous structure that exhibited an increased capacity to adsorb and retain copper ions compared to the UT materials. This behaviour indicates the strong binding between the copper ions and active site of the treated tailings particles with greater capability of this material for preserving heavy metal ions within their structure across a wide pH range (2–10) compared to the UT materials. The results advance the fundamental understanding of how inline flocculation can considerably improve the sorption capacity of high solids/high salinity tailings favouring potential long-term rehabilitation purposes at mine closure and the role of sorption and desorption of heavy metal ions' behaviour play to achieve this goal.

Keywords: inline flocculation; heavy metal; sorption; desorption; high solid tailings; copper ions

1. Introduction

Heavy metal ions released during mining activities are major potential contaminants of soil, water, and vegetation [1,2]. Poorly controlled mine residues are easily dispersed into the environment by surface drainage, wind, and groundwater [3,4]. In particular, mineral processing tailings can have high heavy metal concentrations and consequently are often sources of heavy metal pollution, and owing to the serious toxic effects, it is imperative to prevent and/or mitigate heavy metal pollutions. In line with recommendations and government guidelines, the current inclination is to expand and apply in situ techniques where possible to minimise disturbance of contaminated lands. Therefore, tailings dewatering by an effective technique which would also provide the opportunity for in situ remediation is desirable.

Generally, tailings are comprised of water and a range of materials remaining from the mineral processing including finely sized clay minerals which are the most problematic components in tailings dewatering [5,6]. Among several available techniques for treating fine-particle tailings, polymer-bridging flocculation is an effective approach which achieves solid–liquid separation by increasing the settling rate of individual particles by forming larger aggregated structures [7–9]. In this regard, one of the industrially trialled methods is inline flocculation, where a polymer is added at high dosages to the thickener underflows, resulting in additional aggregation and water release on deposition compared to conventional low-solids thickening alone [10,11]. These highly flocculated materials have potential properties and structure that may be beneficial for removing hazardous metal ions.

Copper (II) is known to be one of the heavy metals widely used in many industries [12], which spreads into the environment through soils and water streams and accumulates along the food chain, resulting in a high risk to human health [13]. Most of the mine tailings, especially the copper sulphide ores tailings, contain copper ions which are generated through the oxidation of sulphide and dissolution/weathering of tailings or the acid mine drainage processes [12]. As copper (II) does not degrade biologically, the control of Cu (II) pollution has special importance for both organisms that live in waters and those that benefit from waters.

Clay minerals have shown a range of sorption capabilities for different heavy metal ions. For example, kaolinite clay was used to remove Cu (II), Pb (II), Zn (II), and Ni (II) in aqueous medium and has been investigated through the process of sorption under a set of varied conditions [14,15]. The removal of metal ions using kaolinite clay is based on ion exchange and sorption mechanism even though kaolinite exhibits the least exchangeability among clay minerals sorption [16,17].

To study the equilibrium sorption of metal ions, it is important to understand the fundamental differences between sorption and ion exchange isotherms. For sorption studies, the isotherm is applicable for a particular system if the temperature remains constant. Though, in the case of ion exchange, the system is not only at a constant temperature but also of constant solution normality and as concluded in a study by Limousin et al. [18], if constant solution normality is not satisfied, then the uniqueness of the isotherm is not acceptable. In this regard, Millar et al. [19] assessed the influence of the bottle-point technique to generate equilibrium ion exchange isotherms. It was found that the constant-mass method in which the concentration of the metal ions was diverse while the sorbent mass was preserved constant seemed to be unreliable in determining the maximum loading of ions on the sorbent surface. Alternatively, the constant-concentration methodology with a solution of constant normality and different sorbent masses, was observed as a more reliable method and emphasised that the ion exchange phenomenon was required to be studied not only at isothermal but also isonormal conditions [20]. In this regard, there are several isotherm statements described in the literature for the fitting of equilibrium data [18] such as Langmuir, Freundlich, Hill models, etc., which apply to the experimental equilibrium data to discover the important scientific observations.

To the best of our knowledge, no previous research has investigated the potential capacity of polymer-treated tailing at high solids in removing heavy metal ions from tailings pore water. Therefore, this study aims to examine the efficiency of polymer-treated tailings for Cu^{2+} sorption/desorption from the aquatic environment and to compare these data with those of untreated materials. The hypothesis was that the porous, non-homogenous structure of polymer-treated materials could allow more Cu^{2+} ions removal and retention. In this regard, Figure 1 presents the tailings profile after flocculation and deposition at the tailings storage facility (TSF) with the potential aggregates for removing the heavy metal ions. Research questions that were addressed include: (1) what are the maximum Cu^{2+} ion exchange capacities of polymer treated and untreated tailings? (2) What equilibrium and kinetic model fits the experimental data more optimally? (3) How does pH impact the sorption and desorption behaviour of materials? (4) Do the polymer-treated sorbents have an acceptable breakthrough time? The investigation comprised three stages. In the first stage, the morphological structure of the polymer-treated and untreated tailings were investigated using the Cryo-SEM imaging and BET analysis. Then the bottle-point method constant concentration method was applied for both the sorption and desorption experiments and studying the impact of pH. The third stage consisted in determining the capacity of materials in removing the copper ions through the column leaching tests and the behaviour of the materials over time in removing the copper ions.

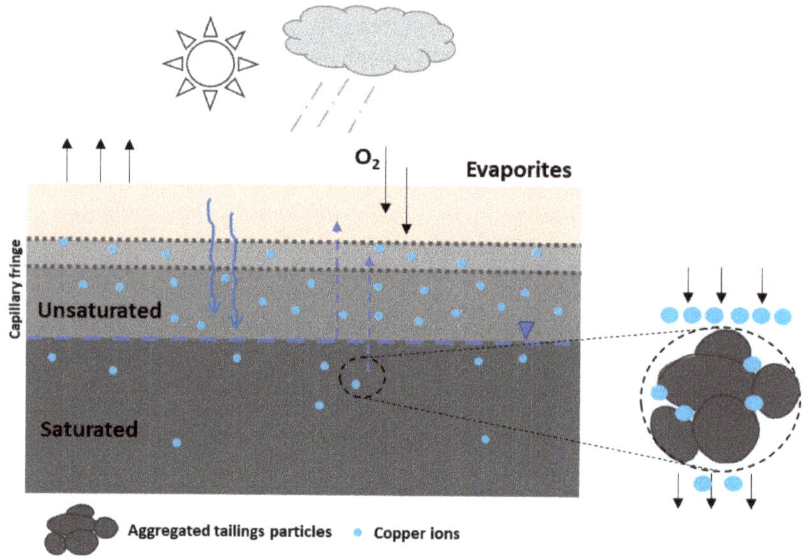

Figure 1. Tailings profile after dewatering and deposition on TSF.

2. Materials and Methods

2.1. Materials

Kaolinite-NY prestige, silica 200 G silt, and silica fine sand were purchased from Sibelco Company, Australia, to prepare the initial slurry. All three slurry components are produced from chemically treated and water-washed high purity materials and the milling process ensures minimal contamination, resulting in high purity products. Table 1 depicts the dominant minerals for each slurry ingredient in this study. Magnafloc® 336 was selected because it gave higher water recovery (more than 60%) with reasonable final solid properties after specific time in preliminary testing under saline conditions. BASF's Magnafloc® products are all acrylamide/acrylate copolymers with their general behaviour considered representative of nominally equivalent products available from other flocculant suppliers. Magnafloc® 336 is a very high molecular weight anionic polyacrylamide flocculant supplied as a free-flowing granular powder. The intrinsic viscosity (IV) of the flocculant, as specified by the supplier, was 24 dL/g, providing a relative indication of its solution dimensions. Higher molecular weight (MW) polymers typically offer a larger hydrodynamic size of chains in solution, with increasing MW producing loops and, more importantly, long extended tails that can form bridges with other particles to create aggregates [21]. At higher ionic strength, which was applied in this study, the polymer chains are expected to be more mobile and flexible on the colloidal surface, which allows the restructuring of aggregated clusters to proceed unhindered [22]. Copper sulfate pentahydrate ($CuSO_4{\cdot}5H_2O$) purchased from Sigma Aldrich was used to prepare the copper ions solution for either the batch or the continuous sorption experiments.

Table 1. Synthetic slurry mineralogy.

	SiO_2 (%)	Al_2O_3 (%)	Fe_2O_3 (%)	TiO_2 (%)
Kaolinite-NY prestige	49.2	35	0.94	0.96
Silica 200 G	99.6	0.17	0.02	0.03
Silica fine sand	98.5	0.1	0.3	0.9

2.2. Slurry and Polymer Preparation

A synthetic homogenous tailings slurry was prepared in 0.6 M NaCl solution to avoid the natural variability that exists with real tailings samples and to enable the preparation of large batches for testing. Slurry was prepared at 50 wt% solids from a combination of 25% kaolin, 45% silica 200 G (93% finer than 53 μm), and 30% silica fine sand (99% finer than 0.3 mm), as described elsewhere [23]. The specific gravity of the slurry solid particles was 2.64. Conditioning of the slurries was conducted at 1200 rpm for five days in a baffled 200 L tank to ensure a stable, reproducible size for the duration of testing which was confirmed by laser diffraction measurements.

Working polymer solutions (0.4 wt%) were prepared in the same saline solution used for slurry make-up. The weighed polymer was added slowly to the vortex produced in salted water under vigorous mixing to fully disperse the powder. Initial strong agitation with a 3-bladed axial flow impeller with blade dimension of each 7 cm × 2.2 cm (Lightnin A310) at 250 rpm for 30 min was followed by gentle agitation (80–100 rpm) for 16 h to produce clear, homogenous solutions. Flocculant solutions were aged for 3 days in a cool room (4 °C) to ensure optimal activity before adding to the slurry and producing polymer amended tailings [24].

2.3. Inline Polymer Addition

Figure 2 shows the lab-scale experimental setup used to simulate the inline flocculation process. To control the polymer dosage (1000 g/t), polymer solution (45 mL/min) and slurry (0.25 L/min) were pumped to the inlet of the pre-conditioning coil (1 m length, 7.7 mm I.D.) before entering a low-shear chaotic mixer for continued mild mixing. The final products were collected from the discharge point for further testing [25].

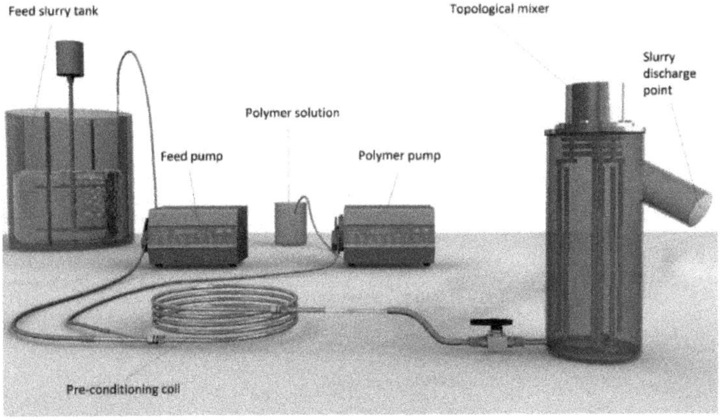

Figure 2. Tapered shear setup for polymer addition to high-solids suspensions [26].

2.4. Cryo-SEM

Morphological properties of the aggregated tailings were investigated using Cryo-SEM. An EM HPM 100 High-Pressure Freezer (HPF) technique was applied to prevent ice crystal formation or soluble aggregation that could provide poor vitrification. Samples were quickly frozen by a jet of high pressure after placing into the HPF and then was fractured with a knife in the cold preparation-chamber to be ready for coating with iridium at 10 mA for 120 s at around 5–10 nm thickness [27]. Scanning electron microscopy (JEOL JSM 7001F FESEM, JEOL, Tokyo, Japan) was operated at different voltages and magnifications to compare the treated and untreated tailings.

2.5. BET

Specific surface area and pore size distribution were defined using Brunauer–Emmett–Teller (BET) and Barrette–Joynere–Halenda (BJH) measurements on a Micromeritics TriStar 3020 analyser (TriStar II Plus, Micromeritics Instrument Corp, Norcross, GA, USA) after degassing under vacuum on a Micromeritics VacPrep061. Multi-point BET and BJH sorption isotherms models were employed to calculate the specific surface areas [15,28].

2.6. Cation Exchange Capacity (CEC) on ICP-OES

For CEC analysis, the collected water samples after each experiment were mixed with silver thiourea (AgTU+) for 24 h. Then, to determine the exchangeable base concentrations, the centrifuged and filtered extracts using a Varian Vista Pro ICP-OES (Inductively Coupled Plasma Optical Emission Spectrometry) (, Varian, Inc, Santa Clara, CA, USA) apparatus were analysed, and CEC was determined by calculating the total of silver ions that were exchanged. ICP-OES analysis was performed using the standard high-concentration wavelengths for each element. 328.0 nm for Ag, 38.215 nm for Al, 317.933 nm for Ca, 766.491 nm for K, 279.553 nm for Mg and 588.995 nm for Na, though other wavelengths can be used if matrix effects make these lines unsuitable [29].

2.7. Methodology for Bottle-Point Method Constant Concentration

For the copper isotherm test, duplicate equilibrium tests were performed by adding an accurately weighed mass of sorbent (1–50 g), i.e., polymer-treated (PT) and untreated materials (UT), into a set of 200 mL Nalgene flasks including copper ions solution with the constant concentration of 200 mg/L. This concentration was chosen to show the equilibrium at the end with attention to the selected range of mass of sorbent after some preliminary works similar to what was shown in previous research investigating the equilibrium sorption of metal ions [13,17,30]. Stock solutions were prepared by dissolving $CuSO_4 \cdot 5H_2O$ in Millipore water. The equilibrium time was previously obtained from dynamic tests and shown to be less than 24 h (details are mentioned in the kinetic study section). The samples were stirred at a low speed (25 rpm) for 24 h at room temperature. Both the untreated and polymer-treated material had a constant pH in the range of 6.5–7.5 [25], while the 200 mg/L copper solution pH was 5.2. Eventually, pH was recorded after the experiment and found to be in the range of 5.5–6.5 for all samples, then the liquid and solid phase were separated by centrifuging (5 min with 2500 rcf), and the liquid phase was syringe filtered. The concentration of Cu^{2+} ion remaining in the solutions after equilibrium, Ce (mg/L), was measured using ICP-OES and the equilibrium loading of copper ions on the UT/PT phase (q_e (mg/g)) deduced from below equation.

$$q_e = V/m \times (C_0 - C_e) \quad (1)$$

where V is the solution volume (L), m is the mass of sorbent (g), and C_o is the initial concentration of the copper ion (mg/L).

2.8. Sorption Isotherm Models

Based on the published studies, between the non-linear least squares (NLLS) and linear least squares (LLS) approaches for fitting isotherm data in both sorption and ion exchange isotherms, the NLLS method is preferable based on the way of estimating the parameters and the error functions [31,32]. Fitting using non-linearised versions of equations was applied to calculate the R^2 values plus the error functions to assess the different models and find the best fit on experimental data by using the Microsoft Solver add-on in Excel.

The mechanism of the sorption can be evaluated based on the theoretical isotherm models and the corelated paraments of each model can describe the properties of the sorbent in the presence of the used metal ions in the solution during the sorption [21]. For example, the Langmuir model which follows the equation 2 estimates that the metal ions sorption take place on a monolayer homogenous surface without any interaction between adsorbed ions.

$$q_e = C_e/(1/bq_0 + C_e/q_0) \tag{2}$$

where q_e is the amount adsorbed (mg/g), C_e is the equilibrium concentration of the metal ions (mg/L), and q_0 (mg/g) and b (L/mg) are Langmuir constants.

Another common isotherm model is the Freundlich model that takes the Equation (3) which is an empirical equation based on sorption on a heterogeneous surface. Freundlich model is not restricted to the monolayer formation in which its application to the multilayer sorption is possible and defines the heterogeneity of the surface as well as the exponential distribution of the active sites and the active sites energies [33]. It proved that the stronger binding sites will be taken and then an exponential decrease in the sorption force will happen upon completing the sorption process.

$$q_e = K_f {C_e}^{1/n} \tag{3}$$

where q_e is the amount adsorbed (mg/g), C_e is the equilibrium concentration of the metal ions (mg/L), and K_f ($mg^{1-1/n}$ $L^{1/n}$ g^{-1}) and n (g/L) are the Freundlich constants and specify the sorption capacity and the sorption strength, respectively [14].

Finally, the Hill isotherm model that supports different species' binding onto homogeneous surfaces, is also a common isotherm model with sorption occurring as a cooperative phenomenon in which the ions have the capability of binding into one location on the sorbent, which could be influencing other binding sites on the same sorbent. Originally, the equation of this sorption isotherm model was developed from the non-ideal competitive sorption isotherm [34]. The non-linear form of this isotherm model is expressed as follows:

$$q_e = (q_{max}C^{\alpha})/(K^{\alpha} + {C_e}^{\alpha}) \tag{4}$$

where q_e is the amount adsorbed (mg/g), C_e is the equilibrium concentration of the metal ions (mg/L), q_{max} (mg/g), and K and α are Hill model constants. Likewise, in this model, if $\alpha = 1$ it indicates that binding is hyperbolic or non-cooperative, if $\alpha > 1$, binding has positive cooperativity, while negative cooperativity happens when $\alpha < 1$ [35].

2.9. Kinetic Studies

The kinetic tests were performed using an initial concentration of 200 mg/L copper with a volume of 50 mL and UT/PT mass of 5 g, at room temperature. Thirty similar bottles in the same condition were placed on a shaker with a speed of 25 rpm for untreated and treated samples. At different time intervals, a bottle was taken, and the supernatant was decanted for transferring to the centrifuge and then analysing by ICP for the residual copper. To understand the dynamic interactions of ions with solid phases and to predict their fate with time, the copper ions sorption constants were calculated using the non-linear form of the pseudo-first-order and pseudo-second-order equations (Equations (5) and (6), respectively).

$$dq_t/dt = k_1(q_e - q_t) \tag{5}$$

$$dq_t/dt = k_2(q_e - q_t)^2 \tag{6}$$

where q_t is copper ions loading at a time (t), q_e is copper ions loading at equilibrium, t represents time, k_1 and k_2 are the pseudo-first-order and pseudo-second-order rate constants. As mentioned earlier, to ensure the accuracy of the results, analysis by NLLS is crucial which requires the use of error functions and the SNE (sum of normalised errors) term should be calculated to introduce the appropriate error function that was studied in detail in a study by Foo and Hameed [36]. Thus, here in this study, based on the calculated data with different error function using Solver function in Microsoft Excel, the suitable fitted model and its parameters were presented.

2.10. Desorption Studies

The same samples that were used and loaded with copper ions during the sorption experiments in different masses of UT/PT samples were used for the desorption tests after decanting their supernatant and replacing it with the constant volume of DI water. Samples were placed on the shaker for 24 h and then analysed for determining the released copper ions to solution.

2.11. Effect of pH on Sorption and Desorption

To investigate the impact of pH on sorption and desorption of copper ions, initially the pH of the solutions was adjusted by using HCl and NaOH for six different points. Then, sorption tests were conducted on 50 mL bottles of 200 mg/L copper concentration containing 5 g of sorbent mass for each test in 24 h shaking conditions. For desorption experiments, 5 g of previously loaded masses of UT/PT samples were transferred to the newly prepared bottles of 50 mL DI water at different pH and were placed on the shaker for 24 h. All the solutions were analysed using ICP for calculating the concentration of copper.

2.12. Column Leaching Experiments

Column tests were performed to investigate the Cu^{2+} removal using untreated and polymer-treated tailings. The Cu^{2+} initial concentrations were 200 ppm, with initial pH values of 6. All experiments were conducted at ambient temperature (25 °C). The fixed bed studies were performed in a 115 mm long acrylic column of 75 mm internal diameter (Figure 3). Untreated and polymer-treated materials were used in the column, which gave a packed bed with a pore volume of 211.2 mL and 261.6 mL, respectively. The copper solution was introduced at a constant volumetric flow rate (Q) of 10 mL/min using a peristaltic pump from the bottom of the column to ensure maximum contacting of the solution with the tailing particles and remove the gravity impact on conducting the copper solution through the porous media. The effluent stream was sampled for ICP analysis after each pore volume. By plotting the exit metal concentration versus time (or pore volume PV), the breakthrough curves were obtained.

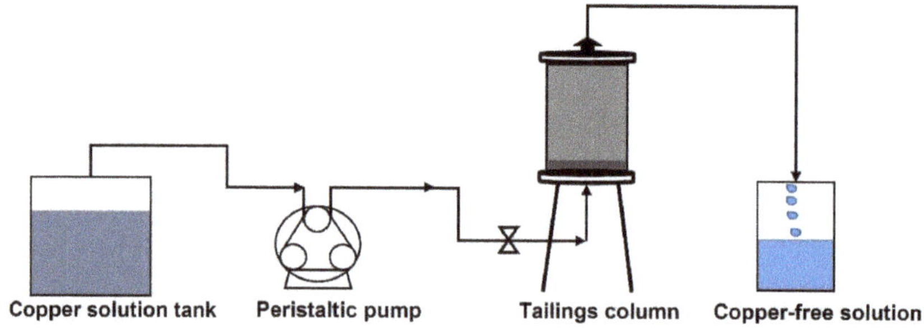

Figure 3. Column leaching sorption experimental setup.

3. Results

3.1. Morphological Structure Characterisations

Cryo-SEM imaging was used to study the effect of polymer addition on the morphological structure of the tailings' particles. Figure 4 compares the morphological structure of untreated and polymer-treated materials in different magnifications through inline flocculation. After adding the polymer, the raw slurry particles were agglomerated and compacted on a configuration that potentially makes the sample with greater porosity as presented in our previous study that showed polymer addition increased the void ratio, and consequently the saturated hydraulic conductivity, of the materials due to producing a porous structure [37]. The pH of the pore solution during flocculation was in the range of 6.5–7.5 which caused the kaolinite particle edges to be positively charged and particle faces to be negatively charged as shown in previous studies [38]. In this configuration, more positively charged sorption sites would be available for the negatively charged polymers and in the presence of a high concentration of NaCl, the van der Waals forces prevailed in drawing particles towards each other. This, together with hydrogen bonding, allowed more contact opportunities for kaolinite edge and face surfaces to form a combination of edge-to-edge (E-E) and edge-to-face (E-F) forms aggregates in high-salinity water [28].

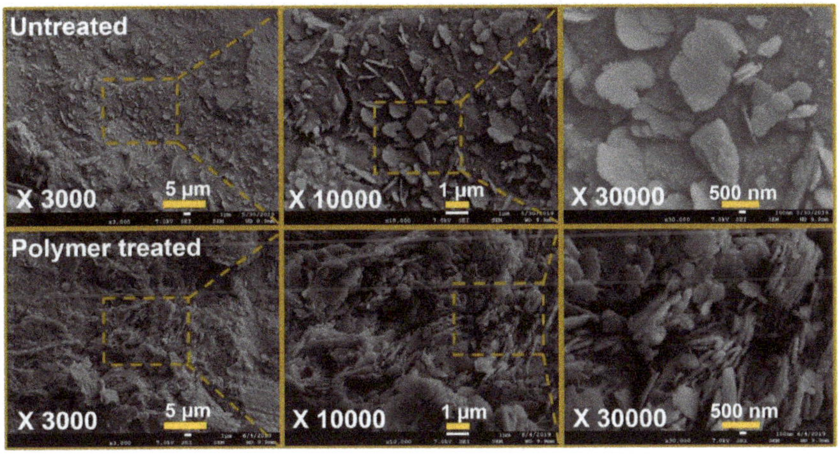

Figure 4. Cryo-SEM imaging of untreated and polymer-treated materials in three different magnifications.

To better understand the morphological structure of untreated and polymer-treated samples, the BET isotherm plots were employed to calculate the specific surface area of tailings before and after polymer addition (Figure 5A,C) [39]. According to the corresponding Barrett–Joyner–Halenda (BJH) pore size distribution curves in Figure 5B,D, the vertical axis (dV/dD) representing the rate of change of pore volume with pore diameter increased with the increasing pore diameter which tended to be flat when the pore diameter approached 450 A°. It seems a significant proportion of the pore areas with diameter less than 50 A° in the untreated sample converted to the larger pores after the polymer addition with the maximum peak around 100–150 A°. It is worth mentioning that the BET testing was done in the same conditions for the UT and PT samples to highlight their differences in structure and pore size distribution and the absolute values do not matter. The calculated sorption percentages for each sample during the designed sorption process are independent of the reported specific surface area measurement of each sample and the interaction of ions with the active sites of the sorbents depends on the applied conditions [40,41].

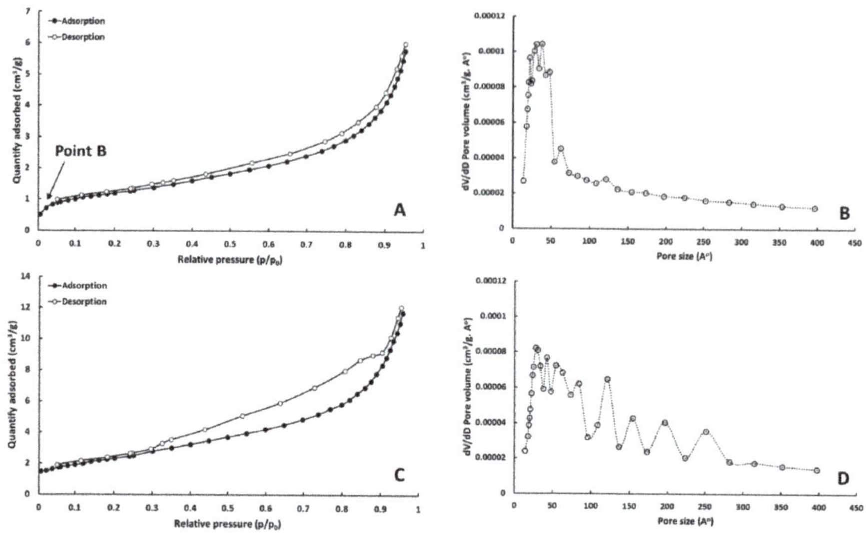

Figure 5. The N_2 gas sorption/desorption isotherm loop for (**A**) untreated and (**C**) polymer-treated tailings plus a histogram of the pore size distribution of (**B**) untreated and (**D**) polymer-treated tailings.

3.2. Sorption Kinetics and Isotherms of Copper Ions

In this study, the pseudo-first-order and pseudo-second-order models were studied to understand the kinetic behaviour of copper ions sorption by UT and PT materials. Figure 6 shows the Cu^{2+} sorption against time for both the untreated and polymer-treated samples with similar trends but a different rate of copper loading on the materials.

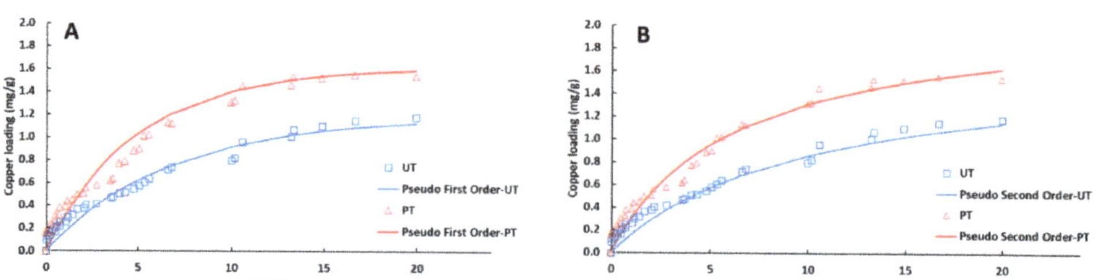

Figure 6. (**A**) Pseudo-first-order and (**B**) pseudo-second-order kinetic models for sorption of copper ions on UT and PT materials.

The copper ions sorption was relatively faster for the polymer-treated sample with over 60% removal during the first 5 h contacting the copper solution with the materials. Similar fast removal of copper ions was also reported in the study of removing heavy metal ions using only kaolinite [42]. The higher sorption degree of both samples at the beginning of the sorption process is because of the higher initial copper concentration, which results in a higher driving force for sorption.

Kinetic constants were verified using NLLS fits of pseudo-first and second-order equations (Equations (5) and (6)) and the calculated constants derived from these models are given in Table 2. From a statistical point-of-view and regression coefficient (R^2) values, the pseudo-first-order fitted the experimental data more accurately for the untreated materials while the pseudo-second-order showed the better fitting on the polymer-treated tailings data.

Table 2. Kinetic constants for the copper ions sorption of untreated and polymer-treated tailings.

Sample	q_{exp} (mg/g)	Pseudo-First-Order				Pseudo-Second-Order			
		q_e (mg/g)	K_1	*SNE*	R^2	q_e (mg/g)	K_2	*SNE*	R^2
UT	1.4	1.34	0.14	4.18	0.97	1.85	0.09	3.96	0.93
PT	2.05	1.32	0.18	4.25	0.91	2.12	0.07	4.02	0.98

In addition, Figure 7 presents the equilibrium sorption data with their fitted models that was performed over 24 h (with kinetic data confirming that equilibrium conditions had been reached). Generally, based on the graphs in Figure 7, copper sorption using polymer-treated tailings was found to be beneficial, meaning the sorption of copper ions occurred even at the low equilibrium concentration [43]. Application of Langmuir expression to the equilibrium Cu^{2+} data for untreated and polymer-treated samples is presented in Figure 7A,B, and did not provide a satisfactory fit to the experimental data for both sorbents.

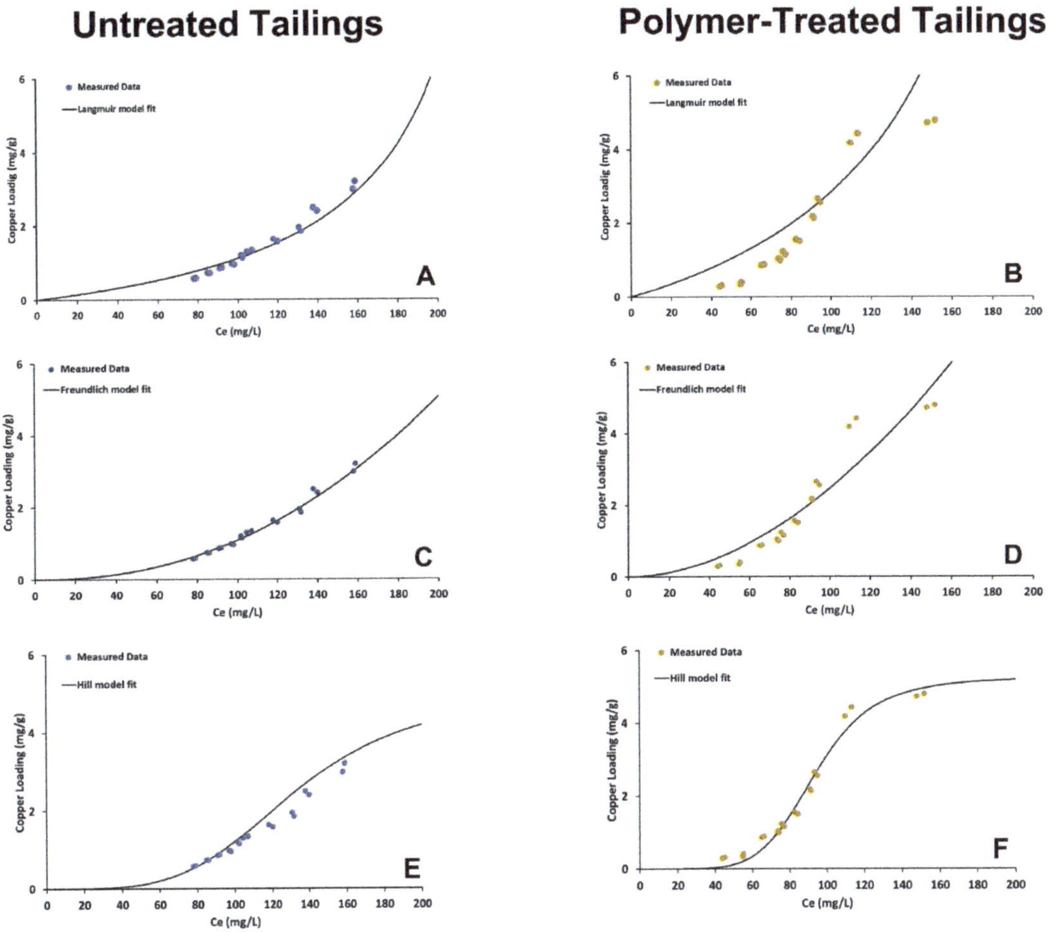

Figure 7. Non-linear least squares (NLLS) model fits for copper ion sorption for untreated and polymer treated materials. (**A**) Langmuir model fit-UT, (**B**) Langmuir model fit-PT, (**C**) Freundlich model fit-UT (**D**) Freundlich model fit-PT, (**E**) Hill model fit-UT, (**F**) Hill model fit-PT.

The Freundlich isotherm only showed good fitting to the sorption data of copper exchange (Figure 7C) with untreated tailings with the maximum copper loading value of 3.43 mg/g. Freundlich model fitted the copper sorption data sufficiently with the Freundlich exponent value of higher than 1 (1.45) which indicates the exchange process between copper ions and the protons on the untreated slurry sites. The model assumption of the heterogeneous surface with not equivalent and independent binding sites was also proven. In contrast, the Freundlich isotherm model was found to be unfavourable for polymer-treated tailings sample (Figure 7D). The empirical factor ($1/n$), which indicates the favourable sorption when sitting within the range of ($0.1 < 1/n < 1$), were 0.67 and 1.9 for the untreated and polymer-treated materials, respectively.

The Hill equation provided a better fitting to the experimental data of copper sorption by polymer-treated sample compared to the untreated sample. Hill equation fitting to equilibrium data of PT sample was satisfactory in the isotherm system at low equilibrium concentrations with R^2 value of 0.98 and maximum copper loading of 5.24 mg/g.

3.3. *Effect of Sorbent Mass on Sorption/Desorption of Copper Ions*

The sorption and desorption percentage of copper ions on untreated and polymer-treated material are presented in Figure 8. As was expected based on the morphological studies and BET data, the polymer-treated sample having a more porous structure and more active sites for ion exchange of copper ions showed higher percentage of sorption over different masses of sorbent with an increasing trend. In fact, by adding the polymers into the tailings slurry, particularly at high salinity (in our study 0.6 M NaCl), the polymer chains are expected to be more mobile and flexible on the colloidal surface, which allows the restructuring of aggregated clusters to proceed unhindered. This interpretation is consistent with previous research examining aggregate size-density relationships in freshwater and seawater for different MW polymers [44]. This kind of behaviour of the polymer chains in flocculating the tailings particles provides the situation to expose more surface of the minerals to the copper ions during the sorption and ion exchange processes.

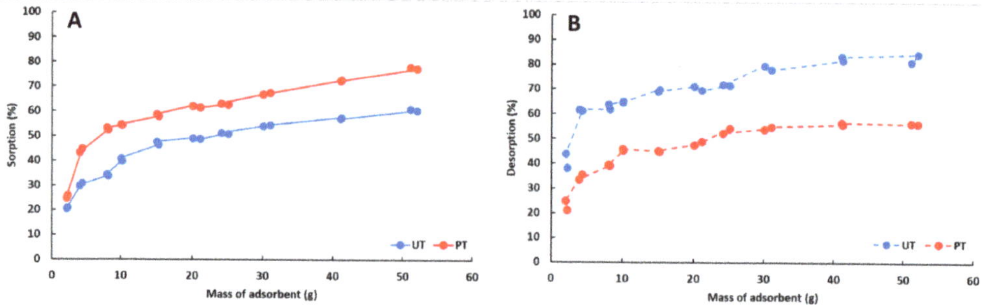

Figure 8. (**A**) Sorption and (**B**) desorption percentage for untreated and polymer treated tailings.

In contrast, the desorption changes by increasing the mass of sorbent of untreated sample was always greater than the polymer-treated sample which indicates that it was hard to desorb the already adsorbed Cu^{2+} on the structure of the polymer treated materials.

The maximum sorption was recorded around 80% at 50 g sorbent sample for the polymer-treated sample in which this amount of the materials at the same condition showed only 56% desorption of copper ions into the DI water. On the other hand, the untreated materials with less capability of copper ion sorption with around 60% removal at the maximum applied sorbent loading (50 g) can readily release the adsorbed ions back into solution (81% after 24 h). This was indicative of the weak binding between the copper ions and active sites of the untreated sample particles and less capability of this material for preserving the adsorbed heavy metal ions within their loosely coagulated structure. In fact, having the constant concentration of metal ions and increasing the mass of the sorbent is like the situation in which the mass of sorbent is constant, and the metal ions concentration is reducing [45]. By increasing the mass of sorbent, the available surface and porous sites to adsorb Cu^{2+} are increasing which contributed to enhancing the driving force at the solid–liquid interface to increase the sorption capacity until the sorbent reached its maximum capacity and plateaued in the removal percentage graphs, similar to what was presented in previous studies and is consistent with the results in Figure 8 [46,47]. It is worth acknowledging the impact of the polymer itself and without the clay materials on the removal of copper ions that was conducted in a separated study and not included here. Therefore, it is a combination of the tailings particles and the polymer impacts that provide conditions in which the copper ions removal will be improved by having more exposure to the polymer-treated material compared to the untreated sample.

3.4. Effect of the Solution pH on Sorption/Desorption of Copper Ions

Changes in pH may alter the surface charges of the sorbents and the behaviour of the metal ions, so the water pH is widely recognised as an important element affecting the sorption properties of heavy metal ions in aquatic conditions [48]. In this study, the sorption and desorption performance of Cu^{2+} onto the untreated and polymer-treated tailings significantly changed with variation of the solution pH (Figure 9). A substantial increase in sorption percentage was observed for both the untreated and polymer-treated samples by increasing the pH with greater impact for the polymer treated sample up to 98% removal at the maximum applied pH. It is suggested that by increasing the pH, electrostatic interactions might engage in important roles in the sorption of Cu^{2+} onto tailings particles [49]. In fact, by deprotonation of functional groups of the sorbents, the formation of the first-order $CuOH^{+}$ species are increased, which are known to be more surface active than the original Cu^{2+} ions and thus adsorb more readily than Cu^{2+} ions and improve the final adsorbed copper ions from the solution.

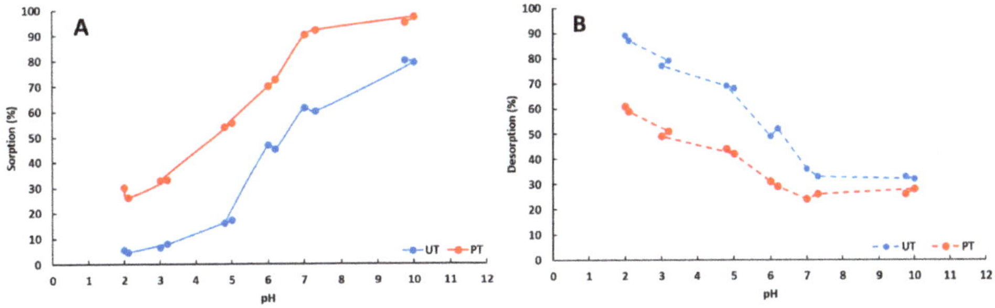

Figure 9. Effect of pH on (**A**) sorption and (**B**) desorption of copper ions on untreated and polymer-treated tailings.

The precipitation of the copper ions onto the surface of the materials was more intense during the desorption process where the increase of the pH resulted in a lower desorption rate of the copper ions into the solution for both samples. For the desorption changes over different pH ranges, opposite to the sorption process, the maximum rate of the desorption was recorded for both samples at lower pH values and was less for the polymer-treated sample, once again demonstrating that the adsorbed copper ions were hard to release. Then, by increasing the pH, the desorption percentage considerably reduced to less than 40% and the two curves tended to converge close to each other.

3.5. Column Leaching Experiments

Batch approach procedure studies have some restrictions and suffer from some undesirable issues such as limited treatment volume [50]. Therefore, sorption of copper ions by untreated and polymer-treated materials was conducted in continuous column experiments and the results are shown in breakthrough curves (Figure 10). As can be seen, untreated and polymer-treated materials performed differently towards Cu^{2+} sorption. The breakthrough happened after 2 PV (pore volume) and 5 PV for untreated and polymer-treated materials, respectively. The concentration of Cu^{2+} at the breakthrough point was found to be 40 mg/L for untreated tailings and for the polymer-treated sample, it was 16 mg/L. Breakthrough concentration can be shown as the minimum detectable or maximum allowable solute concentration in the effluent. As the process advanced for the PT sample, a continuing rise in Cu^{2+} concentration was seen for 15 PV until the saturation point, which took more time compared to the UT sample as the copper ions needed more time to permeate into the porous structure of the PT sample and it took a longer time to get the sorption saturation.

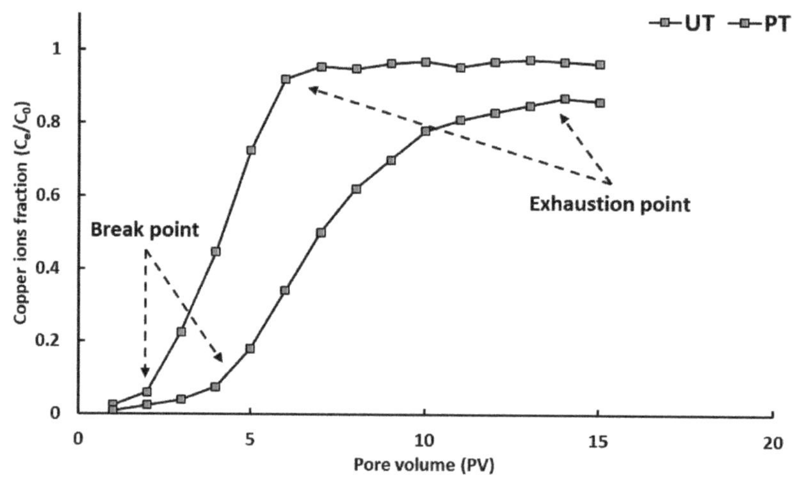

Figure 10. Breakthrough profile for untreated and polymer-treated tailings while treating the copper ions solution (200 ppm).

4. Discussion

In flocculation of mineral particles by inline polymer addition at high dosages, the particles are often bound with suitable cation sorption sites. The highly saline conditions applied in this study (0.6 M NaCl) have practical significance for mineral processing plants utilising seawater in different operations, with mineral surfaces being aggregated through covalent bonding or salt linkages between the anionic groups (e.g., carboxylate) on the polymer chain [22]. Hydrogen bonding may also occur between the polymer's amide groups and OH groups at the mineral surfaces. The resultant highly flocculated clay structures, as presented in the Cryo-SEM images and BET data in Section 3.1, provide further mesoporous media with greater internal voids that can be favourable for sorption of metal ions similar to other applicable mineral materials that are used for the heavy metal ions removal such as copper ions [43,48,51,52]. Moreover, in the presence of high salt, the polymer chains are expected to be more mobile and flexible on the colloidal surface, which allows the unhindered restructuring of aggregated clusters and more contacts with the metal ions to make a new binding to different sorption sites on the mineral particles.

According to IUPAC (International Union of Pure and Applied Chemistry) classification, the sorption-desorption isotherm of untreated materials can be assigned to the combination of Type II and Type III hysteresis while the polymer-treated materials behaviour only followed the Type III hysteresis model [39]. Type II isotherms are assumed by the physisorption on nonporous surfaces and the shape of the graph is related to the unrestricted monolayer-multilayer sorption up to higher p/p_0 values.

Based on the sorption conditions, the copper ions removal from the solution could have happened through the ion exchange with the cations on the tailings particles or via the chemical sorption and establishing bounds with the active sites on the surface of the particles or aggregates [47]. If the first chemical sorption layer begins, then due to surrounding effects such as the ion strength from the existed salts in the system or other minerals forces in the slurry, further layers are formed particularly in case of the polymer-treated sample where polymers chains are more flexible for affecting the sorption mechanism by their functional groups elements [53,54].

In the case of a Type III isotherm, there is no Point B (the PT data in Figure 5C) and consequently no detectable monolayer formation and the adsorbed molecules are clustered around the most favourable sites on the surface of a macro-porous solid [55]. The hysteresis loop on the N_2 gas sorption-desorption isotherm occurred in the range of 0.5 to 1.0 P/P_0 which is greater for the polymer-treated sample, implying the presence of more mesopores [56,57].

To understand the dynamic interactions of metal ions in the solution with solid phases and to predict their variation with time, knowledge of the kinetics of these processes is important [33]. Many researchers have linked the fitting of kinetic data to a specific model as an indication of the type of the sorption process to be either chemisorption or physisorption [58–60]. Table 2 showed that the regression coefficient for pseudo-first-order model of the polymer-treated tailings is low and the projected values of q_e was not reasonably close to experimental q_e values compared to that of untreated slurry, suggesting the insufficiency of pseudo-first-order model to fit the kinetic data of polymer-treated sample. On the other hand, the R^2 values for the pseudo-second-order kinetic model are nearly equal to 1 and the predicted values of q_e are comparable to the experimental ones for the polymer-treated sample compared to that of untreated sample. This implies that the sorption of copper ions onto the polymer treated materials is most likely a chemisorption process involving exchange or allocation of electrons mostly between metal ions and functional groups of the aggregates [61]. Although a simple case (single ion sorption) was studied with a certain concentration of the polymer for creating the aggregates, in presence of the salt, the interaction of the polymers with the minerals surfaces are not simple and provide different options for the copper ions sorption. Interactions of the polymers with mineral surfaces can be expected as a hydrogen bonding of the amide groups to neutral sites, covalent bonding, or salt linkage between the anionic groups. This provides a complex situation and why the kinetic experimental data of PT sample cannot be fitted by pseudo-first-order equation, and it is supposed to be a chemisorption phenomenon.

Regarding the pH impact on the sorption and desorption behaviour of copper ions, the electrostatic interaction between the copper ions and the PT or UT materials determines the extent of attenuation and release of ions. According to Low et al. [62], slight sorption of metal ions at lower pH could be attributed to the hydrogen ions competing with metal ions for exchangeable cations on the surface of the materials. Then, in the pH range from 3 to 7, there is an apparent increase of deprotonation of functional groups, particularly for the polymer-treated sample on the sorbent surface, permitting a change for the better sorption of copper ions [13]. Later, in the pH range from 7 to 10, a significant increase of metal removal is observed for the polymer-treated tailings that might be described by the fact that the sorption sites are no more affected by the pH change. In fact, at a higher pH condition, the copper ions precipitate in the form of $Cu(OH)_2$ and that reduces the rate of sorption observed in both UT and PT samples by reduction of copper ions in the solution after the experiment. This can also be supported by the increasing amount of Na^+ in the solution (attributable to pH adjustment by adding NaOH) which compete with the remaining Cu^{2+} on the exchangeable site. However, in a real situation in which a range of components such as organic flotation agents and other heavy metal ions exist, the selectivity sequence of metal ions by the sorbents varies based on the system employed and mainly depend on the initial concentrations of the metal ions and the pH of the solution (Jiang et al., 2010).

In our previous research on the same artificial tailings that assessed the hydrological properties of the raw slurry and polymer treated materials [37], the addition of polymer altered the physical characteristics of the materials by creating larger macropore volumes to drain more water easily. This material has higher infiltration rates in comparison with the raw slurry, which contains finer particles. Practically, addition of the polymer transforms some microporous media to a macroporous structure with the creation of new flow paths allowing the rapid drainage of run-off water [63]. These modifications along with the larger aggregated sizes are linked with the hydraulic conductivity and provide conditions to move water through the structure of the tailings differently in favour of contacting more with the aggregates. In addition, high salinity conditions are important due to the sorption of cations on tailings particle surfaces and are potentially beneficial to the inline flocculation process; however, there is a lack of published information available on the likely impacts of high saline process liquors on polymer performance. The demonstration of the inline flocculation technique in our previous study (Boshrouyeh Ghandashtani et al., 2022) has shown that it is possible to produce the desired highly aggregated network of tailings solids on deposition even in the presence of elevated concentrations of dissolved salts. This has practical significance for mineral processing plants utilising seawater or other brines.

The polymer treated materials has greater potential for removing the ions in the column experiments. Signs of faster exhaustion were observed for the untreated sample after around 6 PV with no more changes in the copper ions fractions by continuing the experiment. The fast saturation of untreated materials could also be related to the lower specific surface area which explains fewer active sites of this sorbent compared to the polymer-treated sample. These observations are consistent with the results of the batch experiments and revealed that the untreated slurry with lower capacity of copper ions removal saturated significantly faster when a higher volume of water was treated. By integration of the area under the breakthrough curve, it was found that 11 mg Cu^{2+}/g of PT sample had been loaded onto the sample after treating 15 PV. This value was calculated to be 6.5 mg Cu^{2+}/g of UT for the untreated tailings bed.

Further development of the underlying science is needed to determine whether non-homogeneous aggregate structures formed through inline flocculation can be beneficial for tailings rehabilitation. Polymer-amended tailings with altered permeability and distribution of aerated and water-filled pores that have the properties for controlling the metal ions release may favour vegetation establishment and growth at mine closure. These properties as well as the geochemical stability of the tailings in regard to release of potential contaminants as seepage could provide insights into the long-term processes for mine closure activities and rehabilitation plans.

5. Conclusions

Polymer-amended tailings through the inline flocculation process have the potential capability for the attenuation of the heavy metal ions due to their porous structure and the chemistry of the aggregates to adsorb heavy metal ions. To this end, a comprehensive study was conducted to investigate the attenuation and release of the copper ions by polymer-treated tailings and compare it with the untreated materials. It was found from the batch experiments that the maximum Cu^{2+} ions uptake capacity of polymer-treated tailings was 25% more than the raw untreated slurry at the equilibrium state. In addition, the Hill isotherm model was fitted on the data of the polymer-treated sample and the Freundlich model showed better fitting for the untreated tailings. This finding plus the behaviour of the material in the dynamic sorption study suggested that the copper ions removal using the polymer-treated sample was a chemisorption phenomenon while the sorption process on the untreated tailings was an unfavourable and reversible sorption. In contrast to sorption experiments at different pH, the desorption of copper ions decreased by increasing the pH. It also was seen that the desorption of copper ions was more difficult for the polymer-treated sample compared to that of untreated slurry. Finally, the fixed bed column experiments showed higher sorption capacity for the polymer treated with a slower saturation/exhaustion and greater maximum loading of copper ions. Therefore, based on the results in this study, the inline flocculation technique is a promising method for producing highly aggregated, porous structures suited for attenuation of the copper ions released into the solution by the weathering of tailings. This could potentially extend into the removal of the heavy metal ions from tailings pore solutions for rehabilitation activities.

Author Contributions: M.B.G.: conceptualisation, methodology, investigation, visualisation, writing—original draft. M.E.: methodology, validation, visualisation, writing—review and editing, supervision. T.B.: conceptualisation, writing—review and editing, supervision. A.C.: methodology, conceptualisation, writing—review and editing, supervision. S.A.: methodology, visualisation, writing—review and editing. All authors have read and agreed to the published version of the manuscript.

Funding: This research received no external funding.

Conflicts of Interest: The authors declare no conflict of interest.

References

1. Jeong, Y.; Yu, J.; Wang, L.; Lee, K. International Journal of Applied Earth Observations and Geoinformation Bulk scanning method of a heavy metal concentration in tailings of a gold mine using SWIR hyperspectral imaging system. *Int. J. Appl. Earth Obs. Geoinf.* **2021**, *102*, 102382. [CrossRef]
2. Rashed, M.N. Monitoring of contaminated toxic and heavy metals, from mine tailings through age accumulation, in soil and some wild plants at Southeast Egypt. *J. Hazard. Mater.* **2010**, *178*, 739–746. [CrossRef] [PubMed]
3. Jeong, Y.; Yu, J.; Wang, L. Spectral Responses of As and Pb Contamination in Tailings of a Hydrothermal Ore Deposit: A Case Study of Samgwang Mine, South Korea. *Remote Sens.* **2018**, *10*, 1830. [CrossRef]
4. Kemper, T.; Sommer, S. Estimate of Heavy Metal Contamination in Soils after a Mining Accident Using Reflectance Spectroscopy. *Environ. Sci. Technol.* **2002**, *36*, 2742–2747. [CrossRef]
5. Reid, C.; Bécaert, V.; Aubertin, M.; Rosenbaum, R.K.; Deschênes, L. Life cycle assessment of mine tailings management in Canada. *J. Clean. Prod.* **2009**, *17*, 471–479. [CrossRef]
6. Sabah, E.; Cengiz, I. An evaluation procedure for flocculation of coal preparation plant tailings. *Water Res.* **2004**, *38*, 1542–1549. [CrossRef]
7. Qi, C.; Fourie, A.; Chen, Q.; Tang, X.; Zhang, Q.; Gao, R. Data-driven modelling of the flocculation process on mineral processing tailings treatment. *J. Clean. Prod.* **2018**, *196*, 505–516. [CrossRef]
8. Arjmand, R.; Massinaei, M.; Behnamfard, A. Improving flocculation and dewatering performance of iron tailings thickeners. *J. Water Process Eng.* **2019**, *31*, 100873. [CrossRef]
9. Grabsch, A.F.; Fawell, P.D.; Adkins, S.J.; Beveridge, A. The impact of achieving a higher aggregate density on polymer-bridging flocculation. *Int. J. Miner. Process.* **2013**, *124*, 83–94. [CrossRef]
10. Anstey, D.; Guang, R. Relationship between In-line polymer dose and tailings index. In Proceedings of the Tailings and Mine Waste conference, Banff, AB, Canada, 5–8 November 2017; pp. 1–10.
11. Fawell, P.; Adkins, S.; Costine, A. Reagents. In *Paste and Thickened Tailings—A Guide*, 3rd ed.; Australian Centre for Geomechanics: Perth, WA, Australia, 2015; Chapter 6; pp. 87–108. [CrossRef]
12. Paunovic, P.; Dimitrov, A.T.; Jordanov, S.H. Leaching of Mine Tailings in Semi-Industrial Conditions. In Proceedings of the XIth Annual Conference Union Metall, Skopje, North Macedonia, 07–10 May 2007; pp. 36–41.
13. Chen, H.; Dai, G.; Zhao, J.; Zhong, A.; Wu, J.; Yan, H. Removal of copper(II) ions by a biosorbent-Cinnamomum camphora leaves powder. *J. Hazard. Mater.* **2010**, *177*, 228–236. [CrossRef]
14. Jiang, M.; Wang, Q.; Jin, X.; Chen, Z. Removal of Pb(II) from aqueous solution using modified and unmodified kaolinite clay. *J. Hazard. Mater.* **2009**, *170*, 332–339. [CrossRef]
15. Hierro, I.; Fajardo, M.; Sierra, I.; Pe, D. Mesoporous silica functionalized with 2-mercaptopyridine: Synthesis, characterization and employment for Hg (II) adsorption. *Microporous Mesoporous Mater.* **2006**, *89*, 58–68. [CrossRef]
16. Jiang, M.; Jin, X.; Lu, X.Q.; Chen, Z. Adsorption of Pb(II), Cd(II), Ni(II) and Cu(II) onto natural kaolinite clay. *Desalination* **2010**, *252*, 33–39. [CrossRef]
17. Suraj, G.; Iyer, C.S.P.; Lalithambika, M. Adsorption of cadmium and copper by modified kaolinites. *Appl. Clay Sci.* **1998**, *13*, 293–306. [CrossRef]
18. Limousin, G. Sorption isotherms: A review on physical bases, modeling and measurement. *Appl. Geochem.* **2007**, *22*, 249–275. [CrossRef]
19. Millar, G.J.; Outram, J.G.; Couperthwaite, S.J.; Leung, C.W. Separation and Puri fi cation Technology Methodology of isotherm generation: Multicomponent K+ and H+ ion exchange with strong acid cation resin. *Sep. Purif. Technol.* **2020**, *251*, 117360. [CrossRef]
20. Ho, Y.S.; Porter, J.F.; Mckay, G. Equilibrium isotherm studies for the sorption of divalent metal ions onto peat: Copper, nickel and lead single component systems. *Water. Air. Soil Pollut.* **2002**, *14*, 1–33. [CrossRef]
21. Lee, B.J.; Schlautman, M.A. Effects of Polymer Molecular Weight on Adsorption and Flocculation in Aqueous Kaolinite Suspensions Dosed with Nonionic Polyacrylamides. *Water* **2015**, *7*, 5896–5909. [CrossRef]

22. Arinaitwe, E.; Pawlik, M. A role of flocculant chain flexibility in flocculation of fine quartz. Part I. Intrinsic viscosities of polyacrylamide-based flocculants. *Int. J. Miner. Process.* **2013**, *124*, 50–57. [CrossRef]
23. Reid, D.; Fourie, A. Laboratory assessment of the effects of polymer treatment on geotechnical properties of low-plasticity soil slurry. *Can. Geotech. J.* **2016**, *53*, 1718–1730. [CrossRef]
24. Owen, A.T.; Fawell, P.D.; Swift, J.D.; Farrow, J.B. The impact ofpolyacrylamide flocculant solution age on flocculation performance. *Int. J. Miner. Process.* **2002**, *67*, 123–144. [CrossRef]
25. Boshrouyeh Ghandashtani, M.; Costine, A.; Edraki, M.; Baumgartl, T. The impacts of high salinity and polymer properties on dewatering and structural characteristics of flocculated high-solids tailings. *J. Clean. Prod.* **2022**, *342*, 130726. [CrossRef]
26. Costine, A.; Fawell, P.; Chryss, A.; Dahl, S.; Bellwood, J. Development of test procedures based on chaotic advection for assessing polymer performance in high-solids tailings applications. *Processes* **2020**, *8*, 731. [CrossRef]
27. Yu, L.; Yakubov, G.E.; Gilbert, E.P.; Sewell, K.; van de Meene, A.M.L.; Stokes, J.R. Multi-scale assembly of hydrogels formed by highly branched arabinoxylans from Plantago ovata seed mucilage studied by USANS/SANS and rheology. *Carbohydr. Polym.* **2019**, *207*, 333–342. [CrossRef] [PubMed]
28. Liu, D.; Edraki, M.; Berry, L. Investigating the settling behaviour of saline tailing suspensions using kaolinite, bentonite, and illite clay minerals. *Powder Technol.* **2018**, *326*, 228–236. [CrossRef]
29. Clesceri, L.S.; Greenberg, A.E.; Eaton, A.D. *Anionic Surfactants as MBAS*, 20th ed.; Port City Press: Baltimore, MD, USA, 1999.
30. Çay, S.; Uyanik, A.; Özaşik, A. Single and binary component adsorption of copper(II) and cadmium(II) from aqueous solutions using tea-industry waste. *Sep. Purif. Technol.* **2004**, *38*, 273–280. [CrossRef]
31. Mckay, G.; Mesdaghinia, A.; Nasseri, S.; Hadi, M.; Solaimany, M. Optimum isotherms of dyes sorption by activated carbon: Fractional theoretical capacity & error analysis. *Chem. Eng. J.* **2014**, *251*, 236–247. [CrossRef]
32. Lee, I.; Kuan, Y.; Chern, J. Equilibrium and kinetics of heavy metal ion exchange. *J. Chinese Inst. Chem. Eng.* **2007**, *38*, 71–84. [CrossRef]
33. Donald, L. Sparks. In *Kinetics of Soil Chemical Processes*; Elsevier: San Diego, CA, USA, 1989.
34. Koopal, L.K.; van Riemsdijk, W.H.; de Wit, J.C.M.; Benedetti, M.F. Analytical Isotherm Equations for Multicomponent Adsorption to Heterogeneous Surfaces. *J. Colloid Interface Sci.* **1994**, *166*, 51–60. [CrossRef]
35. Al-Ghouti, M.A.; Da'ana, D.A. Guidelines for the use and interpretation of adsorption isotherm models: A review. *J. Hazard. Mater.* **2020**, *393*, 122383. [CrossRef]
36. Foo, K.Y.; Hameed, B.H. Insights into the modeling of adsorption isotherm systems. *Chem. Eng. J.* **2010**, *156*, 2–10. [CrossRef]
37. Boshrouyeh Ghandashtani, M.; Edraki, M.; Costine, A.; Baumgartl, T. Investigation of geotechnical, hydromechanical, and chemical behaviour of polymer-treated tailings. In Proceedings of the Paste 2021: 24th International Conference on Paste, Thickened and Filtered Tailings, Perth, WA, Australia, 21–23 September 2021; Fourie, A.B., Reid, D., Eds.; Australian Centre for Geomechanics PP—Perth: Perth, WA, Australia, 2021; pp. 79–90. [CrossRef]
38. Jeldres, R.I.; Piceros, E.C.; Leiva, W.H.; Toledo, P.G.; Herrera, N. Viscoelasticity and yielding properties of flocculated kaolinite sediments in saline water, Colloids Surfaces A Physicochem. *Eng. Asp.* **2017**, *529*, 1009–1015. [CrossRef]
39. Thommes, M.; Kaneko, K.; Neimark, A.V.; Olivier, J.P.; Rodriguez-Reinoso, F.; Rouquerol, J.; Sing, K.S.W. Physisorption of gases, with special reference to the evaluation of surface area and pore size distribution (IUPAC Technical Report). *Pure Appl. Chem.* **2015**, *87*, 1051–1069. [CrossRef]
40. Rouquerol, F.; Rouquerol, J.; Sing, K.S.W.; Maurin, G.; Llewellyn, P. *Adsorption by Powders and Porous Solids, Second*; Academic Press: Cambridge, MA, USA, 2014.
41. Lowell, S.; Shields, J.E.; Thomas, M.A.; Thommes, M. *Characterization of Porous Solids and Powders: Surface Area, Pore Size and Density*; Springer: Berlin/Heidelberg, Germany, 2004.
42. Shahmohammadi-Kalalagh, S.; Babazadeh, H.; Nazemi, A.H.; Manshouri, M. Isotherm and Kinetic Studies on Adsorption of Pb, Zn and Cu By Kaolinite. *Casp. J. Environ. Sci.* **2011**, *9*, 243–255.
43. Hui, K.S.; Chao, C.Y.H.; Kot, S.C. Removal of mixed heavy metal ions in wastewater by zeolite 4A and residual products from recycled coal fly ash. *J. Hazard. Mater.* **2005**, *127*, 89–101. [CrossRef]
44. Jeldres, R.; Yang, T.; Costine, A.; Fawell, P.; Bellwood, J. The impact of high salinity and seawater on aggregate structures in clay tailings flocculation. In Proceedings of the 23rd International Conference on Paste, Thickened and Filtered Tailings—PASTE, Santiago, Chile, 10–12 June 2020; pp. 1–8. [CrossRef]
45. Ouyang, D.; Zhuo, Y.; Hu, L.; Zeng, Q.; Hu, Y.; He, Z. Research on the adsorption behavior of heavy metal ions by porous material prepared with silicate tailings. *Minerals* **2019**, *9*, 291. [CrossRef]
46. Wang, W.; Tian, G.; Zhang, Z.; Wang, A. A simple hydrothermal approach to modify palygorskite for high-efficient adsorption of Methylene blue and Cu(II) ions. *Chem. Eng. J.* **2015**, *265*, 228–238. [CrossRef]
47. Duan, P.; Yan, C.; Zhou, W.; Ren, D. Development of fly ash and iron ore tailing based porous geopolymer for removal of Cu(II) from wastewater. *Ceram. Int.* **2016**, *42*, 13507–13518. [CrossRef]
48. Melo, D.Q.; Neto, V.O.S.; Oliveira, J.T.; Barros, A.L.; Gomes, E.C.C.; Raulino, G.S.C.; Longuinotti, E.; Nascimento, R.F. Adsorption equilibria of Cu^{2+}, Zn^{2+}, and Cd^{2+} on EDTA-functionalized silica spheres. *J. Chem. Eng. Data* **2013**, *58*, 798–806. [CrossRef]
49. Zhou, Y.; Yang, Y.; Liu, G.; He, G.; Liu, W. Adsorption mechanism of cadmium on microplastics and their desorption behavior in sediment and gut environments: The roles of water pH, lead ions, natural organic matter and phenanthrene. *Water Res.* **2020**, *184*, 116209. [CrossRef]

50. Shahbazi, A.; Younesi, H.; Badiei, A. Functionalized SBA-15 mesoporous silica by melamine-based dendrimer amines for adsorptive characteristics of Pb(II), Cu(II) and Cd(II) heavy metal ions in batch and fixed bed column. *Chem. Eng. J.* **2011**, *168*, 505–518. [CrossRef]
51. Milićević, S.; Vlahović, M.; Kragović, M.; Martinović, S.; Milošević, V.; Jovanović, I.; Stojmenović, M. Removal of copper from mining wastewater using natural raw material—comparative study between the synthetic and natural wastewater samples. *Minerals* **2020**, *10*, 753. [CrossRef]
52. Esmaeili, A.; Mobini, M.; Eslami, H. Removal of heavy metals from acid mine drainage by native natural clay minerals, batch and continuous studies. *Appl. Water Sci.* **2019**, *9*, 1–6. [CrossRef]
53. Benzaoui, T.; Selatnia, A.; Djabali, D. Adsorption of copper (II) ions from aqueous solution using bottom ash of expired drugs incineration. *Adsorpt. Sci. Technol.* **2017**, *36*, 114–129. [CrossRef]
54. Scheufele, F.B.; Módenes, A.N.; Borba, C.E.; Ribeiro, C.; Espinoza-Quiñones, F.R.; Bergamasco, R.; Pereira, N.C. Monolayer–multilayer adsorption phenomenological model: Kinetics, equilibrium and thermodynamics. *Chem. Eng. J.* **2016**, *284*, 1328–1341. [CrossRef]
55. Xu, L.; Zhang, J.; Ding, J.; Liu, T.; Shi, G.; Li, X.; Dang, W.; Cheng, Y.; Guo, R. Pore Structure and Fractal Characteristics of Different Shale Lithofacies in the Dalong Formation in the Western Area of the Lower Yangtze Platform. *Minerals* **2020**, *10*, 72. [CrossRef]
56. Asadullah, M.; Kabir, M.S.; Ahmed, M.B.; Razak, N.A.; Rasid, N.S.A.; Aezzira, A. Role of microporosity and surface functionality of activated carbon in methylene blue dye removal from water. *Korean J. Chem. Eng.* **2013**, *30*, 2228–2234. [CrossRef]
57. Ekundayo, J.M.; Rezaee, R. Volumetric Measurements of Methane-Coal Adsorption and Desorption Isotherms—Effects of Equations of State and Implication for Initial Gas Reserves. *Energies* **2019**, *12*, 2022. [CrossRef]
58. Ismail, L.; AbdelKhalek, M. Adsorption Study of Some Heavy Metal Ions on Modified Kaolinite Clay. *Int. J. Adv. Eng. Technol.* **2016**, *3*, 152–163.
59. Hashim, M.A.; Mukhopadhyay, S.; Sahu, J.N.; Sengupta, B. Remediation technologies for heavy metal contaminated groundwater. *J. Environ. Manag.* **2011**, *92*, 2355–2388. [CrossRef]
60. Ogbu, I.; Akpomie, K.; Osunkunle, A.; Eze, S. Sawdust-kaolinite composite as efficient sorbent for heavy metal ions. *Bangladesh J. Sci. Ind. Res.* **2019**, *54*, 99–110. [CrossRef]
61. Barka, N.; Abdennouri, M.; El Makhfouk, M.; Qourzal, S. Biosorption characteristics of cadmium and lead onto eco-friendly dried cactus (Opuntia ficus indica) cladodes. *J. Environ. Chem. Eng.* **2013**, *1*, 144–149. [CrossRef]
62. Low, K.S.; Lee, C.K.; Lee, K.P. Sorption of copper by dye-treated oil-palm fibres. *Bioresour. Technol.* **1993**, *44*, 109–112. [CrossRef]
63. Rai, R.K.; Singh, V.P.; Upadhyay, A. Soil Analysis. In *Planning and Evaluation of Irrigation Projects*; Academic Press: Cambridge, MA, USA, 2017; Chapter 17; pp. 505–523. [CrossRef]

minerals

MDPI

Review

Practical Experience of Filtered Tailings Technology in Chile and Peru: An Environmentally Friendly Solution

Carlos Cacciuttolo Vargas [1,*] and Giovene Pérez Campomanes [2]

[1] Civil Works and Geology Department, Catholic University of Temuco, Temuco 4780000, Chile
[2] Facultad de Ingeniería, Universidad Continental, Huancayo 12001, Peru; gperezc@continental.edu.pe
* Correspondence: carlos.cacciuttolo@gmail.com or ccacciuttolo@uct.cl

Abstract: In the last 20 years many mining projects around the world have applied a tailings deposition technology named "dry stacking of filtered tailings" at tailings storage facilities (TSFs). This technique produces an unsaturated cake that allows storing this material without the need to manage large slurry tailings ponds. The application of this technology has accomplished: (i) an increase in tailings water recovery, (ii) a reduction of the TSF footprint (impacted areas), and (iii) a decrease in the risk of physical instability, being TSFs self-supporting structures under compaction (such as dry stacks), and (iv) a better regulator and community perception satisfying the need of stable TSFs. This paper presents the main features, benefits, and advances in filtered tailings technology applied in Chile and Peru with emphasis on: (i) filtering technology evolution over the last decade: description of main equipment, advantages, and disadvantages, (ii) design considerations for main TSF geometrical configurations, tailings transport and placement systems, TSF water management, TSF operational and emergency plans, and TSF progressive closure, (iii) operation experiences at site-specific conditions, (iv) technology acceptance in regulatory frameworks, (v) lessons learned and advances, and (vi) new trends and future developments, considering technical, environmental, regulatory frameworks and cost-effective manners.

Keywords: filtered tailings; water recovery; compaction; technology performance

1. Introduction

Dry climate, water scarcity, community issues, and environmental constraints around the globe, make the efficient use of water an important aspect of mining [1]. For this reason, equipment manufacturers (cyclones, thickeners, filters, pumps, etc.), and consulting engineering firms are innovating to develop new solutions and appropriate technologies to address these challenges [2].

In recent years, the improvements in tailings dewatering technologies (thickening and filtering), have allowed increased water recovery. These technologies have been successfully applied for production rates up to 50,000 mtpd (metric tonnes per day); showing performance improvements on large-scale projects with high ore production rates [1]. In this scenario, there is still a need for more reliable equipment for filtering processes on a large scale, focusing on the tailings water recovery enhancement for its reuse in mining processing. Figure 1 shows the tailings continuum concept [3].

In current Chilean and Peruvian large-scale mining in dry climate areas, most typical tailings disposal schemes consist of conventional or slightly thickened at modest levels of tailings solids weight concentration (Cw 48–52%). Conventional TSFs have dams built of the coarse fraction of tailings obtained by hydrocyclones or have slightly thickened tailings deposits with dams built of borrowed material. Conventional tailing dams may have water recoveries as high as 65–75% in very well-operated TSFs, which means they have appropriate tailings distribution, good control of the pond (volume and location), and adequate seepage recovery. In conventional dams, water at the settling pond is decanted by floating pumps, or decant towers, and dam seepages are collected by a drainage system and cutoff trench systems. However, a high seasonal evaporation rate can substantially reduce water recovery from the pond area, and infiltration from the pond in contact with natural soil can produce water losses. Some mining operations with this technology are: (i) Cerro Verde (Peru), Cuajone and Toquepala (Peru), Los Pelambres (Chile) and Los Bronces (Chile) [4,5]

Thickened Tailings Disposal (TTD) technology requires more background data than conventional tailings disposal. In the conventional approach, the properties of tailings are fixed by the concentrator plant, whereas in a TTD impoundment, the properties of the tailings and their placement are "engineered" to suit the topography of the disposal area [6]. The behavior of tailings in the two approaches is entirely different. In conventional disposal, tailings segregate as they flow and settle out to an essentially flat deposit, whereas in TTD technology, a sloping surface is obtained. The principal difference is that, in TTD technology, tailings are thickened before discharge to a homogeneous heavy consistency that results in laminar non-segregating flow. In this way, TTD produces high water recovery (80 percent of tailings water recovery) and a self-supporting deposit with sloping sides, requiring small dams. Some mining operations with this technology are: (i) Toromocho (Peru), Constancia (Peru), Centinela (Chile), and Sierra Gorda (Chile) [4,5].

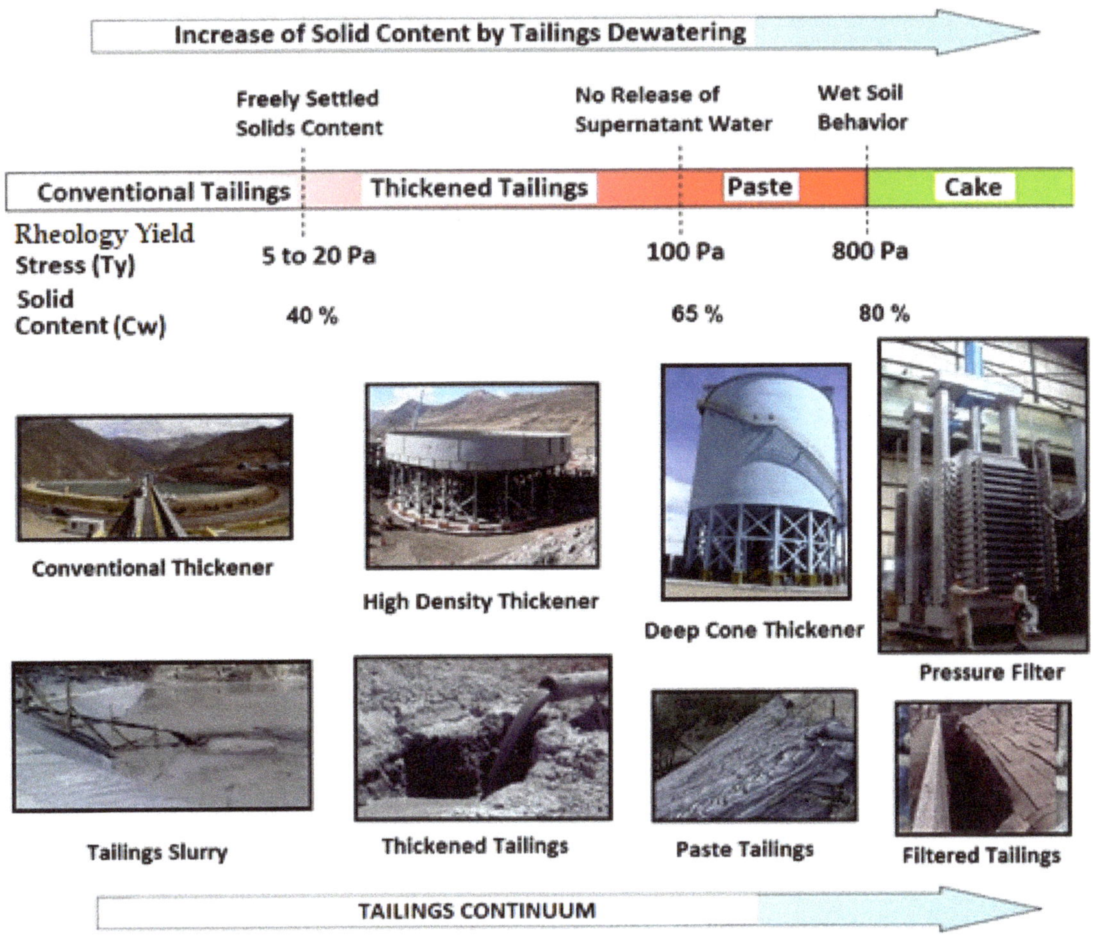

Figure 1. Dewatering Tailings Technologies–Tailings Dewatering Continuum [11].

Paste Tailings Technology has been applied on a small production scale because a limitation of equipment manufacturing ability exists. This method permits obtaining a medium make-up water requirement. However, in some cases, there are difficulties in tailings transportation requiring the use of positive displacement pumping, resulting in the highest capital/operating costs. The main advantage of this method is that large dams are not required, only small dams are needed. Some mining operations with this technology are: (i) Chungar (Peru), Cobriza (Peru), Las Cenizas (Chile), and Alhue (Chile) [4,5].

Finally, in the last 20 years, many mining projects around the world have applied a tailings disposal technology called dry stacking of filtered tailings. This technique produces an unsaturated cake that allows storage of this material without the need to manage large slurry tailings ponds. The application of this technology has accomplished: (i) an increase in water recovery from tailings (90 percent), (ii) a reduction of TSF footprint (impacted areas), and (iii) a decrease in the risk of physical instability because TSFs are self-supporting structures under compaction (such as dry stacks), and (iv) a better community perception. Some mining operations with this technology are: (i) Cerro Lindo (Peru), Catalina Huanca (Peru), El Peñon (Chile), and Mantos Blancos (Chile) [4,5].

2. Tailings Filtering Equipment Development and Advances

Filter suppliers have gained experience carrying out projects during the last decades, learning that each particle size distribution (PSD) and mineralogy of tailings exhibit their own unique filtering behavior, making efficient and reliable solid/liquid separation units [6–10].

2.1. Vacuum Filters

Coarse tailings with low percentages of clay minerals can be handled by vacuum filters with low flocculant consumption. The main characteristics of commercial vacuum equipment are presented below [8,11,12]:

2.1.1. Horizontal Belt Filters

- Equipment Functionality Principle: Vacuum, a filter cloth belt receives tailings and by vacuum suction pressure the filtrate (water) is obtained, and a tailings cake is formed. See Figure 2.
- Equipment Operation Mode: Continuous with shorter dewatering and washing cycles.
- Equipment Capacity Range: 1000–2000 mtpd.
- Equipment Filtration Area/Solids Loading Range: 30–200 m^2/300–1000 kg/h/m^2.
- Cake Moisture Content Typical Range: 15–25%.
- Application: Gold tailings, cake washing high efficiency, and CN/Au high recovery.
- Flocculant Consumption: 20–40 g/t.
- Sensitivity: Low performance at high elevations (masl), and with high percentages of <74 μm clay minerals.

2.1.2. Ceramic Disc Filters

- Equipment Functionality Principle: Vacuum and capillary action, a ceramic disc with microscopical pore structure rotates on a tank with tailings, and by vacuum suction pressure the filtrate (water) is obtained, and on the disc surface a tailings cake is formed by drying. See Figure 3.
- Equipment Operation Mode: Continuous with shorter dewatering and washing cycles.
- Equipment Capacity Range: 1000–2000 mtpd.
- Equipment Filtration Area/Solids Loading Range: 45–145 m^2/575–1500 Kg/h/m^2.
- Ceramic Disc Features: Diameter range of 1–5 m, with a maximum of 15 units by tank.
- Cake Moisture Content Typical Range: 12–18%.
- Application: Copper tailings, low energy consumption, high equipment availability.
- Flocculant Consumption: 40–70 g/t.
- Sensitivity: Low performance at high elevations (masl), and with high percentages of <74 μm clay minerals.

Figure 2. Typical Tailings Horizontal Belt Filter [13].

2.2. Pressure Filters

Cases where tailings have a fine PSD and require high throughput of tailings are preferentially dealt with by filter press. The main characteristics of commercial industrial filter press equipment are presented below [11,14]:

Vertical Plate and Frame Automatic Filter Press

- Equipment Functionality Principle: Pressure by hydraulic mechanisms, a series of plates are covered with a filter cloth when the plates are pressed and clamped, forming chambers. Tailings are pumped into the chambers, and then filtrate (water) is retired by compression air application. Finally, the filter is opened, the cake is removed and the procedure is repeated. See Figure 4.
- Equipment Operation Mode: Batch, with larger dewatering and washing cycles.
- Equipment Capacity: 1000–14,000 mtpd.
- Equipment Filtration Area/Solids Loading Range: 25–2000 m^2/50–1000 $Kg/h/m^2$.
- Cake Moisture Content Typical Range: 10–15%.
- Flocculant Consumption: Not needed, but required filter cloth maintenance per plate.
- Sensitivity: Low equipment availability, which requires stand-by equipment.
- Application: Copper and gold tailings, not sensitive at high elevation (masl).

Figure 3. Typical Tailings Ceramic Disc Filter [15].

Figure 4. Typical Tailings Vertical Plate and Frame Automatic Filter Press [9].

3. Filtered Tailings Conveyance System Improvements

The design of a dry stack TSF needs to consider a filtered tailings conveyance system that needs to be compatible with the dry stack construction sequence/plan, using conventional conveyance/haulage and mechanical placement equipment [7,9,16]. The main filtered tailings conveyance types are presented below.

3.1. Truck Haulage

Filtered tailings cake is piled at the discharge of the filter plant in a temporary stockpile from where haul trucks (25 tons or 60 tons of capacity depending on the project) are loaded using front-end loaders (Figure 5) to be sent to the tailings deposit [17]. To have safe truck transport it is necessary to build dual-lane roads that allow connecting the filter plant with the TSF, and provide adequate access roads within the TSF [17].

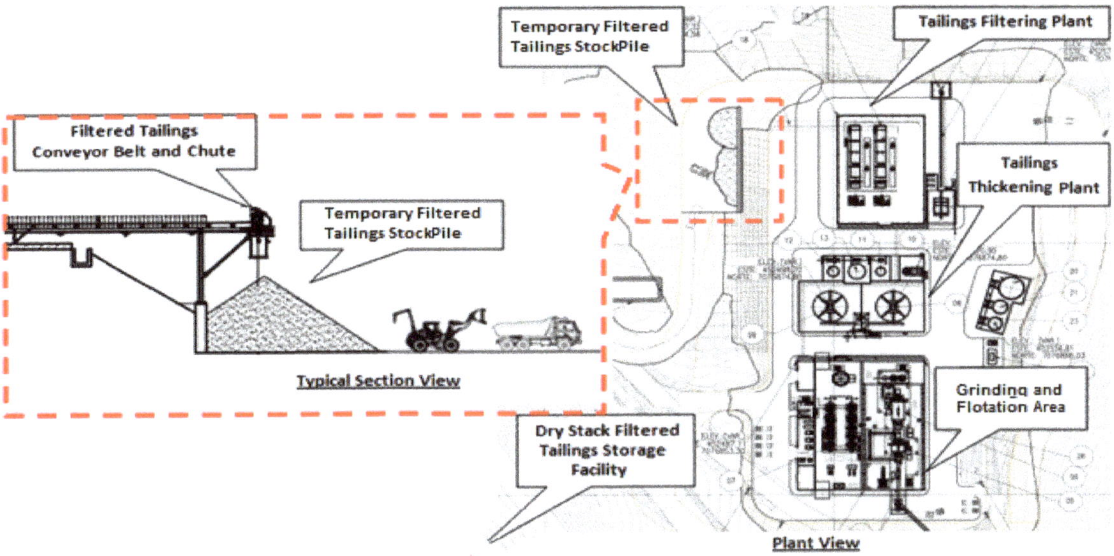

Figure 5. Typical Layout and Profile of Filtered Tailings Temporary Stockpile [18].

The main issue associated with the placement of the filtered tailings by truck is usually trafficability. The filtered tailings are generally produced above the optimum moisture content for compaction as determined by the compaction Proctor Standard Test (ASTM D698) [19]. This means that a construction/operating plan is required to avoid trafficability difficulties. Typically, strategies to overcome the low tailings bearing capacity involve some combination of strengthening, use of very small equipment (D6R Dozers), and running the equipment on thin lifts of capping material rather than directly on the tailings [20].

Mining operations with low throughput and filtering technology implemented has the TSF located quite close to the filter plant, and in this case, tailings are usually hauled by truck to the TSF (Figure 6) [7,20–22].

3.2. *Conveyor Transport*

Filtered tailings cake is discharged from the filter plant to a fixed conveyor belt to be sent to the tailings deposit. Then at the TSF, this conveyor belt transfers the cake by a chute or hopper to a mobile conveyor belt system which has the function of disposal of the filtered tailings at the TSF [22,23]. The main typical conveyor systems used to transport filtered tailings are:

- Grasshopper conveyors and spreaders, or shiftable conveyors and radial stackers.
- Mobile stacking conveyor (MSC) with translating tripper boom.

In order to provide adequate conveyor belt mechanical availability and not to have tailings spillage, it is recommended that the moisture content of filtered tailings cakes be lower than about 20% [23,24]. The main issues associated with the placement of the filtered tailings by the conveyor system are usually bearing capacity and conveyor system alignment and positioning control.

MSC offers versatility and flexibility which means that it can move linearly, radially, up or down a slope without having to follow a horizontal line. For the horizontal movement, each MSC bridge is equipped with crawlers to maintain alignment, and hydraulic cylinders to maintain level assisting the passage of the tripper car (Figure 7) [23,24]. These alignment functions are controlled via a set of field instruments coupled to a computerized control system. Control of movement, as stated previously is achieved via instrumentation and a PLC control system.

In general mining operations with a throughput of over 5000 mtpd implement dry stack TSF with conveyor transport, considering: (i) trafficability constraints, (ii) necessity for use of more truck fleet and increased costs for fuel, and (iii) necessity for a system that allows disposal filtered tailings at high production rates, accomplish the construction program of TSF dry stack (Figure 8) [11,19,20,25].

Figure 6. Typical View of Temporary Filtered Tailings Stockpile [20].

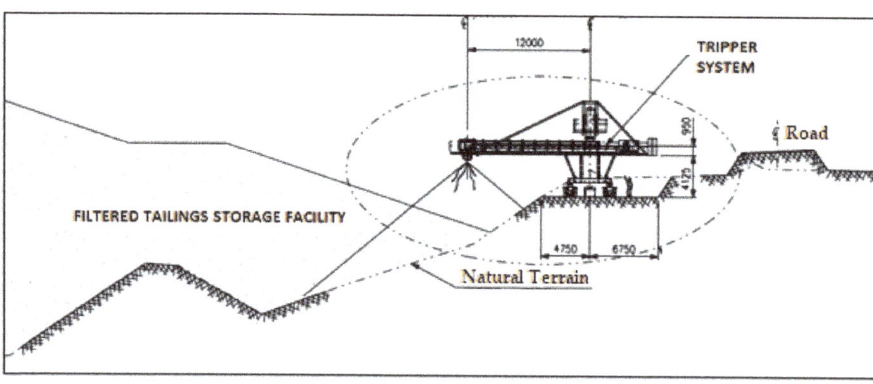

Figure 7. Typical View of Filtered Tailings Facility using Conveyor Belt and Tripper System [23].

Figure 8. Typical View of Filtered Tailings Transported by a Conveyor Belt [23].

4. Filtered Tailings Disposal Scheme and Placement Procedures

The design of dry stack filtered TSF geometry needs to be compatible with how the dry stack can be practically constructed using the selected haulage/conveyance and placement equipment. Mechanical re-handling of tailings by dozing and/or compaction with compactors needs to allow the construction of stackable geometries under the land topography of the TSF site [11]. An adequate placement strategy needs to be applied to accomplish the filtered tailings disposal scheme in different zones of the TSF and coordinate the works for progressive TSF closure and land reclamation [7,9].

4.1. Filtered TSF Geometry Configuration and Disposal Scheme

The dry stack filtered TSF consists of the placement of successive compacted filtered tailings lifts that form a stable platform, conformed by terraces and berms. The typical design values for the characteristic geometrical elements of dry stack TSF in seismic areas are: (i) local side slopes on order to H:V = 3.0:1.0, (ii) 5 m bench width to provide enough space for vehicle traffic and geotechnical instrumentation installation, (iii) terraces at a maximum height of 5 m, obtaining overall side slopes on order to H:V = 3.5:1.0. Steeper side slopes need a review of the filtered tailings geotechnical parameters to assure the stack stability [11,22].

4.1.1. Flat Topography Configuration

The dry stack TSF in a flat terrain is restricted by the surrounding mine facilities. The construction of this type of dry stack TSF is relatively easy, either on flat terrain or terrain contouring on gentle alluvial fans such as in the Chilean and Peruvian Atacama desert [11]. The filtered tailings are stacked in relatively thin lifts (0.3 m typically). The successive placement of compacted lifts forms the TSF terraces or pads. The construction of the TSF terraces is from the bottom area to the top area, decreasing the filtered tailings placement surface when the dry stack filtered TSF is raising [11].

In some cases, the perimeter slopes of the TSF are designed to provide a perimeter buttress dam, constructed with borrow or waste rock materials. This buttress dam provides physical stability and dust control and is part of the progressive reclamation activities [7,22]. In these cases MSC are preferred for transportation purposes, bulldozers are used for spreading, and smooth drum vibrating compactors for cake compaction (if required).

Figure 9 presents a typical arrangement and disposal scheme of dry stack TSF on flat terrain.

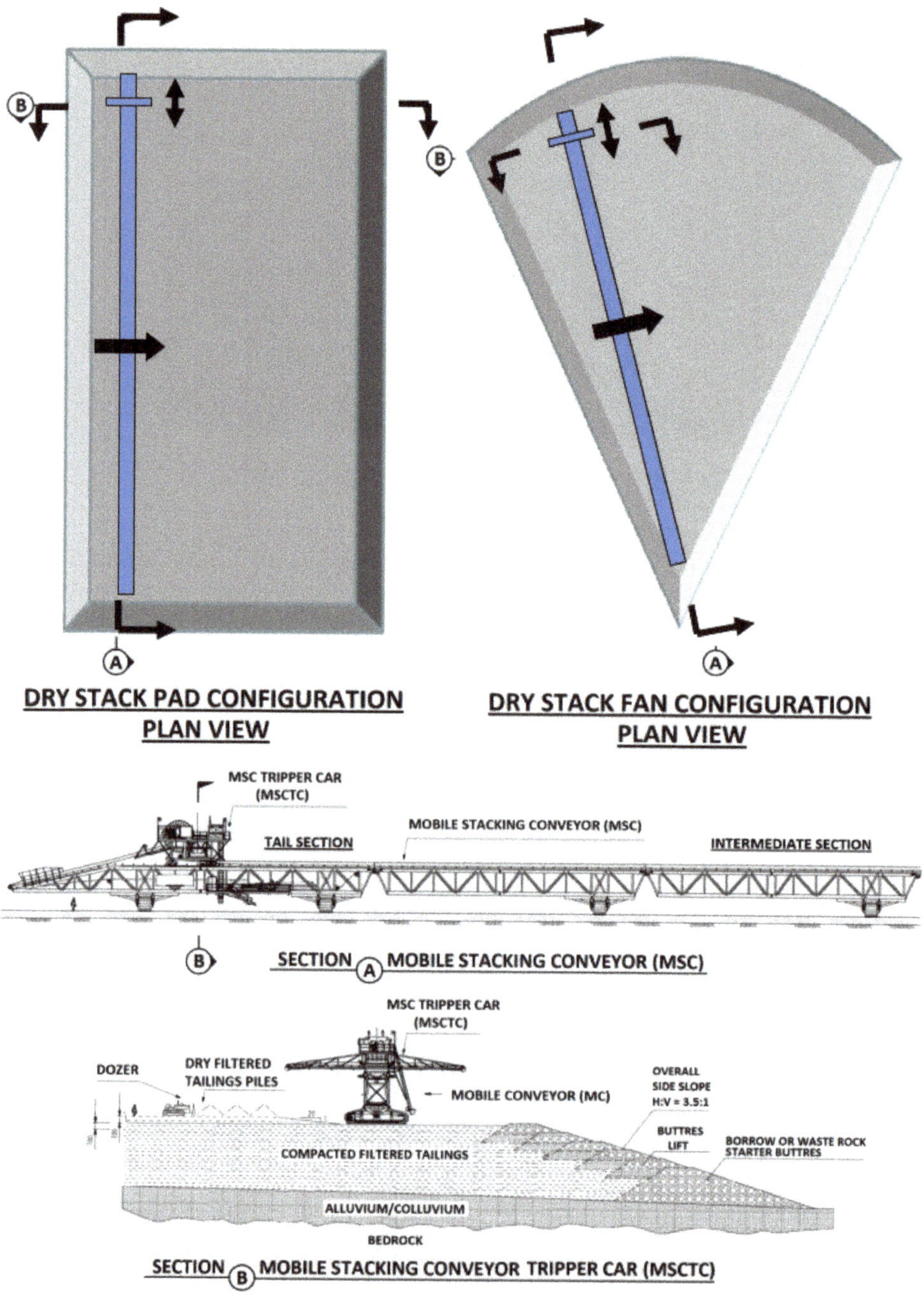

Figure 9. Dry Stacked TSF Disposal Scheme with MSC–Construction Method [11].

4.1.2. Valley Topography Configuration

The dry stack TSF in an abrupt terrain is restricted by the surrounding land topography. The construction of this type of dry stack TSF needs a strategic plan. In these cases down valley placement or up valley placement methods can be applied [11,20–22]. The up valley placement method considers that filtered tailings are stacked in relatively thin lifts (0.3 m typically). The successive placement of compacted lifts forms the TSF terraces on the valley landform. In these cases, a perimeter buttress dam is provided and has the same function described above [22]. Down valley placement can be applied at upset conditions to disposal in a temporary TSF [11]. Figure 10 presents the arrangement plant view and growth scheme of the filtered tailings dry stack TSF. The dry stack TSF is constructed in the upstream direction of the typical Andes valley topography. The construction of the TSF terraces is from the lower sector to the higher sector of a valley basin, according to the typical geometric characteristics, in order to facilitate an adequate construction with haul trucks, bulldozers, and smooth drum vibrating rollers, assuring the TSF physical stability [11,20–22]. Figure 11 shows an example of filtered tailings dry stack TSF that is a mass composed of seven terraces and berms.

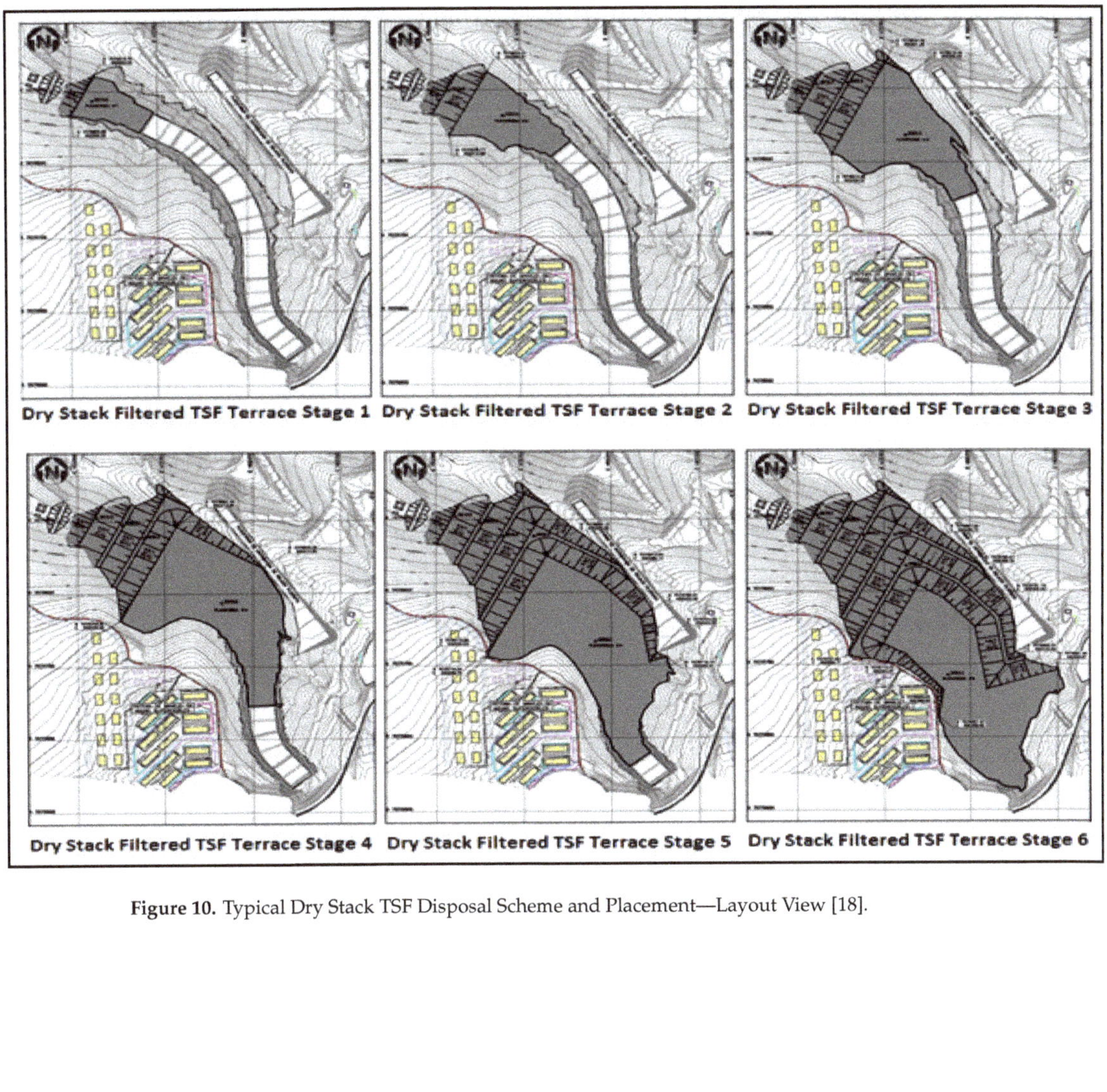

Figure 10. Typical Dry Stack TSF Disposal Scheme and Placement—Layout View [18].

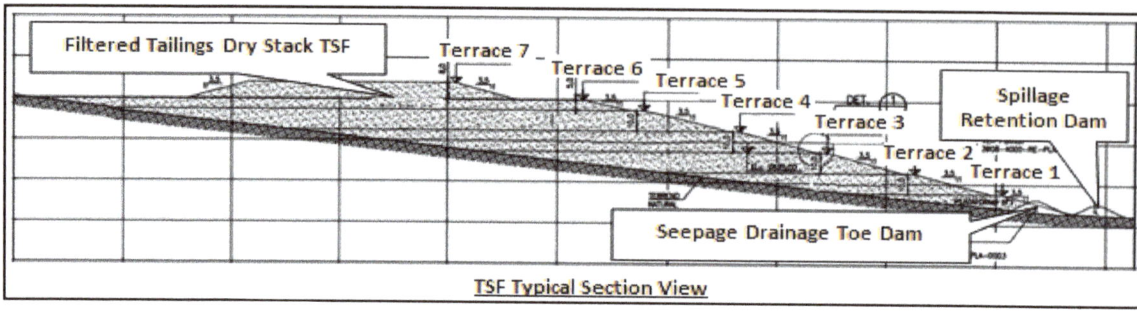

Figure 11. Typical Dry Stack TSF Geometrical Configuration—Typical Cross Section View [18].

4.2. Filtered Tailings Placement Issues

TSF construction and operating quality assurance and quality control (QA/QC) programs are required to avoid trafficability difficulties, especially at high raising rates of TSF, because trafficability drops as moisture content rises [3,7,9,22].

The filtered tailings are deposited in layers, of 30 cm thickness typically, which are exposed to reach their range of optimum moisture content of 12–14%, and are then compacted to reach 95% Proctor Standard (ASTM D698) [3,7,9,19,22].

Moreover, in high seismic areas, there is often a design requirement to compact the filtered tailings to a higher density at the TSF. QA/QC is applied at the mining operation units [20,25].

4.2.1. Spreading Operation

- Filtered tailings are dumped in piles by haul trucks or MSC tripper boom. Typical moisture content is below 20%, several percent below tailings saturation.
- Dozers or graders spread the filtered tailings piles into 0.3 m loose lift (Figure 12).
- For adequate MSC movement, a good practice is to ensure no excessive temporary filtered tailings piles are laid in front of the tracks crawlers, and improved operations will be obtained by running a grader or dozer in front of the MSC prior to movement.

Figure 12. Filtered Tailings Spreading Operation–Construction Method.

4.2.2. Drying Activities

- Drying is relevant, different placement zones are required, wet (saturated) zones, and dry (unsaturated) zones. The effects of sun, wind, and the aid of earthmoving machinery allow filtered tailings to dry and reach high compaction densities.
- Filtered tailings should be spread by a dozer for some days after the placement (Figure 13). The drying time depends on cake moisture content, climate, and day or night conditions drying needs to reach a density value near optimum moisture content.

Figure 13. Filtered Tailings Drying Operation–Construction Method.

4.2.3. Compaction Field Trial

- Smooth Drum Vibratory Compactor equipment is adequate to provide compaction for TSF access roads, filtered tailings lifts, and bearing capacity for dozer, haul trucks, and MSC [11,20].
- After the drying period is accomplished, the compaction (if required) proceeds with passes of a smooth drum vibrator compactor (Figure 14). Compaction field trials are recommended to determine the optimal number of compaction passes to obtain the density on order to 95% of Proctor Standard (ASTM D698) [19], providing, in this manner, adequate bearing capacity for equipment trafficability.

Figure 14. Filtered Tailings Compaction Field Trial—Construction Method.

4.3. Progressive Rehabilitation/Closure Liability Issues–Dust Emission Control

The lack of a tailings supernatant pond, very low seepage from the unsaturated tailings, and high degree of structural stability, allow dry stacks TSFs to develop progressive reclamation in many instances. A closure cover material is provided to manage runoff erosion, and create an appropriate ground surface for project reclamation (Figure 15) [3,7,11]. Dust emission controls will play an important role as good design and proper implementation will provide the primary control mechanism for dust in accordance with regulatory air quality requirements. Some dust control alternatives are: soil cover, top soil/revegetation cover, binder material, or chemical agglomeration [11].

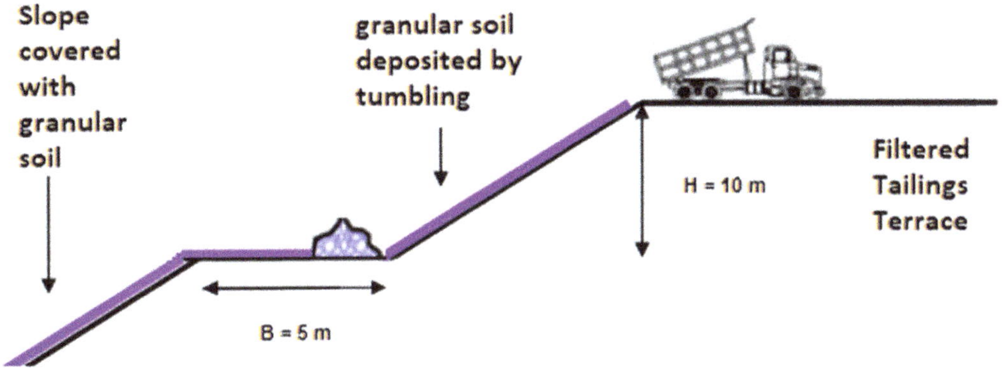

Figure 15. Filtered Tailings Storage Facility Covered with Granular Soil–Progressive Closure Activities.

4.4. Contingency Plan–Upset Conditions

An Operating, Maintenance, and Surveillance (OMS) Manual is necessary to address adequate procedures to operate, maintain, and monitor the performance of the dry stack TSF, ensuring its operations are in accordance with its design, meet regulatory framework, and provide an emergency and response plan to act during upset conditions [11,18,26]. Some emergency plans and design features are mentioned below.

- Thickener capacity: Thickeners need to be designed with over capacity to constantly feed the filtering plant.
- Contingency TSF site: When the filter plant is under an upset condition event, thickened tailings are diverted to an emergency pond. Once this event is overcome the

thickened tailings are returned to the filtering plant in such an amount to not exceed the filtering design capacity.

- Stand by filter units: Additional filter units are needed to support changes in potential tailings feed upset conditions by ore PSD variability, or mineralogy changes. These additional units are needed to maintain high mechanical availability of the filter plant.
- Stand-by conveyor: It is anticipated that a secondary conveyor system consisting of a bypass diverter or radial stacker conveyor is needed to allow temporary disposal of the filtered tailings stack, for placement with dozers while the primary MSC is inactive due to relocation, maintenance, or upset conditions. Alternatively, additional mobile conveyor systems might be implemented, such as grasshopper conveyors, or spreaders.
- Optimal moisture content handling: The tailings moisture content in the filters discharge can sometimes exceed the maximum moisture value admissible to be placed in the stack. In these cases, treatment moisture reduction is applied by spreading the tailings in thin layers and passing plows to aerate them. This treatment can be performed on the stack footprint provided sufficient room or in a special area in the vicinity of the facility.

5. Filtered TSF Physical and Hydrological Stability

To ensure the physical stability and protect the structure of a dry stack TSF, it is necessary to have proper water management at the construction, operation, and closure stages. This means to address the management of (i) the precipitation and runoff (natural water), and (ii) the excess process water [18,22].

5.1. Geometric Design Criteria, Tailings Geotechnical Properties, and Physical Stability

The filtered tailings are commonly deposited in layers of 30 cm thickness, which are exposed to reach their optimum moisture content such as 12%, and are then compacted to reach 95% Proctor Standard (ASTM D698) [19]. The filtered dry stack TSF is a mass composed of a number of terraces (or benches) and berms, which needs to be subjected to static and pseudo-static stability analysis (Figure 16). The dry stack TSF is constructed in the upstream direction of a valley topography [11,18–21]. Table 1 describes the main geometric design criteria and geotechnical properties for filtered copper tailings storage facilities commonly registered in mining operations:

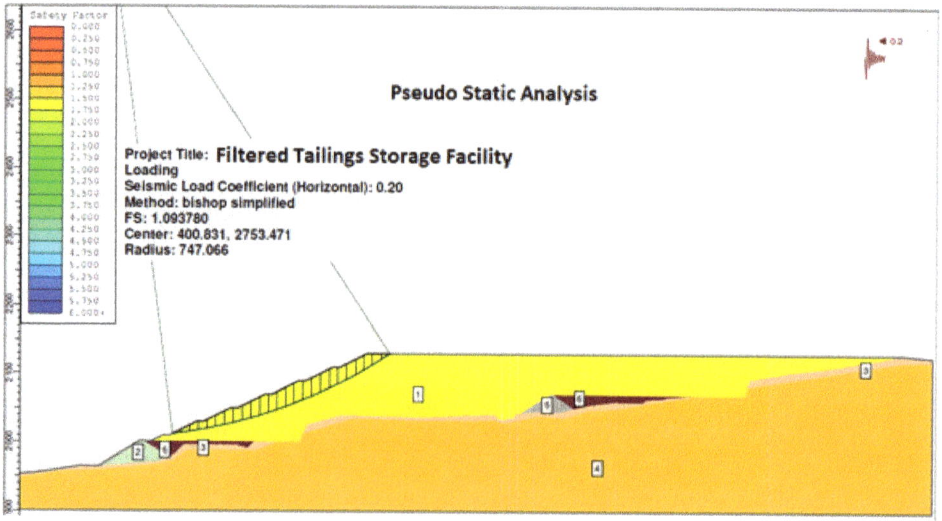

Figure 16. Typical Dry Stack TSF Stability Analysis.

Table 1. Geometric Design Criteria and Geotechnical Properties of Dry Stack TSF [18,22].

Geometric Design Criteria of Filtered TSF			Geotechnical Properties of Filtered Tailings		
Characteristics	**Range**	**Unit**	**Characteristics**	**Range**	**Unit**
Number of terraces (benches)	3–8	-	Particle size distribution P_{80}	50–90	(μm)
Maximum terrace height	7–10	(m)	Solid Specific Gravity	2.6–4.0	-
Minimum berm width	2–5	(m)	Fines Content	30–60	(%)
TSF terrace local slope	H:V = 3.0:1.0	-	Permeability	10^{-5}–10^{-6}	(cm/s)
TSF global slope (all terraces)	H:V = 3.5:1.0	-	Cohesion/ friction angle–c/ϕ	0/25–35	$(t/m^2)/(°)$
TSF maximum height (all terraces)	70–80	(m)	Residual Shear Strength Su/p′	0.15–0.30	-

5.2. Water Management and Hydrological Stability

- Considering the use of filtered tailings technology, the physical and hydrological stability of the TSF has to be assured, which implies designing civil works where the surface and underground water flows are managed adequately [3,22]. It is important to maintain these tailings with low moisture levels, close to optimum moisture content, avoiding saturation, to be able to achieve high dry densities and, therefore, adequate shear resistances assuring the stability of the pile [3,22]. The following paragraphs present projected civil works to carry out proper water management at a filtered tailings storage facility.
- Spillage Retention Dam: This civil work is a dam of 5 m in height typically, located downstream of the TSF aimed to contain potential downstream spills of filtered tailings, as a result of potential rainfalls.
- Underdrainage System: It is the base drainage system, formed by an excavated trench with unwoven geotextile, gravel, and filter materials, located at the bottom stream of the valley site, in order to capture potential seepages from the filtered tailings. This system is important due to the collection of potential seepage from tailings or contact waters. Figure 17 presents a typical drain section and hauling road to handle of filtered tailings.
- Seepage Collection Sump: The drainage sump is located typically approximately 5 m downstream of the spillage retention dam, allowing collection of potential seepages from the filtered tailings. This sump is entirely lined with a geotextile and geomembrane. To remove the accumulated water from this sump, one option is to install a submergible pump and recover it to a tank truck for its removal. In case of an excessive accumulation of water by an extreme rainfall event, the sump must have a safety spillway to eliminate excess water.
- Rainfall Diversion Ditches: The filtered TSF projects must include the construction of perimeter channels, one for the right margin, and another for the left margin of the valley. These channels are formed by ditches of trapezoidal section lined, such as concrete cloth ditch liner, with average slopes of 1%, and using corrugated steel pipelines in lengths with slopes over 10%. Both channels collect water runoffs, restricting their entrance to the filtered tailings dry stack deposit, and discharging these flows downstream of the TSF. A collection and diversion system for non-contact water with the TSF (runoff water), consisting of perimeter ditches (lined with geoweb/concrete or precast concrete cloths) (Figure 18).
- An interception system for contact surface water and any impacted groundwater or seepage that may result from the dry stack TSF. "On stack" water should be managed by routing flows to engineered temporary channels by a terrace, sized with grades and sections to convey water and to control erosion and settling. The seepage collection system usually consists of a cutoff and an underdrain system.
- Both systems convey the contact water to collection ponds where water quality is controlled. If the water quality meets the standards, it is possible to discharge it to

natural courses. If it is not, water is treated or pumped back to be reused in the process. Monitoring wells are installed downstream, to control the water quality periodically.

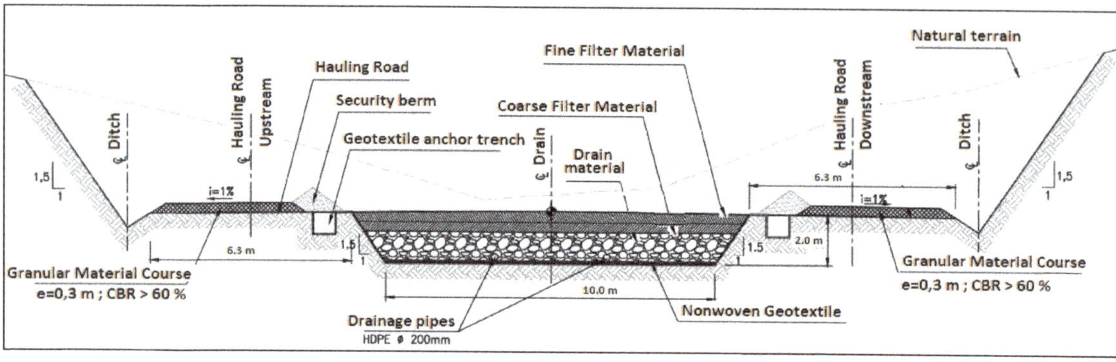

Figure 17. Filtered TSF Seepage Underdrainage System—Typical Cross Section [18].

Figure 18. Collection and diversion system for non-contact water–perimeter ditches [27].

5.3. *Operational and Constructability Issues*

To construct a stable filtered dry stack tailings deposit, assurance controls of the quality of the construction procedures and materials used are required [20,21]. Operators must control and perform follow-up geometric, geotechnical, and compaction features of deposited filtered tailings, to assure that design specifications are strictly followed during the operation of the project [3,22]. Table 2 shows controls that must be complied with in a QA/QC (Quality Assurance and Quality Control) plan.

Table 2. Measurements and Controls of Geometry, Compaction and Geotechnical Properties for Filtered TSFs (QA/QC) [20,25].

Measurements and Controls of Filtered TSF Geometry			
Type of Control	**Value**	**Unit**	**Control Frequency Recommendation**
Local slope between terraces	3.0:1.0	H:V	Monthly
TSF Global slope	3.5:1.0	H:V	Every 2 constructed terraces
Minimum berm width	2-5	m	Per terrace
Maximum terrace height	5-10	m	Per terrace
Measurements and Controls of Filtered Tailings Compaction Parameters			
Type of Control	**Value**	**Unit**	**Control Frequency Recommendation**
Recommended tailings thickness lift	30–35	cm	Per compacted tailings lift
Maximum dry compacted tailings density	1.9–2.6	t/m^3	Per 5.000 m^3 of compacted tailings
Optimal tailings moisture content	12–14	%	Per compacted tailings lift
Measurements and Controls of Filtered Tailings Geotechnical Properties			
Type of Control	**Value**	**Unit**	**Control Frequency Recommendation**
Grain size distribution	ML-CL	-	Monthly
Solids specific gravity	2.65–3.75	-	Monthly
Casagrande piezometer lecture	Phreatic line lecture	m	Monthly per terrace

6. Tailings Filtering Experiences—State of Practice

The following table summarizes experiences of operations that have implemented the dry stack filtered tailings technology. The main characteristics and data of cases are provided in Table 3.

Table 3. Filtered Tailings Storage Facility Global Experiences.

TSF Name	Country	Tailings Throughput (mtpd)	PSD d_{50} (μm)/Ore Type	Gravimetric Moisture Content w (%)/PSD	Type of Filter/Number of Filter/Filtration Area per Filter (m²)	Transport/ Spread/ Compaction Method	Reference
Metales (*)	Mexico	120,000	60/(Au)	14 (TT)	Pressure Filter/(14)/860	MSC/D/SDVC	[28]
Media Luna	Mexico	14,000	65/(Au)	14 (TT)	Pressure Filter/(15)/100	MSC/D/SDVC	[28]
El Sauzal	Mexico	5300	60/(Au)	16 (TT)	Belt Filter/(03)/73	HT/D/SDVC	[28]
Alamo Dorado	Mexico	3500	65/(Au)	15 (TT)	Belt Filter/(02)/73	HT/D/SDVC	[28]
Greens Creek	USA	1500	55/(P*)	14 (TT)	Pressure Filter/(02)/70	HT/D/SDVC	[17,29]
Efemçukuru	Turkey	1500	50/(Au)	15 (TT)	Belt Filter/(02)/23	HT/D/SDVC	[15,30]
Pogo	USA	2500	55/(Au)	18 (TT)	Pressure Filter/(03)/90	HT/D/SDVC	[26]
Chingola	Zambia	50,000	90/(Cu)	17(TT)	Belt Filter/(26)/80	CB-RS/D/NA	[31]
Karara	Australia	50,000	85/(Fe)	15 (TT)	Pressure Filter/(30)/100	MSC/D/SDVC	[23]
Rosemont (*)	USA	75,000	65/(Cu)	18 (TT)	Pressure Filter/(14)/860	MSC/D/SDVC	[32]
Casposo	Argentine	1000	85/(Au)	14 (TT)	Belt Filter/(02)/60	HT/D/SDVC	[6,33–35]
Cerro Lindo	Peru	5000	65/(P*)	12 (SL)	Belt Filter/(02)/73	HT/D/SDVC	[20,21,25]
Poderosa	Peru	700	95/(Au)	15 (TT)	Pressure Filter/(2)/100	HT/D/SDVC	[36,37]
Curaubamba	Peru	2000	79/(Au)	15 (TT)	Pressure Filter/(2)/100	HT/D/SDVC	[38,39]
Catalina Huanca	Peru	1850	74/(Cu)	17 (TT)	Pressure Filter/(2)/100	HT/D/SDVC	[40,41]
Tambomayo	Peru	1500	85/(Au)	14 (TT)	Pressure Filter/(2)/100	HT/D/SDVC	[42,43]
Chungar	Peru	4200	70/(Cu)	14 (SL)	Pressure Filter/(3)/100	HT/D/SDVC	[44]
Potrerillos	Chile	1300	60/(Cu)	12 (SG)	Ceramic Disc Filter/(02)/45	HT/D/SDVC	[45,46]
Mantos Blancos	Chile	12,000	95/(Cu)	18 (UF)	Belt Filter/(03)100	CB-RS/D/NA	[13,19,46]
La Coipa	Chile	20,000	85/(Au)	18 (TT)	Belt Filter/(12)/100	MSC/D/NA	[13,46]
El Peñon	Chile	3500	60/(Au)	20 (TT)	Belt Filter/(04)/54	HT/MG/T	[13,46]
El Gato	Chile	5500	74/(Cu)	16 (TT)	Ceramic Disc Filter/(02)/60	HT/D/SDVC	[46,47]
Tambillos	Chile	3000	84/(Cu)	16 (TT)	Ceramic Disc Filter/(03)/60	HT/D/SDVC	[46,48]
Tambo de Oro	Chile	750	85/(Au)	14 (TT)	Pressure Filter/(1)/100	HT/D/SDVC	[46,49]
Salares Norte	Chile	5500	80/(Au)	15 (TT)	Pressure Filter/(3)/100	HT/D/SDVC	[46,50]
El Espino	Chile	20,000	84/(Cu)	15 (TT)	Pressure Filter/(15)/100	MSC/D/SDVC	[46,51,52]
El Indio	Chile	3000	80/(Au)	20 (TT)	Pressure Filter/(2)/100	HT/D/SDVC	[46,53]
Huasco	Chile	5000	75/(Fe)	20 (TT)	Pressure Filter/(4)/100	HT/D/SDVC	[46,54,55]

Nomenclature: PSD: Particle Size Distribution, TT: Total Tailings, UF: Cycloned Tailings Sand, SL: Slimes, SG: Slag Copper Tailings, P*: Polymetallic Ore (Cu-Pb-Zn), CB-RS: Conveyor Belt–Radial Stacker, MSC: Mobile Stacking Conveyor, HT: Haul Truck, D: Dozer, MT: Motor Grader, T: Tractor, SDVC: Smooth Drum Vibratory Compactor, and NA: Not Applied, compaction by a D7R Dozer self-weight. (*): project under regulator environmental/construction permits process.

7. Successful Cases in Chile and Peru

The development of filtering technologies has changed the criteria used in evaluating the benefits of increased tailings density. Nowadays, considering a more stringent regulatory framework, more concentrator plants apply tailings filtering technologies together with optimized tailings disposal schemes to recover water, minimize TSF footprints, and comply with regulations. The following paragraphs present various successful cases in South America specifically in mining projects in Chile and Peru.

7.1. La Coipa Filtered TSF–Valley Topography Configuration–Atacama Desert–Chile

La Coipa gold and silver mine treated 20,000 tonnes of ore per day at the processing facilities, however, nowadays this mine is under closure stage. The mine is located in Chile´s Region III, in the Atacama Desert area of the Chilean Andes, at 3800 m above sea level, roughly 130 km from Copiapó City. The climate is typical of Andes mountain region conditions. Annual precipitation is approximately 50 mm with essentially this entire amount falling in the winter months (May–September) as snow [56,57].

Tailings were processed in a filtration plant by 12 vacuum belt filters, obtaining a cake with 18% moisture content, and then transported by conveyor belt system to the main TSF or auxiliary TSF [56,57].

The conveyor belt system had a 1700 m long fixed conveyor belt and an 825 m long MSC, which delivered the filtered tailings at the main TSF. It is important to mention that the conveyor belts cited had a negative slope that allows power generation, thus contributing to global savings for the tailings transport system energy consumption [56,57].

The main TSF which approximately parallels the original topography is developed with an up valley placement method where the MSC rotates around a fixed point (pivot) in one direction and in the other and allowed the MSC tripper to dispose of the tailings along the slope, forming a fan shape TSF. Dozers carried out spreading and compaction activities on filtered tailings lifts of average 20–30 cm in thickness (Figure 19) [56,57].

Individual lifts include an inwardly sloping bench which creates zero lift thickness and intersection with the natural ground. Each lift creates an inter-ramp slope (step-in) to create the overall slope angle prescribed by geotechnical design. Until the surface of the TSF lift is uniformly tilted about 5% upstream, the MSC rotation point is changed upstream, shortening the transfer belt approximately 75 m to start the construction of a new TSF lift [13]. The following figure shows the disposal of filtered tailings with MSC at the main TSF.

Figure 19. La Coipa Mobile Stacking Conveyor (MSC) at Rako TSF [24,58].

The auxiliary TSF is developed by the down valley placement method applied at MSC upset conditions, where saturated filtered tailings flow along roughly repose angle slopes, buttressed at the toe by a containment dyke and downstream supported by a sedimentation collection pond (Figure 20) [56,57].

Nowadays, the main TSF and auxiliary TSF are developing closure and post-closure plans, and carrying out reclamation programs allowing for the re-establishing of native vegetation and landforms [56,57].

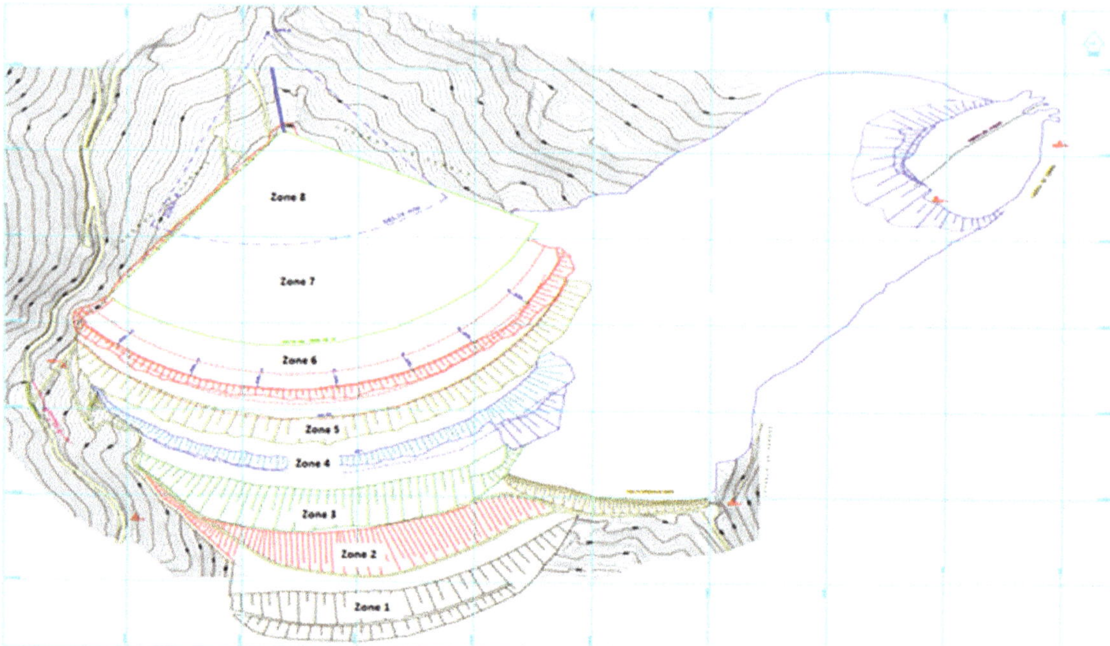

Figure 20. La Coipa Rako TSF Layout [24,58].

7.2. El Indio–Dry Stack of Filtered Tailings–Andean Region of Chile

El Indio copper-gold mine is located in the Chilean Andes Area approximately 180 km east of La Serena City. Actually, the El Indio mine is in post-closure phase. The mine processed 3000 tonnes per day (tpd) of ore for obtaining polymetallic minerals gold, silver, and copper. The climate is typical of the Andes mountain region of dry conditions, with variable winter precipitations (May–September), mainly as snow. During the operation phase, the mine had an open pit and underground activities, with two waste rock dumps, three process/metallurgic plants, and three tailings storage facilities.

Final tailings from the mill are thickened to 50% to 60% solids by two 30.5 m diameter thickeners. Approximately 50% of the thickened tailings are filtered by two Edwards and Jones pressure filters. Filter cake containing approximately 18% to 22% moisture is conveyed by belt conveyors and front end loader to deposition modules where they are stabilised by compacting. The remaining final tailings, which are not filtered, flow by gravity to a tailings pond. Water reclaimed from the tailings pond and thickener overflows is pumped to a reservoir for reuse in the mill [53,59].

Tailings and waste facilities managed in El Indio Mine were: (i) Pastos Largos TSF, (ii) El Indio TSF, (iii) dry tailings modules (filtered tailings), and (iv) sedimentation pond (polishing pond). Figures 21 and 22 show the filtered dry stack tailings modules during the operating phase, located at Malo river stream [53,59].

The mining company negotiated a voluntary agreement with the Chilean Region IV regulatory authorities after 20 years of mine lifetime to carry out the closure stage of the El Indio Mining project, as there was no legislation yet in place in Chile focused on mine closure. One of the key components of the closure plan was surface water management works developed with the overall objective to "establish a physically and chemically stable drainage system with minimal maintenance and monitoring requirements" (Robledo and Meyer, 2007). As a consequence of mining activities, the Malo River stream was modified in several sectors with diversion civil works. The main closure activities were: (i) restoration of the Malo River drainage system in the process plant area and on TSFs by the construction of engineered lined channels (lined with rockfill and cobblestones), and (ii) abandonment of the existing Malo river diversion system [53,59].

Tailings and waste facilities managed in El Indio Mine were: (i) Pastos Largos TSF, (ii) El Indio TSF, (iii) dry tailings modules (filtered tailings) (Figure 22), and (iv) sedimentation pond (polishing pond) (Figure 23). Specific closure works at the TSFs were considered including surface grading, placement of a cover to prevent hydraulic and wind erosion, and the construction of spillways to manage storm flows. The sedimentation pond is located at Malo River downstream of TSFs and stores approximately 66,000 m^3 of sediments (As, Pb, and cyanide among others). Sediments were transported by haul trucks to Pastos Largos TSF for disposal. The pond was removed and the Malo River was restored to its natural stream. Tailings were processed in a filtration plant by press filters, obtaining a cake with 18% of moisture content [53].

Figure 21. El Indio Filtered TSF Front View [53].

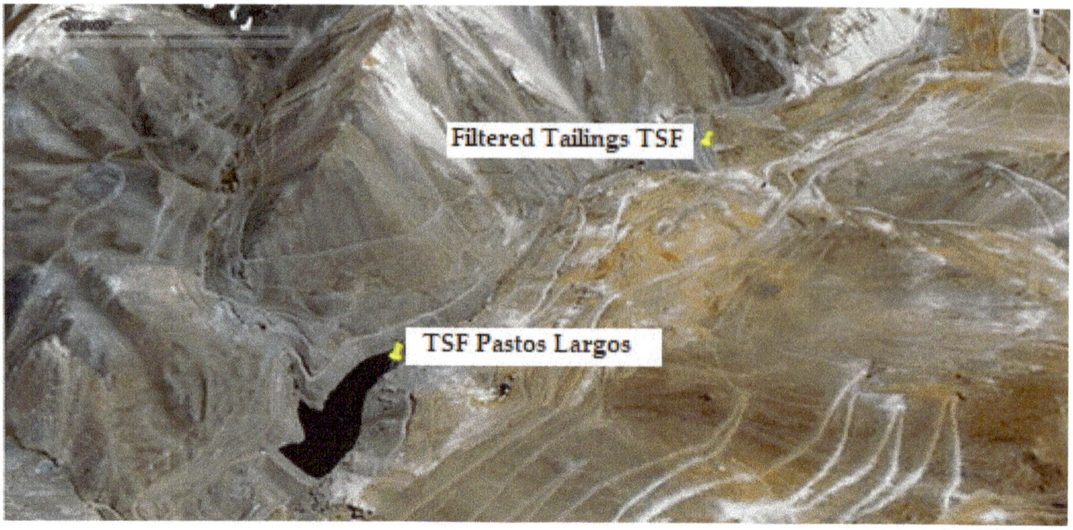

Figure 22. El Indio Filtered TSF Layout [53].

Figure 23. El Indio Filtered TSF Overview–Haul Trucks and Dozers works [53].

7.3. El Espino Filtered TSF–Flat Topography Configuration–Atacama Desert–Chile

El Espino is a copper-gold project that considers the construction of a LX-SX-EW plant for the treatment of 6.3 ktpd of leachable mineral, and a concentrator plant for the treatment of sulfide ore at a rate of 20,000 mtpd. The estimated oxide reserve is 7.2 ktpy of copper cathodes, while the sulfides production is 32,800 tonnes per year of fine copper and 25,000 oz per year of copper contained in concentrates, for a useful life of almost 18 years [51,52,60].

El Espino mine has incorporated a high dewatering technology of tailings, consisting of thickener/horizontal pressure filter plants with a 20,000 mtpd throughput, which produces filtered tailings composed predominantly of copper-gold. Tailings were processed in a filtration plant by press filters, obtaining a cake with 15% of moisture content [51,52,60].

Then the filtered tailings with a moisture content of 15% are delivered to the TSF area by a conveyor, disposed of by a MSC, spread by dozers in 0.3 m lift, and finally compacted by smooth drum vibratory compactors (Figure 24). The TSF has a flat topography configuration and will have a final footprint area of 113 Ha, with a lifetime of 18 years. The filtered tailings are stacked at the TSF in three stages, with a TSF final height of 45 m (Figure 25). To further assist the stacking of filtered tailings, the tripper car itself is equipped with a translating boom which can run 370 m along the MSC and allows for accurate placement of the ground in front or behind the MSC. The translating boom is programmed to detect the side slope of the TSF terrace via a laser probe and is able to lift (layer) the filtered tailings accurately along this terrace level. The crawlers are advanced together using limits to ensure alignment between bridges [51,52,60].

This tailings management technology was chosen due to two main aspects: (i) total cost savings over the mine lifetime (construction + operation + closure liability), and (ii) efficient water recovery. The following figure shows the MSC disposal filtered tailings at El Espino TSF.

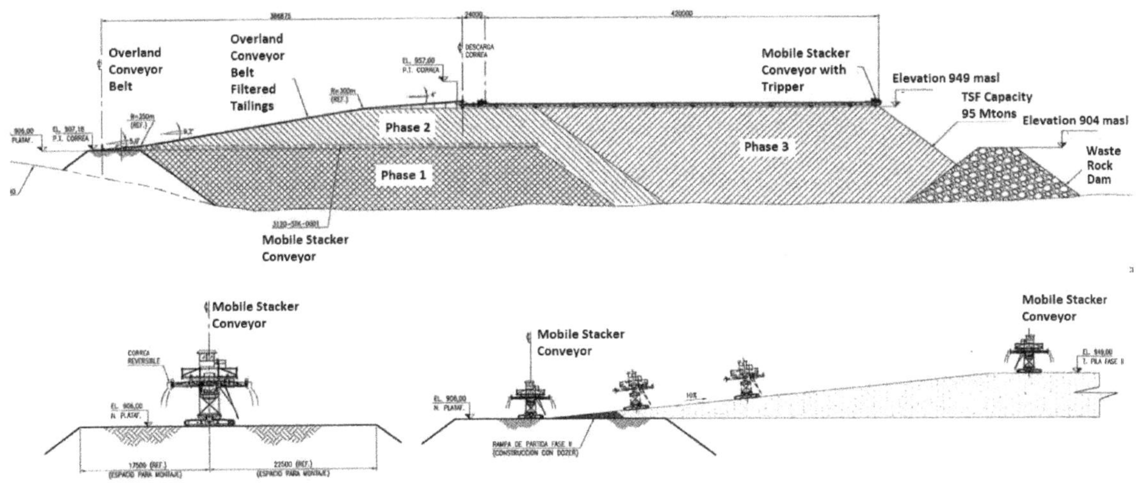

Figure 24. El Espino Mobile Stacker Conveyor with Tripper–Typical Cross Section [51,52,60].

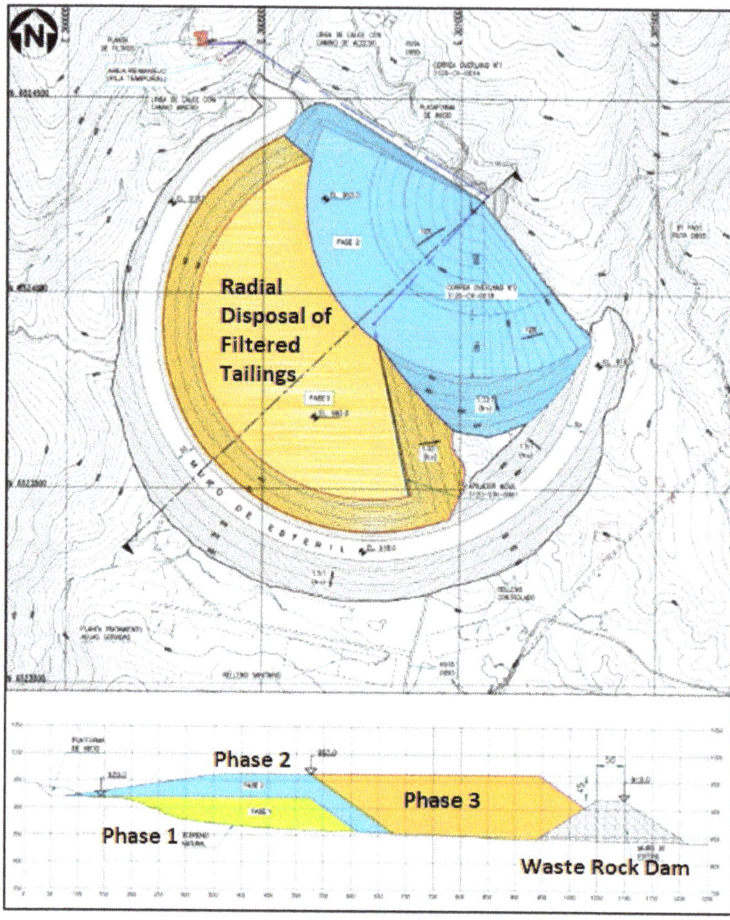

Figure 25. El Espino Filtered Tailings Storage–Layout and Typical Cross Section [51,52,60].

7.4. El Gato Creek Filtered TSF–Flat Topography Configuration–Atacama Desert–Chile

The S.C. Minera Atacama Kozan copper mine is located approximately 15km southeast of the town of Copiapó in northern Chile. The owners, a joint venture consisting of Nittetsu Mining Co. Ltd., a Japanese firm, and Inversiones Errazuriz Ltd., a Chilean company, have been mining copper at this location since 1998 [47,61].

Due to the expansion of reserves consigned in the update of available resources, prepared by an expert through drilling campaigns (period 2011–2016), the Atacama Kozan Copper Project for the production of tailings of 5500 mtpd considers the formation of a Deposit of Filtered Tailings with a total capacity to store 12 Mm^3 of tailings (Figure 26). The new tailings deposit will use the filtered tailings methodology, since it is environmentally more favorable and with the competitive advantage of maximizing water recovery [47,61].

The extension in the useful life of the Company will not imply a modification in the unit operations or variation in the level of treatment, both Mine and of the Plant. This increase in useful life meant developing a project to deposit more tailings since the old El Gato tailings dam ends its useful life in 2020 [47,61].

Tailings are transported and deposited at a nominal production average of 5500 mtpd. Considering that the concentrator and filtered tailings plant will have a continuous operation, with a mechanical availability of 85% and operational utilization of 90%, an annual production equal to 1,425,000 tonnes of dry tailings is estimated. The Filtered Tailings Deposit has a total capacity to store 12 Mm^3 of tailings, equivalent to 24,456,736 tonnes of filtered tailings (Figure 27). Then, based on the average production of filtered tailings, the operating life of the Filtered Tailings Deposit is 17 years [47,61].

The filtering plant allows the solid-liquid separation process of the tailings to be carried out, using a vacuum ceramic disc filter system (Figure 28). In this process, a filtered cake with a 16% moisture content on a wet basis is obtained. The plant has been designed for a nominal average daily capacity of 5500 mtpd [47,61].

From the filtered tailings stockpile, located near the filtering plant in the upper southwest sector of the current tailings dam, a path is planned for the transit of trucks that will transport the filtered tailings to the deposit.

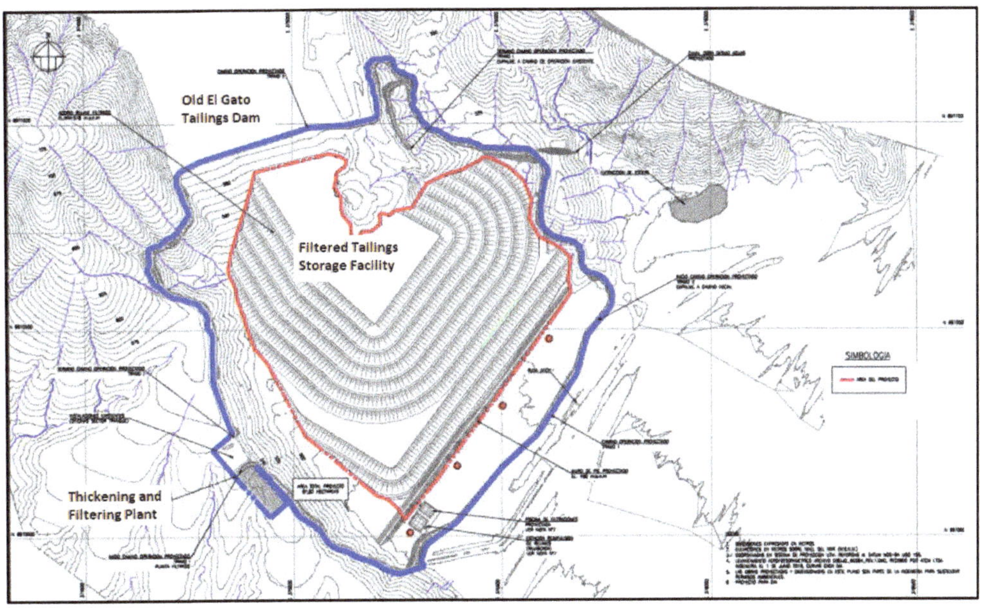

Figure 26. El Gato Creek Filtered Tailings Storage Facility—Layout [27,47,61].

The filtered tailings deposit system corresponds to a mechanized method, consisting of transport from the temporary storage area to the filtered tailings deposit, in 25-tonne hopper trucks, allowing the placement of the tailings in the storage areas using a bulldozer, forming layers of filtered tailings, compacted by 10-ton compacting rollers until reaching the maximum dry compacted density, equivalent to 95% of the Proctor Standard [19].

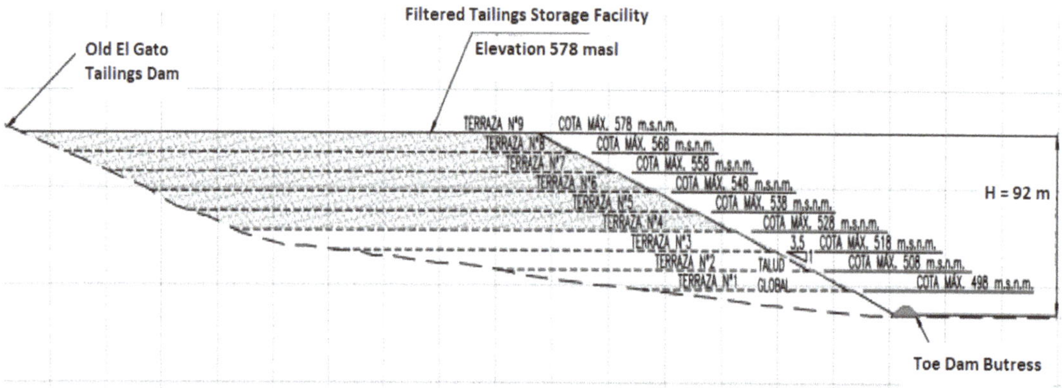

Figure 27. El Gato Creek Filtered Tailings Storage Facility–Typical Cross Section **[27,47,61]**.

Figure 28. El Gato Creek Filtered Tailings Plant [27,47,61].

7.5. El Peñon Filtered TSF–Flat Topography Configuration–Atacama Desert–Chile

Yamana Gold owns 100% of the high-grade gold and silver mine of El Peñón, located in the Atacama desert, in the II Region of Chile and 160 kilometers southeast of the port city of Antofagasta and 1,800 m above sea level. The El Peñón Deposit is exploited underground, by a Bench and Fill. Its annual movement is 2.5 million tonnes of ore and 58,000 m of tunnels are built. The water supply is made from deep wells 16 kilometers away from El Peñón [62,63].

The metallurgical process of its minerals is carried out in a plant with a capacity of 4200 mtpd, with a crushing-grinding circuit and agitated cyanidation leaching in tanks, obtaining the metal doré through a Merrill Crowe process and precipitate smelting. The tailings are filtered and deposited in a dry tailings storage facility [62,63].

The process contemplates the washing stages of the ore pulp before being deposited in the tailings deposit, carried out in the CCD circuit with a sterile solution and in the belt filters with a sterile solution. The purpose of washing the solids is to displace the contents of gold and silver towards the washing solutions. However, washing with industrial water in the belt filter (2nd washing) allows the NaCN content to be reduced from 2 g/lt to 200–300 ppm in the liquid impregnated with the tailings [62,63].

The process of depositing tailings in the filtered tailings storage facility (Figures 29–31) begins with the discharge of the filtered material, corresponding to a cake between 18 and 20% humidity, which is sent to a temporary tailings storage facility (in a material transport belt 55CVR01) that has a weight gauge (55WIB01), a wet tailings sampler (55AM01), and a chute (55CH04) for unloading and distribution to the stockpile site. The base of the tailings stockpile is made up of a concrete slab and a retaining wall to protect the belt, with a capacity of 2500 tonnes of tailings and has a floor pump (55PPS04) to collect solutions and cleaning water that are recycled to the Filtering Surge Tank (55TNK01) [62,63].

The stacked tailings are loaded onto trucks (50 tonnes) with a front loader (CAT 966) and transported to the tailings deposit (Figure 32). Once the tailing has been deposited, it is spread in layers of 15 cm, by means of a motor grader [62,63].

To complete the process, the tailings are removed by means of a disc plow and irrigated with industrial water three times a day to achieve the degradation of the cyanide contained in the solution that accompanies the solid (18–20% humidity) (Figure 33). This aeration and irrigation process is repeated for 72 hours, a process with which a tailing with cyanide concentrations of less than 2 ppm is obtained [62,63].

Figure 29. El Peñon Filtered Tailings Storage Facility Overview [62,63].

After 72 hours of treatment, having reached a residual cyanide concentration of 2 ppm and once the tailings are dry, a second 15-cm layer of material is deposited and so on until four layers (50~60 cm) are completed. The tailings are compacted by means of a 10-tonne roller [62,63].

The management of the tailings deposit follows a cyanide degradation, and tailings compaction program, to achieve a mechanically stable deposit area, free of contamination and with access roads that allow the circulation of vehicles involved in the operation [62,63].

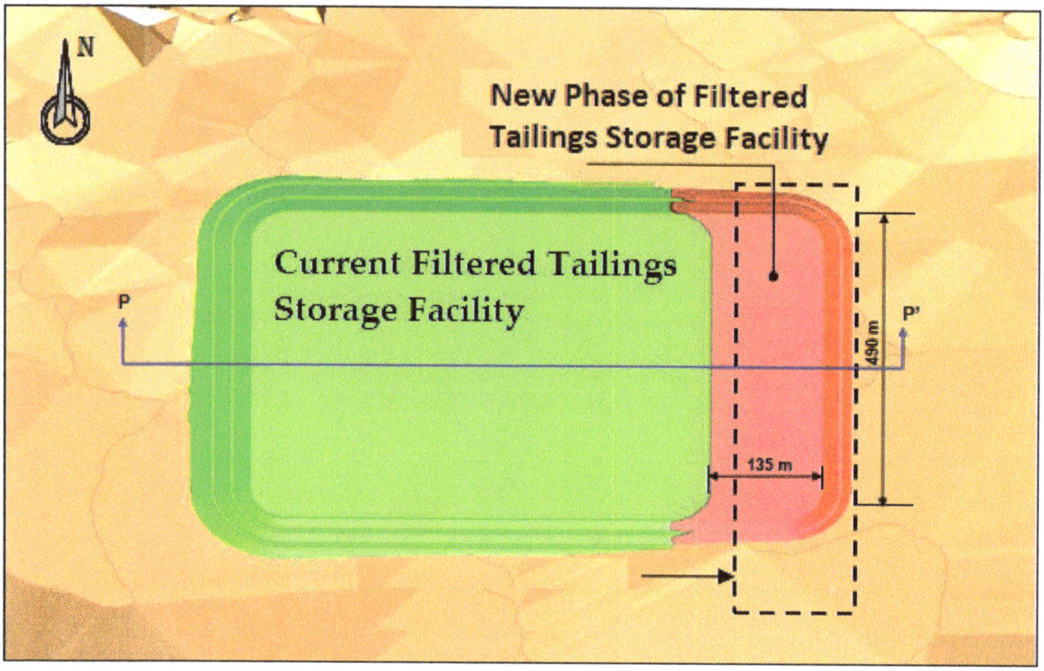

Figure 30. El Peñon Filtered Tailings Storage Facility—Old and New Phase Layout [62,63].

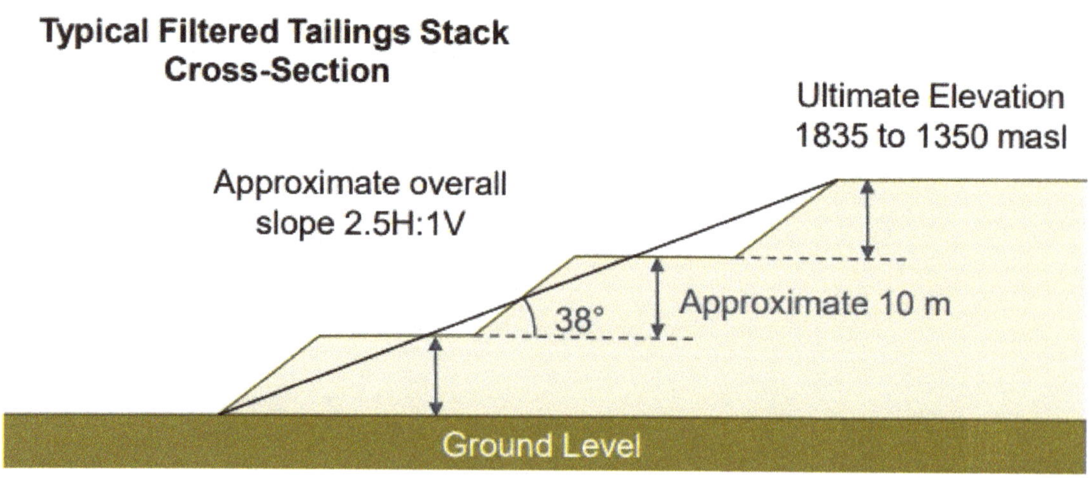

Figure 31. El Peñon Filtered Tailings Storage Facility–Old and New Phase Cross Section [62,63].

Figure 32. El Peñon Filtered Tailings Storage Facility—Haul Truck with Tails Disposal [62,63].

Figure 33. El Peñon Filtered Tailings Storage Facility–Tailings are removed by a disc plow [62,63].

7.6. Huasco Filtered TSF–Flat Topography Configuration–Atacama Desert–Chile

Valle de Huasco is a mining complex owned by the Chilean steel producer Compañía de Acero del Pacífico (CAP) with operations close to the city of Vallenar, approximately 650 kilometers north of Santiago. The area is comprised of the operating Los Colorados mine and the now-closed El Algarrobo mine. The Huasco Pellet Plant is a mining facility that dates back to 1978, whose objective is to process and add value to the product of iron mining, in the Huasco Commune. Its process consists of receiving pre-concentrated iron minerals from mines located in the Chilean iron strip and submitting them to grinding and concentration through physical processes (gravimetric and magnetic) to generate a commercial product called iron concentrate or pellet feed, with a concentration of magnetite (Fe_3O_4) over 90%. Alternatively, this product can be agglomerated and subjected to a thermal process to obtain iron pellets [54,55].

The total tailings from the grinding process are made up of the overflow of the hydro separators and the rejection of the magnetic batteries of each Grinding and Concentration line. Tailings generated are transported to the two Tailings Thickeners, where a flocculant product is added in order to increase the sedimentation rate of solids and optimize water recovery. From this point, the project under study contemplates the installation of a tailings filtering plant and a conveyor belt to connect this plant with a new filtered tailings deposit located 1.2 km south of the Pellet Plant (Figure 34) [54,55].

The start of tailings management corresponds to the discharge from the existing regulating agitator ponds, which will feed the filtering plant with tailings. From these ponds, the tailings are pumped to the filtering plant where the filters with the capacity to process a nominal amount of 5000 mtpd of tailings are fed. The filter plant will have up to four press-type filters available for filtering [54,55].

The Filter Plant has four vertical plate press filters available to filter the tailings until reaching the desired humidity. The process generates filtered tailings with up to 20% humidity, which are disposed of in the new deposit. For its part, the filtered water is returned to the Pellet Plant process [54,55].

At the Filter Plant, through collection belts, the filtered tailings are conveyed to a hopper located in the tailings transfer building, which will transfer it to an overland belt that will convey it to a second transfer station next to the Tailings Deposit. At the end of the belt, a transfer station is located, where there is a closed unloading building, the purpose of which is to allow tailings to be loaded onto trucks with a high load capacity, thus reducing transportation cycles [54,55].

The transfer building will have an alternative discharge system in case of failure of the overland belt. In case of emergency, trucks are located under the alternative exit and will transport the filtered tailings through the existing road of the Pellets Plant to an emergency stockpile located 200 m from the Pellets Filters Plant (Figure 35) [54,55].

At the work front of the filtered tailings deposit, the trucks will unload the tailings, which are spread out in layers using heavy machinery. These layers are compacted. Once the tailings layers have been compacted, they are irrigated with a solution of dust suppressant with water. In turn, the entire path traveled by the trucks, inside and outside the tailings storage facility, is stabilized with a dust suppressant to minimize emissions of particulate dust [54,55].

The filtered tailings deposit is delimited at its base by a toe dam of variable height, between 5 and 10 m high, to reach a height of 40 m, with respect to sea level (Figure 36). The dam is built with borrow material, and its objective is to contain the starting slope of the deposit, ensuring its physical stability and providing support to the first bank. The reservoir is filled in five successive growth stages (terraces) [54,55].

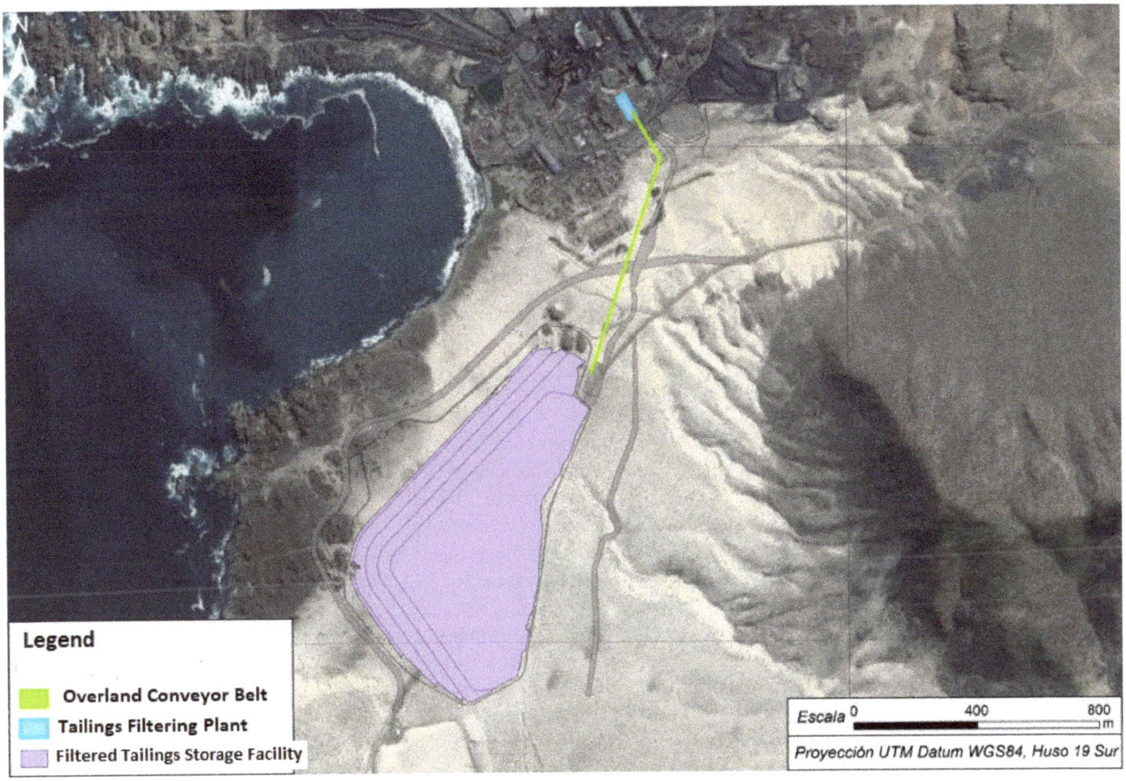

Figure 34. Huasco Filtered Tailings Storage Facility—Layout [54,55].

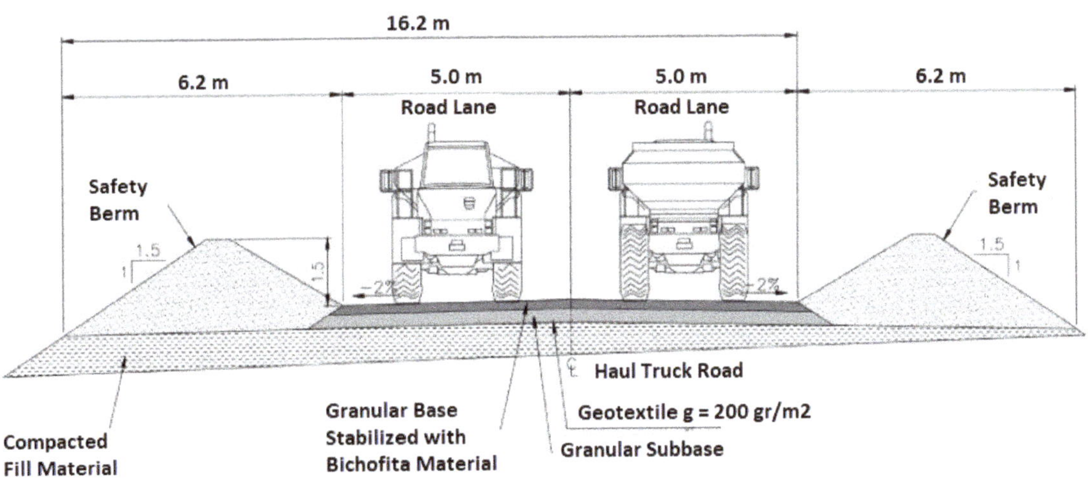

Figure 35. Huasco Haul Truck Road at Filtered Tailings Storage Facility—Typical Cross Section [54,55].

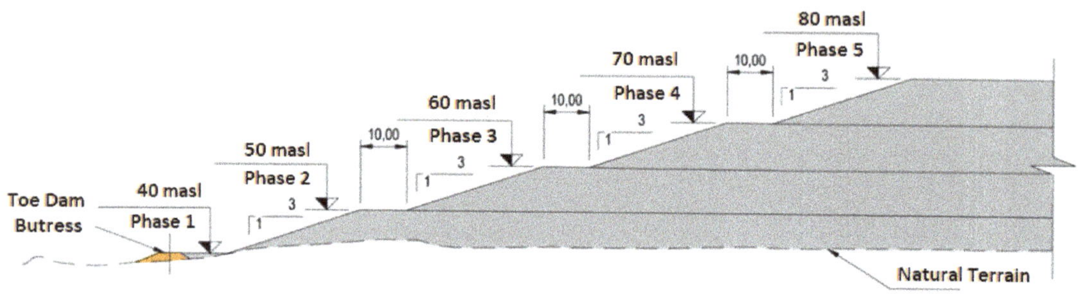

Figure 36. Huasco Filtered Tailings Storage Facility—Typical Cross Section [54,55].

7.7. Salares Norte Filtered TSF–Flat Topography Configuration–Atacama Desert–Chile

Salares Norte is located between 3900 and 4700 m above sea level, in the middle of the highest peaks of the Andes mountain range and the driest desert in the world. Its height and geographical environment are some of the challenges faced by this project that seeks to process 2 million tonnes of ore per year for the production of metal doré [50,64,65].

The exploitation of gold and silver is carried out in the open pit. The ore is processed by a crushing and grinding system. The extraction is through a hybrid scheme of conventional cyanide leaching -Merrill Crowe- and carbon in pulp for the recovery of gold and silver and deposit of filtered tailings [50,64,65].

The system called wad cyanide detoxification system will occupy an approximate area of 190 m^2. Its objective is the detoxification of tailings (reducing the concentration of wad cyanide to less than 15 ppm, before being stockpiled) and the conditioning of moisture in the tailings to its final disposal in the filtered tailings deposit. Sodium metabisulfite is used to detoxify the wad cyanide in the tailings from the coal-in-pulp stage [50,64,65].

The plant, regarding the treatment of cyanide, has been designed taking into account the requirements of the international cyanide code, a non-binding regulation in Chile, which contemplates the highest world standard in the handling of this substance [50,64,65].

Following the detoxification of the tailings, the pulp is taken to a tailings thickener, this thickener is 30 m in diameter and will occupy an area of 1100 m^2 inside the plant, where the envelope flow or overflow is sent to the plant for incorporation into the process, while the pulp is conducted through pumps to the tailings filtering area, located approximately 1.2 km from the gold process plant (Figure 37) [50,64,65].

The tailings filtering sector will have an approximate area of 1000 m^2, which includes: a thickening sector that includes a flocculant plant and recovered water management ponds; and a filtering area with auxiliary services and a system for returning the water recovered to the process [50,64,65].

The equipment considered in the tailings filtering stage is: (i) A 30 m diameter tailings thickener; (ii) A filter tank of 350 m^3 (process water tank); (iii) Three vertical plate type press filters (two in operation and one on standby), and (iv) Two tanks of 1500 m^3 for feeding the filter [50,64,65].

The tailings deposit that is on the intermediate platform of the southern waste rock dump, consists of a tailings storage facility previously filtered, self-supporting stockpile of tailings, which are built in layers compacted by means of a vibrating roller, with humidity around the optimum according to the results of Standard Proctor compaction tests, in order to reach the compaction levels established in the Technical Specifications of the Project, which corresponds to a range between 85% and 90% of the Standard Proctor test, which are carried out periodically during the operation of the deposit (Figure 38) [50,64,65].

The filtered tailings deposit is filled at a tailings production rate equivalent to 5500 mtpd, it will also be made up of two platforms with a bank height of 20 m, occupying an area of 54 ha and a volume of 16 million m^3 (Figure 39) [50,64,65].

Both the base of the tailings deposit (intermediate platform of the southern waste rock dump) and the slopes of the hills and inclined surfaces of the dump on which the tailings will rest are waterproofed by means of a geomembrane [50,64,65].

The filtered tailings deposit has a system for capturing contacted water consisting of a basal drainage system, collection of water at the foot of the slope, a collection pool downstream of it, and a storage pool located on the south side of the filter plant. It also has a safety parapet downstream to contain potential spills [50,64,65].

Additionally, as the tailings deposit grows, fill material or mine waste is placed on its slope to reduce wind erosion on the said surface [50,64,65].

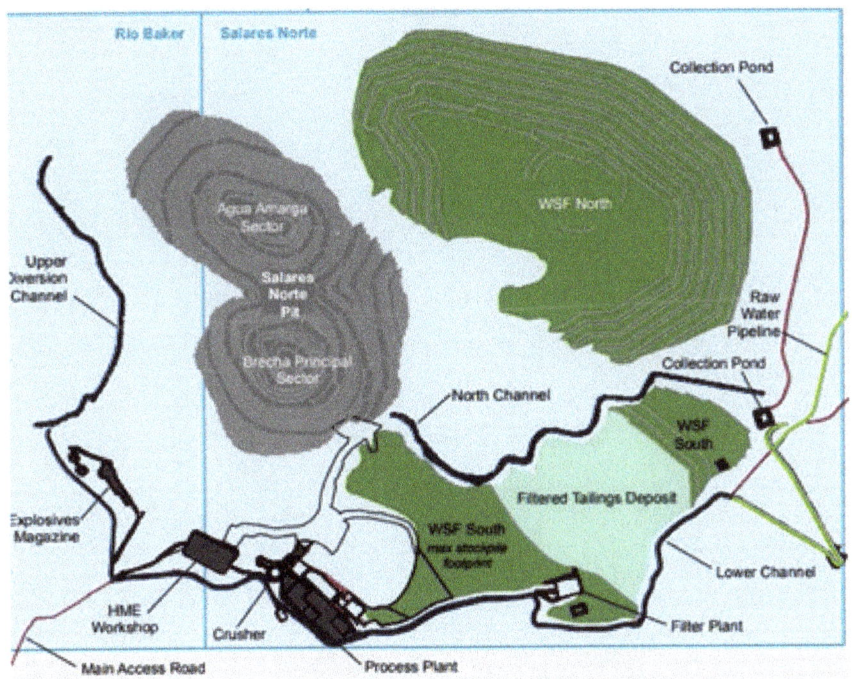

Figure 37. Salares Norte Filtered Tailings Storage Facility—Layout [50,64,65].

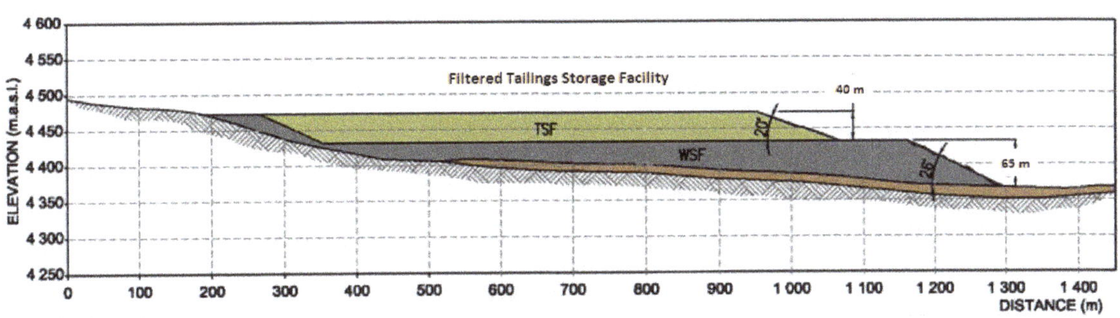

Figure 38. Salares Norte Filtered Tailings Storage Facility–Typical Cross Section [50,64,65].

Figure 39. Salares Norte Filtered Tailings Storage Facility–Dozer forming filtered tailings layers [50,64,65].

7.8. Tambo de Oro Filtered TSF–Valley Topography Configuration–Atacama Desert–Chile

The Tambo de Oro Mining Site is located in the Punitaqui Mining District, 5 km southeast of the town of Punitaqui and 29 km southeast of the city of Ovalle, commune of Punitaqui, Province of Limarí, Region of Coquimbo. The Tambo de Oro Mining Site corresponds to underground exploitation of high-grade gold ore, environmentally authorized for extraction of 15,000 t/month -using the Bench and Fill (B&F) extraction method- and processing, via flotation, of 22,500 t/month, whose final product is gold-copper concentrate and thick gold concentrate obtained by gravitational methods. In this regard, it should be borne in mind that the processing capacity considers the purchase of 7500 t/month of ore from third parties, which, added to the extraction of 15,000 t/month, completes the processing capacity processing of 750 mtpd [49].

Tailings are the fraction of the flotation that is mainly made up of gangue and other species that have no commercial interest because they have low grades of gold, silver, and copper. In this area, the installation of a thickener and a filter system has been considered.

The tailings from the gold flotation are thickened from an initial concentration of 37% by weight to a final concentration of a minimum of 55% and a maximum of 60%. The thickening of the tailings to 55% solids will allow the recovery of around 9.4 m^3/h of water that returns to the process and the solids are used for tailings filtration. The tailings thickening circuit is made up of a thickener and an accumulator/conditioning tank that collects the underflow from the thickener and is also the feed point for the tailings filtrate [49].

The process tailings from the underflow of the tailings thickener are accumulated and/or conditioned in a tank equipped with an agitator, from where an automatic filter press with 190 plates of 1200 mm × 1200 mm is fed (Figure 40). With the capacity to process 4.41 m^3 per cycle (11 minutes per cycle), this guarantees production with 20 h of operation per day. The product of the tailings filtration will generate flotation tailings with average humidity between 12 and 16%. The filtered material is transported in trucks to the tailings deposit [49].

For the transport of tailings, trucks with a capacity of 14 m^3 are used, which will move from the processing plant to the tailings deposit, approximately 500 m east of the processing plant, with a frequency of 40 trips/day in the highest processing stage of the plant (Figure 41) [49].

The filtered tailings deposit will have a capacity of approximately 864,000 tonnes, compacted at a density of 1.85 tonnes / m^3 (Figure 42). The deposit will have a total height of 60 m and covers an area of 4.7 ha. The base of the deposit is prepared in order to ensure a good anchoring of the tailings material to the ground. The tailings disposal slope is H:V = 2.5:1.0, with banks every 15 m and berms 3 m wide (Figure 43). The material is spread out in horizontal layers that do not exceed 30 cm in loose thickness and is compacted until it reaches a minimum dry density of 95% of the maximum compacted density dry from the Standard Proctor test [19,49].

Figure 40. Tambo de Oro Tailings Vertical Plate Filter Press [49].

Figure 41. Tambo de Oro Filtered Tailings Storage Facility Overview and Front Loader Works [49].

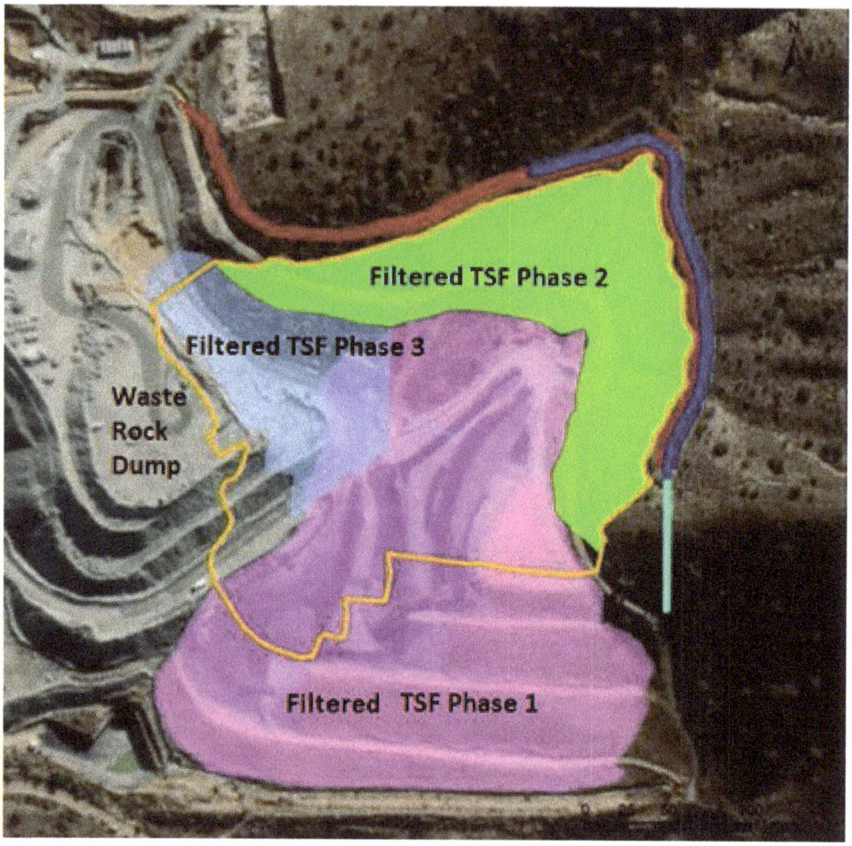

Figure 42. Tambo de Oro Filtered Tailings Storage Facility—Layout [49].

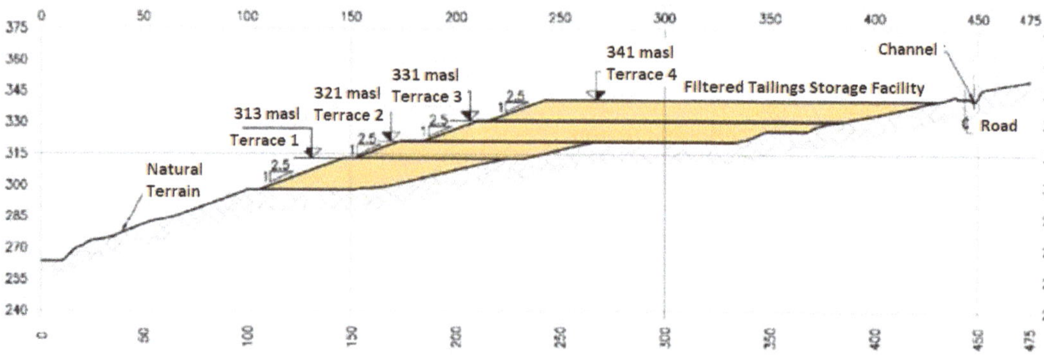

Figure 43. Tambo de Oro Filtered Tailings Storage Facility–Typical Cross Section [49].

7.9. Tambillos Filtered TSF–Flat Topography Configuration–Atacama Desert–Chile

The Tambillos Mining Site is located in the Coquimbo Mining District, 28 km south of the city of La Serena, Region of Coquimbo, Chile. From Mina Florida, ore is supplied to the Beneficiation Plant for the production of copper concentrate, located four km to the Southeast (SE) of the town of Tambillos [48,66].

In this regard, it should be noted that the useful life of the Project increases by seven years the useful life of the mining operations carried out by Florida Mining Company at its Tambillos Site. The foregoing is a consideration of the existence of sufficient copper ore reserves in Mina Florida to continue processing ore at 3000 mtpd [48,66].

The tailings are transported from the concentrator plant. It will originate in the tailings of the flotation plant, which are driven using the existing caisson and pumps (which currently drive towards the tailings impoundment in operation). The pipe is HDPE 250 PN6 PE100 and will continue to the thickener feed box and projected filtrate.

The tailings thickening unit will consist of a high-efficiency thickener (HRT) with a diameter of 18 m, which has a feed elevator box and will serve as a compact flocculant preparation and conditioning unit (Flocculant C1590). The pulp discharged from the thickener is pumped to the upper loading drawers of each of the filters (three in total), through HDPE 160 PN6 PE100 pipes with a nominal diameter of 160 mm with 50 HP centrifugal pumps (1 in operation and 1 on standby) [48,66].

The filtering unit is enabled on a platform of approximately 2000 m^2, where the plant is mounted, as well as the tailings control and loading systems. This unit comprises three ceramic disc filters installed in series, with a filtering area of 120 m^2 each (Figure 44). The tailings that have passed the filtering process are unloaded using a front loader, which in turn will load 40-tonne trucks for transport to the deposit area, where they are distributed using bulldozers and compacted using a pneumatic roller (Figure 45) [48,66].

It is important to point out that the location of the filtering unit is optimal from the operational and environmental point of view since it allows for optimizing the travel times of loaded trucks from the filters to the tailings disposal area (distance of 600 m), reduces fuel costs and the operational risk associated with the displacement of a large volume of tailings [48,66].

The civil work to be projected for the management of mining waste consists of a tailings deposit made up of terraces 7 m high each, arranged in such a way as to allow the existence of 5 m berms and a general slope of H:V = 3.0:1.0 (Figures 46 and 47) [48,66]. In specific terms, the filtered tailings facility will have the following geotechnical characteristics:

- Average geotechnical humidity of the tailings cake: 16%.
- Construction method: terraces.
- Platform height: 7 m.
- Berm width: 5 m.
- Local terrace slope: H:V = 2.5:1.0.
- Global Slope: H:V = 3.0:1.0.
- Maximum dry compacted density (95% Standard Proctor): 2.05 t/m^3.

Figure 44. Tambillos Tailings Ceramic Disc Filters [48,66].

Figure 45. Tambillos Tailings Thickening and Filtering Plant [48,66].

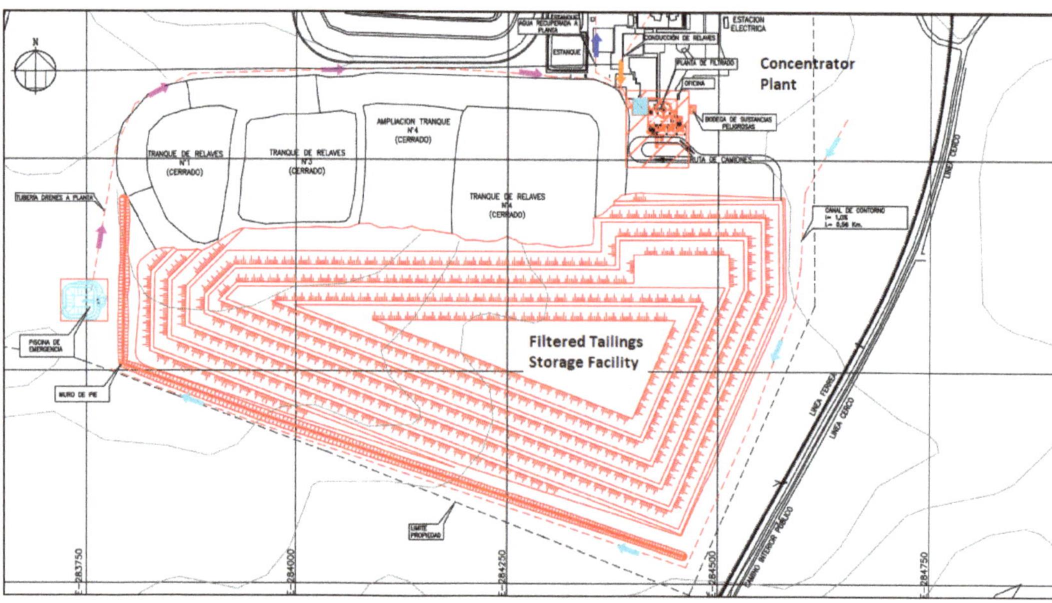

Figure 46. Tambillos Filtered Tailings Storage Facility Layout [48,66].

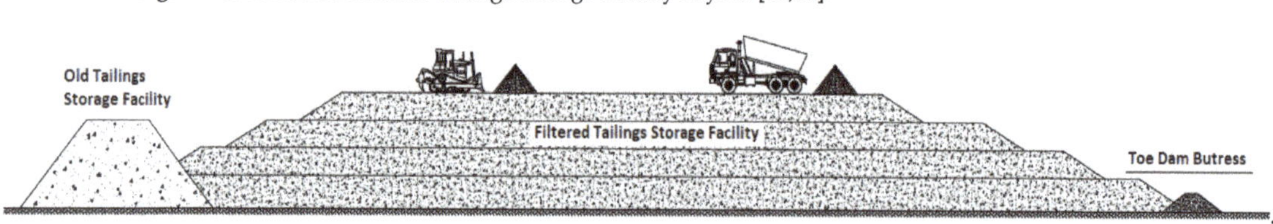

Figure 47. Tambillos Filtered Tailings Storage Facility Typical Cross Section [48,66].

7.10. Mantos Blancos Filtered TSF–Flat Topography Configuration–Atacama Desert–Chile

The Mantos Blancos copper mine is located 48 km from the Chilean port city of Antofagasta and is 800 m above sea level. Anglo American is the owner of the mine; which is one of the company's six operating divisions in the region. The Mantos Blancos division includes an open-pit mine, crushing plants, and installations for processing oxide and sulfide ore. The mine mainly produces copper. The Sulfide ore plant at Mantos Blancos Division of Anglo American Chile Ltd, which started its operations in 1981, has had a tailings treatment plant from the beginning. Basically, it consists of a classification circuit, slimes thickening, and filtration of the coarse fraction. As this plant is located in one of the most arid lands in the world, coarse tailings are filtered in order to maximize water recovery. Disc filters were used at the beginning. Later in 1984, during the expansion of the concentrate plant to 12,000 mtpd, horizontal belt filters were incorporated and are still being used today. Tailings are finally disposed of in a dam for thickened slimes and in a dam for the coarse fraction, in a 40/60% ratio, respectively, and a solids concentration of 60% and 80%, respectively [18,67,68]

The tailings from the concentrator plant are classified in a station of 500 mm Hydrocyclones, functioning by gravity, in a conical configuration, with four operating units and two on stand-by. This classification system produces a solids distribution of 50–55% and a cut size of d50 between 50–60 microns. The overflow of the cyclones, CW 24–26% of solids, is transported to three thickening units (Larox of 60 m diameter, Dorr Oliver of 44 m diameter, and Eimco of 44 m diameter) for the process of sedimentation, using a dosage of flocculent of 5 gr/ton, that later feeds with a solids content of Cw 55–60% the filtration units and the fine tailings dam in a ratio restricted by the capacity of absorption of fines in the filters. The thickeners recover approximately 63 to 67% of water, which is returned to the concentrator plant (Figure 48) [18,67,68].

The underflow of the cyclones, with a solids content of Cw 65–67%, is sent to three filtration units (band filters of 100 m^2 each) and are previously mixed with parts of thickened fine tailings. Tailings cakes are obtained with a 17% moisture content considering a wet basis (equivalent to a 23% moisture content considering a dry basis), which are later transported to the dry stockpile of coarse material by conveyor belts (Figure 48). The filtered water, plus the wash water of the filtering cloth, is recirculated to the thickeners to sediment solids in suspension [18,67,68].

The Filtered Coarse Tailings Storage Facility is developed by down valley placement method applied at Mobile Stacker Conveyor with tripper forming a fan shape, where saturated filtered tailings flow along roughly repose angle slopes, buttressed at the toe by a containment dyke and downstream supported by a sedimentation collection pond. Dozers carried out spreading and compaction activities on filtered tailings lifts of average 20–30 cm in thickness (Figures 49 and 50) [18,67,68].

Figure 48. Mantos Blancos Filtered Tailings–Belt Filters and TSF overview [18,67,68].

Figure 49. Mantos Blancos Disposal of Filtered Coarse Tailings with Mobile Stacker Conveyor and Tripper [18,67,68].

Figure 50. Mantos Blancos Aerial View of Disposal of Filtered Coarse Tailings [18,67,68].

7.11. Potrerillos Filtered TSF–Valley Topography Configuration–Atacama Desert–Chile

The Potrerillos Smelter is located in the third Region of Atacama, northern Chile at 3000 m above sea level, in the Andes Mountains. Access is through a highway with the city of El Salvador 40 km away. The closest commercial port is Antofagasta, a distant 300 km to the northwest. The city of El Salvador has the El Salvador Bajo commercial airport.

The "Slag Flotation Potrerillos Foundry Teniente Converter" Project includes the elimination of the HLE Slag Cleaning Furnaces, implementing a series of facilities that allow processing of the slag generated in the Teniente CT5 converter, which has an average Cu Law between 8 and 11% (Figure 51) [18,69].

For this, the following processes are implemented:

- Cooling of slag from CT in pots.
- Grinding in SAG mill and Balls.
- Rougher and Scavenger flotation.
- Thickening of concentrate and tailings.
- Filtering of concentrate and tailings
- Transport of filtered tailings to tailings deposit.
- Transportation of concentrate to the concentrate handling area in the smelter.

The final product is copper concentrate, which is recirculated to the Smelter. The daily treatment design capacity of the Flotation Plant is 1700 tonnes of slag, of which the nominal processing capacity corresponds to 1530 tonnes. As a result of the processing, 412 mtpd of copper concentrate and 1288 mtpd of filtered tailings are generated [18,69].

The main target is increasing the water recovery of tailings, and for this, the solid/liquid separation of tailings considers two process unit operations, mainly: (i) thickening (high rate or high-density thickeners), and (ii) filtering with two Ceramic Disc vacuum filters. The tailings are received from the flotation process with solids content by weight (Cw) around 27–30% range, to be thickened to a Cw of approximately 60–65% range, to be fed to the filters. These units provide a cake-type product typically with a Cw between 85–90% range, with a cake thickness in the order of 10–12 mm, and with moisture content (wet basis) of approximately 10–15% [18,69].

Once the tailings cake is obtained, it needs to be stored in a temporary stockpile to transfer to a filtered tailings conveyance system (Figures 52 and 53). The design of a dry stack TSF needs to consider a filtered tailings conveyance system that needs compatibility with the dry stack construction sequence/plan, using conventional conveyance/haulage and mechanical placement equipment. The main filtered temporary stockpile and transport system used typically is: (i) conveyor belt/chute for temporary filtered tailings stockpile at the discharge of the filter plant, (ii) transfer and loading by a front end loader to haul trucks of filtered tailings, and (iii) haul truck access and operation roads to deposited filtered tailings at TSF [18,69].

The project considers the transport of filtered tailings through 25 tonnes capacity haul trucks (15 m^3), which allows for the transportation of approximately 750–2500 m^3/day range of generated tailings.

The filtered tailings are commonly deposited in layers of 30 cm thickness, which are exposed to reach their optimum moisture content such as 12%, and are then compacted to reach 95% Proctor Standard (ASTM D698) [19]. The filtered dry stack TSF is a mass composed of several terraces (or benches) and berms. The dry stack TSF is constructed in the upstream direction of a valley topography (Figure 54) [18,69].

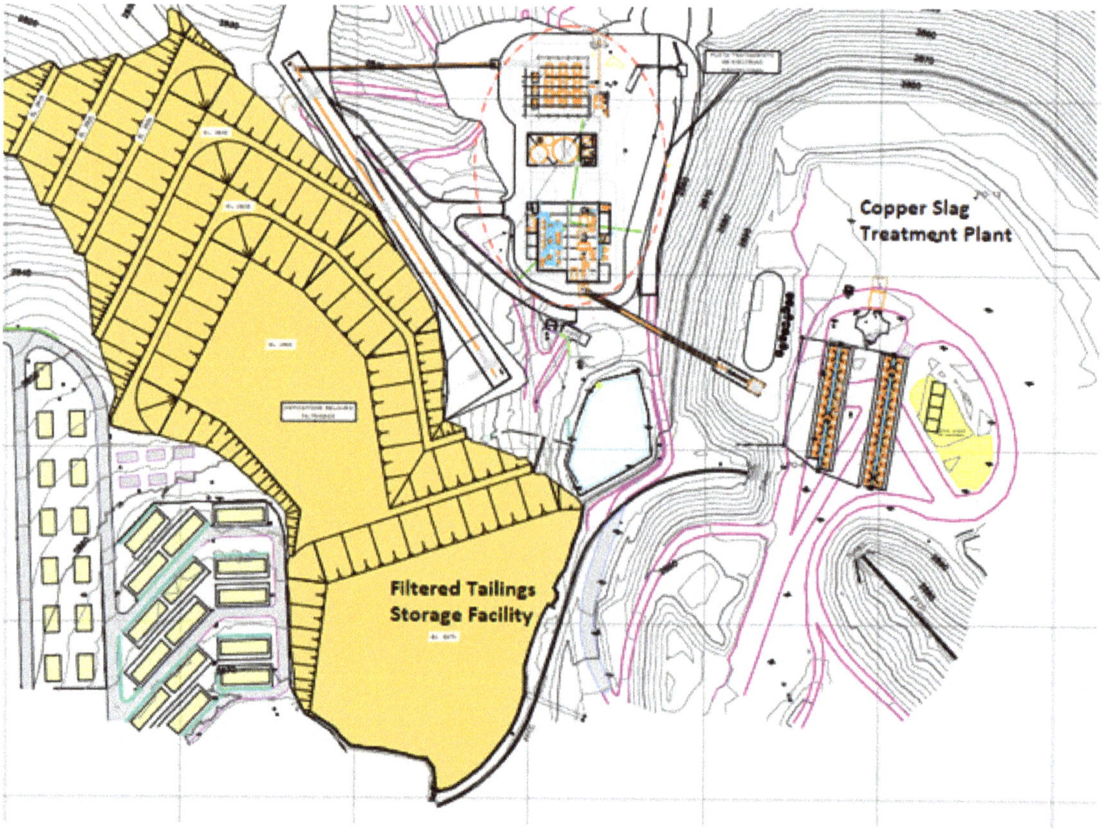

Figure 51. Potrerillos Filtered Tailings Storage Facility Layout [18,69].

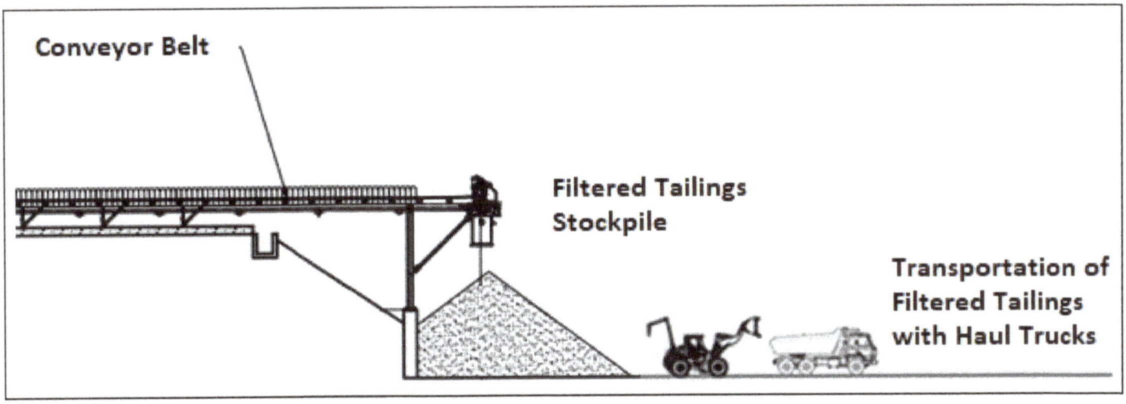

Figure 52. Potrerillos Filtered Tailings Storage Stockpile Facility [18,69].

Figure 53. Potrerillos Copper Slag Treatment Plant Facility [18,69].

Figure 54. Potrerillos Copper Slag Treatment Plant and Filtered Tailings Storage Facility [18,69].

7.12. Cerro Lindo–Dry Stack of Filtered Tailings–Andean Range–Peru

The Cerro Lindo Project, property of Compañía Minera Milpo S.A. (Milpo), is located in the province of Chincha, department of Ica, 200 km south of Lima-Peru city. The mine currently processes 7000 mtpd for obtaining polymetallic minerals zinc, lead and copper, using the method of sub-level stoping exploitation. Concentrates produced, in the order of 10% of the processed mineral, are transported by trucks to Lima-Peru city for its commercialization. From the generated tailings, 55% is used for paste filling of the underground mining works, and 45% is deposited on the surface as filtered tailings. The filtered tailings to be deposited in the TSF site called Quebrada Pahuaypite (Figure 55) require compaction in layers of the tailings, due to the elevated site topography and the high seismic activity of the site [20,21,25].

Tailings generated in the process plant are discharged to the surface as filtered tailings and are used as paste fill to be used in underground mining works, 55% of the time tailings are discharged to be used as backfill for underground mining, and 45% of the time they are discharged on the surface [20,21,25].

Tailings are released from the process plant with 30% of solids and are conveyed to a 30-m diameter High Compression Thickener. Underflow tailings from the High Compression Thickener reach 76–78% solids content and are sent to the filter plant. There are two filter plants, one for surface tailings deposition and the other for underground mining. Both of them are located less than 500 m from the above-mentioned facilities [20,21,25].

In the filter plant, tailings are dewatered with belt filters to reach a solids content between 83 and 85% for a proper mix design with cement which varies between 0 and 5% by weight depending on the stope to be filled in the underground mining (elevation 1850 masl). In the filter plant for surface deposition, the filtered tailings have a solids content (Cs) between 87 and 88%, which is equivalent to a moisture content (w%) between 14 and 15%. These tailings at this moisture content can form stacks with slopes of H:V = 1.3:1.0. Tailings are handled with wheel loaders to trucks with a capacity of 25 tonnes.

Trucks discharge tailings into the tailings deposit by sectors to obtain the dewatering and compaction cycles [20,21,25].

Likewise, the topographic conditions and the narrowness of the Quebrada Pahuaypite made it advantageous for the tailings transport to be by means of trucks instead of conveyors. An additional advantage of compacting tailings in layers was that it created a medium of low permeability, diminishing the risks of infiltration of rainwater, and inhibiting the entrance of oxygen, thus minimizing the generation of acid drainage [7].

Tailings are processed in a filtration plant by vacuum belt filters, obtaining a cake with 18% moisture content. The filtered tailings deposit is composed of two desiccation platforms of approximately 2250 and 14,050 m^2 each, located at elevations 2028 and 2082 masl, respectively. In these platforms, filtered tailings are discharged, and spread by sectors to obtain desiccation and to allow further compacting [20,21,25].

The tailings deposit has been designed to store 14 Mt of tailings and is built as a large fill which will have an overall slope of H:V = 2.8:1.0 (Figure 55). The tailings storage construction has benches with minimum width of 10 m each and 20 m high (Figure 56). As part of the tailings deposit, a basal drainage system has been provided for seepage collection from the tailings deposit. The installation of inclinometers and underground water monitoring wells is also considered. They are progressively installed in berms that will form the deposit [20,21,25].

To obtain tailings that achieve a minimum initial dry density, that is seismically stable, and not below 2.8 t/m^3, the tailings need to be first dewatered until reaching a moisture content in the range of 5–7% and then compacted to 95% standard proctor density. To reach a dry density for compaction, operation cycles from 3–5 days are applied depending on the period of the year. In summer (December to March), the wet season, cycles could last five days. For the rest of the year, cycles could last approximately three days. Operation cycles consist of the following activities: discharge, spreading, drying–compacting (Figure 57).

The layer thickness to form tailings is 30 cm, and after reaching moisture from 5–7%, tailings are compacted through vibrating rollers of 10 tonnes until reaching the required density, which is obtained generally with two cycles of roller passes [20,21,25].

Figure 55. Cerro Lindo Filtered TSF Overview [20,21,25].

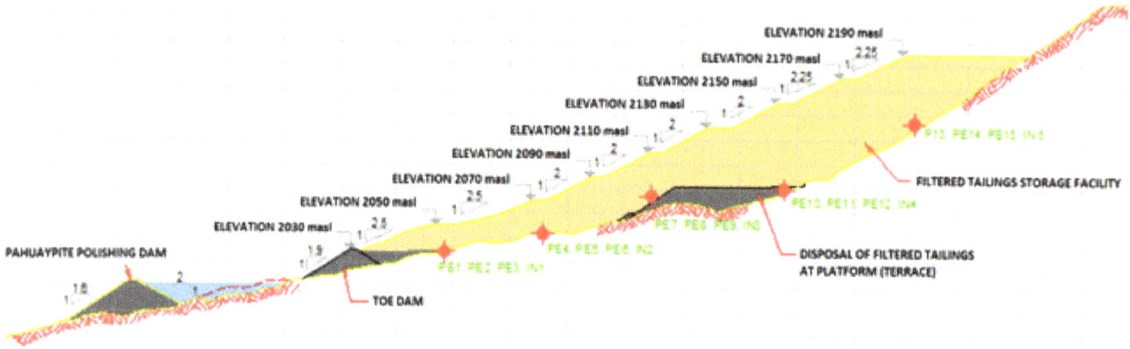

Figure 56. Cerro Lindo Filtered TSF Typical Cross Section [20,21,25].

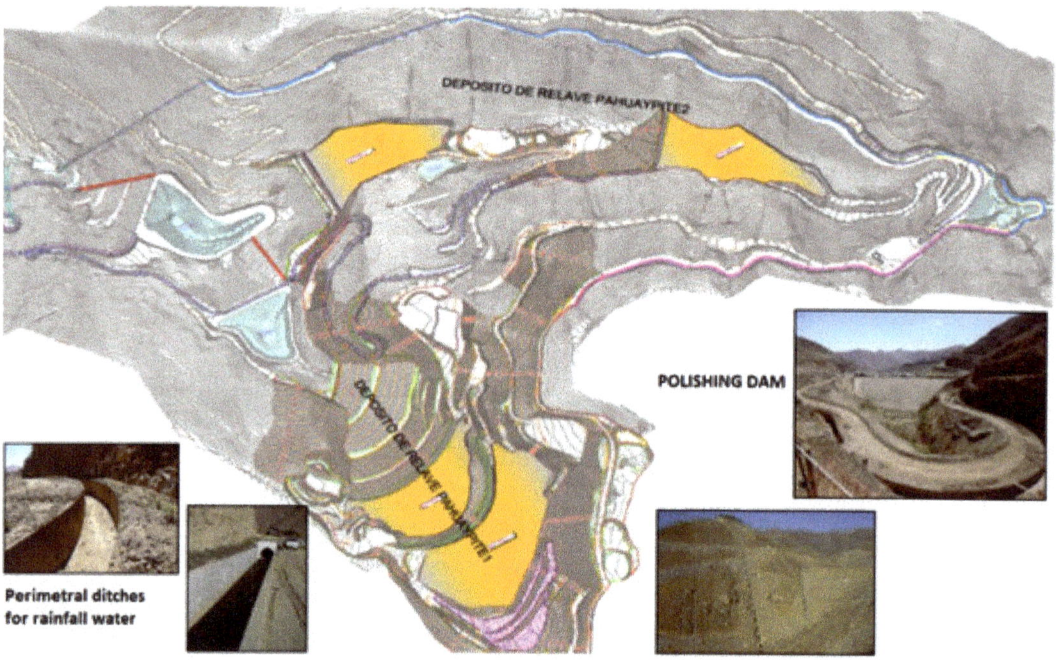

Figure 57. Cerro Lindo Filtered TSF Layout—Pahuaypite 1 TSF and Pahuaypite 2 TSF [20,21,25].

7.13. Poderosa Filtered TSF–Valley Topography Configuration–Andean Range–Peru

The project area, called Pataz Batolito specifically, the Marañón tailings deposits are located on the right bank of the Marañón river valley, in the district and province of Pataz, department of La Libertad, at an average altitude of 1180 to 1280 masl in the Unit Production Marañón, which in turn is part of the Compañía Minera Poderosa.

At the Marañón metallurgical plant, gold is recovered through the cyanide leaching or direct cyanidation process. The total treatment capacity of the plant is 700 tonnes per day. Once the gold dissolves, it is precipitated through the Merrill Crowe process and melted down to obtain the gold bars that are traded. The recoveries of cyanide solution obtained in the plant are above 92%, the residues are stored in filtered tailings deposits, where all the effluents that can be generated are controlled to avoid contamination [36,37].

The tailings filtration plant has a treatment capacity of 700 mtpd. Thickener No. 4 discharges tailings in a slurry with a density of 1650 tons/m^3, passing through an automatic sampler. The slurry is driven by gravity to the agitator, from which it is pumped to two press-type filters. There, the tailings, with a humidity of less than 15%, are deposited in a temporary storage pile by means of a conveyor belt. Finally, it is transported with dump trucks to the Asnapampa tailings deposit for its compaction and final disposal (Figures 58 and 59). The recovered filtered liquid is deposited in the recovery tank, then with the activation of a pump, the solution returns to Thickener No. 3 and the solution is recirculated to the metallurgicalplant [36,37].

With the change of disposal to filtered tailings with 13% moisture equivalent to the optimum moisture content (±2% tolerance), physical stability has been improved for static and pseudo static conditions, as well as ensuring hydrological and hydrogeochemical stability and compliance with these requirements [36,37].

The filtered tailings are deposited in terraces that are adapted to the topography of the place. There are banks 10 m high and berms 2 m wide, with local slopes of H:V = 1.5:1.0 (Figures 60 and 61).

Figure 58. Poderosa Filtered Tailings Plant [36,37].

Figure 59. Poderosa Haul Trucks with Filtered Tailings [36,37].

Figure 60. Poderosa Filtered Tailings Storage Facility Layout [36,37].

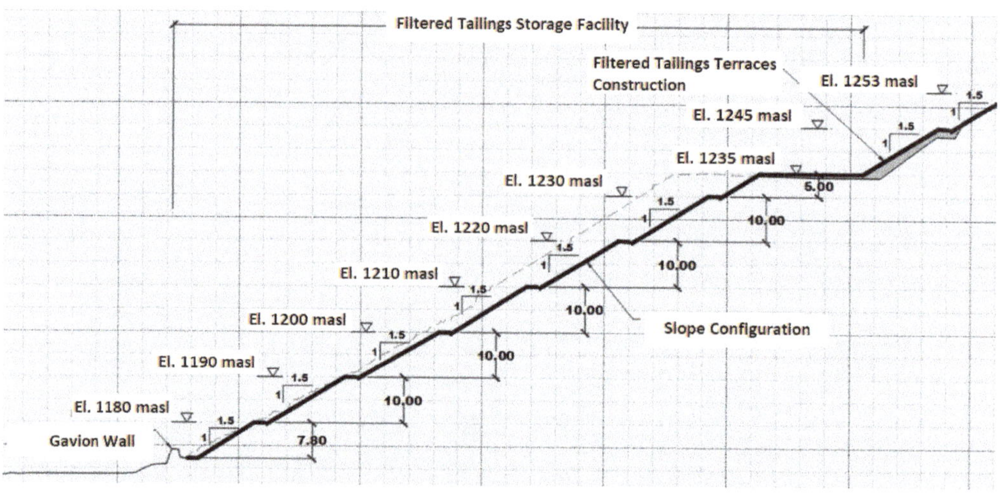

Figure 61. Poderosa Typical Cross Section Filtered Tailings Storage Facility [36,37].

7.14. Curaubamba Filtered TSF–Valley Topography Configuration–Andean Range–Peru

Consorcio Minero Horizonte (CMH) exploits a gold deposit in the district of Parcoy, province of Pataz, department of La Libertad. As a consequence of this exploitation, mining residues such as tailings and waste are produced, which have to be deposited in suitable places, and it is in this sense that CMH has decided to expand the storage capacity of the Curaubamba tailings and waste deposit [38,39].

The project area is located in the Curaubamba basin, on the left bank of the Parcoy River; adjacent to CMH's Chilcapampa tailings deposit.

The new tailings filtering plant considers the processing of tailings from the Parcoy plant from its "overflow" process with a projected capacity of 2000 mtpd, receiving tailings with a 30% solid content by weight, until it is thickened and filtered, obtaining a "cake" with approximately 85% solids content by weight and a humidity of 15%. The filtration equipment corresponds to press-type filters [38,39].

The project for the regrowth of the tailings deposit consists of the use of the Curaubamba creek, for the storage of the tailings and waste generated by the mining activity in the plant. The project will have two deposits, one for mine waste material and another for filtered tailings (Figures 62 and 63) [38,39].

The storage capacity of the regrowth of the tailings and waste deposit has been estimated at 22 years, to take advantage of the maximum capacity of the Curaubamba creek [38,39].

The design of the regrowth of the tailings and waste deposits will consist of raising the existing deposits. In other words, in the case of the filtered tailings deposit, it is founded on the final elevation of the existing deposit at the 2605 masl level and grows by means of a 30-m-high compacted embankment until it reaches the 2635 masl level. In the case of the waste rock material deposit, this regrowth is carried out by progressively raising the eight existing banks of waste rock, by a maximum height of 30 m for each bank (Figures 64–66). The following characteristics are present in the filtered tailings deposit:

- Material of the embankment: filtered tailings.
- Maximum elevation: 2635 m above sea level.
- Starting level: 2605 m above sea level.
- Number of TSF: 01.
- TSF height: 30.0 m.
- Deposit slope: H:V = 2.5:1.0.
- Storage Volume: 1.15 M m^3.
- Operation time: 22.0 years.

For the transport of tailings, trucks with a capacity of 15 m^3 are used, which will move from the processing plant to the tailings deposit, approximately 200 m east of the processing plant, with a frequency of 40 trips/day in the highest processing stage of the plant [38,39].

The layer thickness to form tailings is 30 cm, and after reaching a moisture content from 10–12%, tailings are compacted through vibrating rollers of 10 t until reaching the required density, which is obtained generally with two cycles of roller passes [38,39].

Figure 62. Curaubamba Filtered TSF–Terraces View [38,39].

Figure 63. Curaubamba Filtered TSF—Layout [38,39].

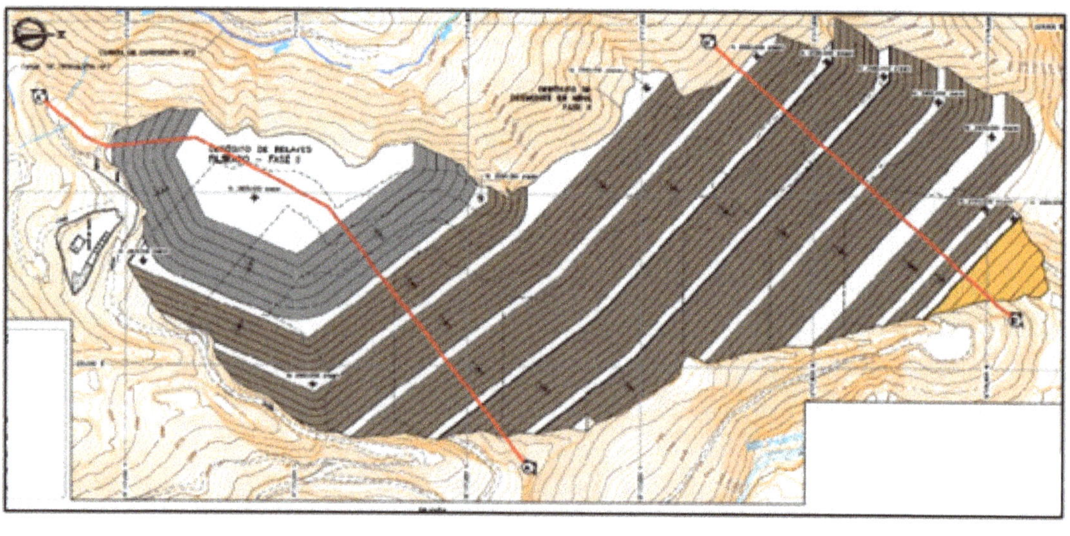

Figure 64. Curaubamba Filtered TSF—Terraces Overall View [38,39].

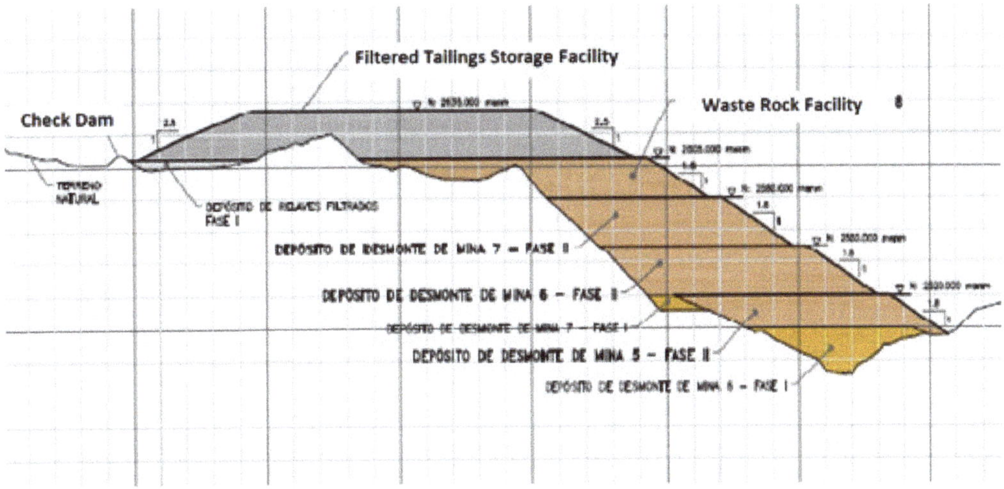

Figure 65. Curaubamba Filtered TSF–Typical Cross Section [38,39].

Figure 66. Curaubamba Filtered TSF–Benches and Terraces [38,39].

7.15. Catalina Huanca Filtered TSF–Valley Topography Configuration–Andean Range–Peru

Catalina Huanca Sociedad Minera S.A.C (CHSM) is a mining company that is dedicated to the exploitation and processing of polymetallic minerals; where minerals such as zinc, lead, and silver are extracted through underground mining and transferred to the plant in 3-axle and 4-axle dump trucks; being processed with a capacity of 2000 mtpd.

There is an area for the tailings deposit and auxiliary facilities in the Ramahuayco project area, located in the district of Canaria, province of Víctor Fajardo, department of Ayacucho at an altitude of 3500 m above sea level [40,41].

The extracted ore is processed by conventional flotation at the San Jerónimo Beneficiation Plant, which has a capacity of authorized treatment of 2000 mtpd, in which concentrates are obtained that are transported by highway via Nazca to the port of Callao. The Catalina Huanca Mining Unit currently has 5407 hectares of mining concessions.

The Ramahuayco tailings deposit project is located to the west from the mine, in the upper sector of the Sacllani ravine, 150 m upstream from the mine offices, with respect to the concentrator plant, which is 12 km away (Figure 67) [40,41].

The project is for ore processing with a current nominal production capacity of 2000 mtpd, an average concentrate rate of 15%, and a filtered tailings production of 1850 mtpd. The ore processing is carried out using the conventional flotation system. The generated tailings pass to a thickener and is then transported through pipes to the filtering plant [40,41].

The filtering plant is conformed by a press filter that receives the tailings from the concentrator plant with 65% solids content, where the tailings are transformed into pulp to filtered tailings, with 88% solids content (Figure 68). After that, the tailings are transported by trucks to the filtered tailings deposit area [40,41].

For stability reasons, filtered tailings require compaction. In this case, in situ tailings dewatering is required. Tailings production, rainfall-evaporation of the site, drying time and areas, and operational flexibility can limit the application of filtered tailings, and the key parameter is the water content in the deposited tailings (Figure 69) [40,41].

Located in the Sacllani basin, the filtered tailings are transported by trucks, from the filtering plant to the tailings deposit facilities, where they are unloaded, spread, dried, and compacted to 95% of the Standard Proctor [19], for which the maximum humidity of tailings compaction must be less than 12% (Figure 70) [40,41].

Figure 67. Ramahuayco Filtered TSF—Terraces Overview [40,41].

Figure 68. Ramahuayco Filtering Plant [40,41].

The tailings deposit, in its first stage, is made up of the main dam and a structure of boulders arranged at the base. The entire structure is supported on limestone rock, the project plans the construction of a dam up to a level of 3584 m above sea level, the formation of a base embankment with borrowed material, and the placement of a foot structure made up of a dam of plain concrete [40,41].

Tailings drying can be done in two ways; the first with a bulldozer which removes the tailings and leaves them in a furrowed form in a certain area; the second is by means of excavators, this is, using its spoons to remove the tailings, passes the shovel, and vents it in such a way that the tailings lose the initial humidity that it has of 17 dropping to 12% to be ready to be mixed with the borrow material [40,41].

Figure 69. Ramahuayco TSF Overview [40,41].

Figure 70. Ramahuayco TSF Civil Works [40,41].

7.16. Tambomayo Filtered TSF–Valley Topography Configuration–Andean Range–Peru

The Tambomayo project is politically located in the annex of Punachica and Tocallo, district of Tapay, province of Caylloma, in the department of Arequipa. It is located 54 km southwest of the Caylloma district and 100 km east of the Paula mine, at an altitude between 4700 and 4900 m above sea level [42,43,70].

Tambomayo is an epithermal-mesothermal polymetallic deposit in veins of gold and silver with base metals in quartz gangue, emplaced in Tertiary andesitic volcanics. Gold is found in its native state and electrum, while silver is found in sulfides and sulfosalts accompanied by galena and sphalerite. There are two main vein systems: Mirtha, NW–SE oriented, and Paola, N–NE oriented. There are also nearby exploration areas [42,43,70].

The Tambomayo metallurgical process consists of primary crushing, fine grinding, gravimetry, and cyanidation in tanks to obtain a solution rich in gold and silver to be processed at the Merrill Crowe plant (zinc precipitation). The precipitate is dried and melted obtaining doré bars (gold and silver). The cyanidation tailings enter the flotation process, from which lead-silver and zinc-silver concentrates are obtained (Figure 71) [42,43,70].

The average production of tailings to be thickened, filtered, and deposited is 1500 mtpd from the discharge of the tailings thickener from the process plant.

The tailings thickener feeds the filtering plant that considers two press filters (it comes from the cyanide destruction system and goes to the tailings distributor). The filtering plant considers the cloth washing systems, the membrane compression system and the transport by trucks of the filtered tailings to the tailings deposit (includes the truck tire washing system) at the exit of the filtering plant (Figure 72) [42,43,70].

The Filtered Tailings Deposit 1, through a drainage system, collects and directs the waters that are generated through the drying of the tailings that were previously filtered through the filter presses. The percentage of moisture that remains in the remainder of this operation varies between 8 to 14% humidity, being a percentage of filtered water that accounts for between 80 to 140 liters of solution for each metric tonne, and if this is the case, this amount of water will filter through the drainage system and will be stored in this pool for storage and subsequent treatment (Figure 73) [42,43,70].

The filtered tailings are transported by trucks to the tailings deposit area which is lined with geosynthetic geomembrane materials throughout its base. Once the filtered tailings are placed in the deposit, they are spread in 30 cm layers and subsequently compacted with a 10-tonne drum-type vibratory roller (Figure 74) [42,43,70].

Figure 71. Tambomayo Concentrator Plant [42,43].

Figure 72. Tambomayo Tailings Filtering Plant [42,43].

Figure 73. Tambomayo Filtered TSF [42,43,71,72].

Figure 74. Tambomayo Filtered TSF Normal Operation and Contingency Operation Overview [42,43,71,72].

7.17. Chungar Filtered TSF–Valley Topography Configuration–Andean Range–Peru

The Chungar Mine is located in the Department of Pasco, Province of Cerro de Pasco, and District of Huayllay, Peru. It is a polymetallic deposit that contains copper, silver, zinc, and lead, an underground working mine, and whose workings are at the 4100 masl.

The process begins with the pumping of the flotation tailings towards the hydraulic fill hydrocyclone battery, which is carried out with a pumping line of 02 pumps in series with 150 HP of power and another line is on standby. The hydrocyclone battery is composed of 05 hydrocyclones, the underflow of the hydrocyclones is stored in 02 silos of 220 and 240 m^3 capacity, to later be sent to an underground mine to be used as hydraulic fill. The hydrocyclone overflow is diverted to transfer tank "A" where it is mixed with water from the underground mine. The overflow mixture plus underground mine water (slurry) is sent to the DCT Deep Cone thickener 17 m in diameter × 21 m in height through 01 pumping line (12″ diameter HDPE pipe) composed of 03 pumps in a series of 200 HP power each [44].

The thickened fine tailings or slimes are transported by pipeline to a tailings filtering plant composed of three press-type filters. In this facility, the filtered tailings produced at

4200 mtpd will reach a moisture content of the order of 16% (Figure 75). The tailings are temporarily stockpiled in the filter plant area. The loading of the tailings is by means of a front loader, and for transport, trucks of 15 m^3 and 30 tonnes of load capacity are used [44].

Figure 75. Chungar Tailings Filtering Plant.

The tailings deposited with a natural humidity of 16% are spread and will go through a minimum desiccation cycle until reaching a humidity content equal to or better than 14%. The tailings are mixed with mine waste in terraces, then the mixed material is compacted to a minimum density of 95% of the Standard Proctor [19] and humidity of 12%. The surface of the final platform and the sidewalks of the tailings deposit are built with a 1% slope towards both abutments of the deposit, so that rainwater on the platform can run off to an area where it can be quickly evacuated from the reservoir tailings deposit area (Figures 76 and 77) [44].

Filtered tailings disposal requires a suitable climate in which to operate. In wet climates like Chungar Mine, care must be exercised to ensure that the "dry stack" does not resaturate upon exposure to the elements (atmospheric precipitation). It is common practice to design dry stack facilities to contain filtered tailings that are less dense or have a high moisture content in a separate, non-structural area (Figure 76) [44].

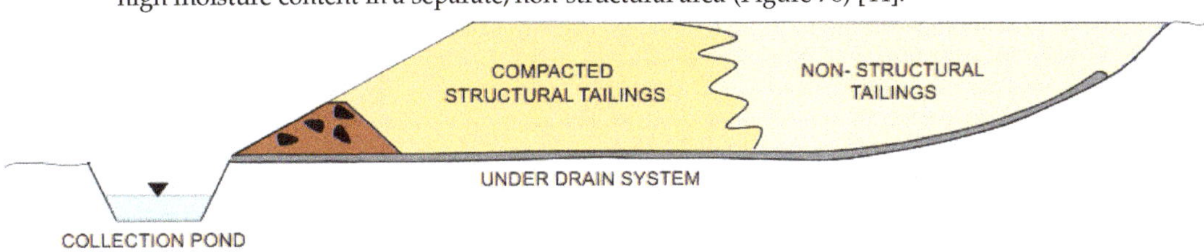

Figure 76. Chungar Filtered TSF Disposal Strategy.

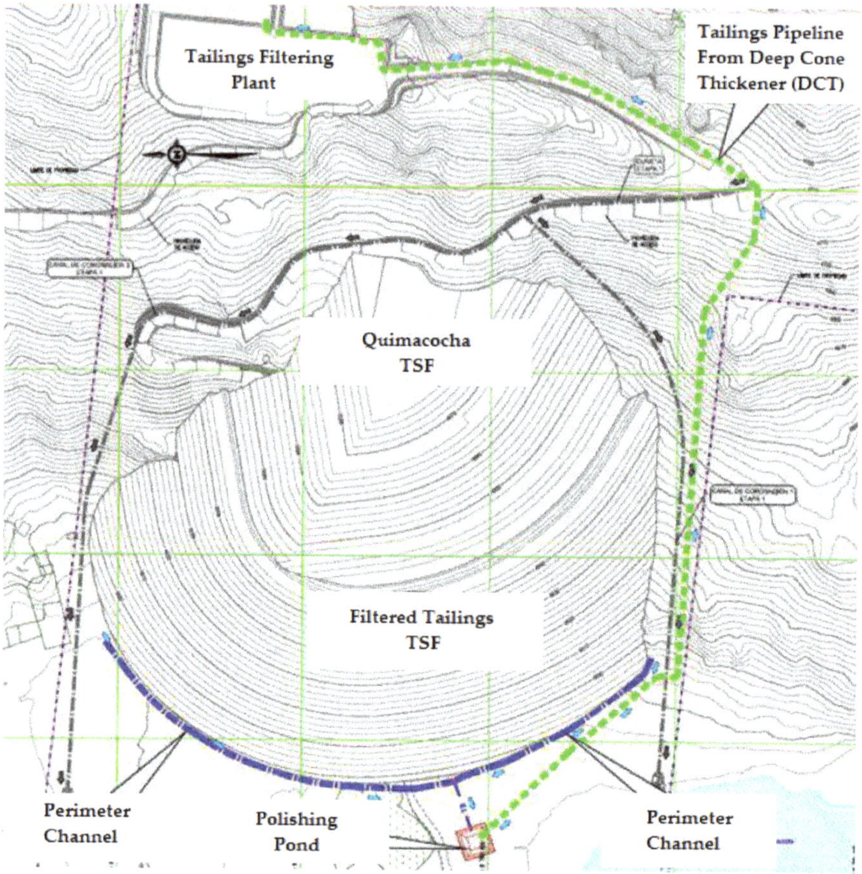

Figure 77. Chungar Filtered TSF Layout View.

Filtered tailings is a technology that is environmentally friendly, since it generates negligible seepages, decreasing the risk of transportation of contaminants significantly [12,14,16]. However, a relevant feature to consider is to avoid dust or fine particulate entrainments carried by local winds throughout the life of the TSF. Dust emission controls will play an important role as good design and proper implementation will provide the primary control mechanism for dust in accordance with regulatory air quality requirements [73]. Some dust control alternatives to implement are: soil cover (borrow material), topsoil/revegetation cover, phytostabilization, binder material, or chemical agglomeration [11].

7.18. Geochemical Stability

The lack of a tailings supernatant pond, very low seepage from the unsaturated tailings, and high degree of structural stability, allow dry stacks TSFs to develop progressive reclamation in many instances, a closure cover material is provided to manage runoff erosion, and create an appropriate ground surface for project reclamation with a proper construction machinery and trucks trafficability [3,22].

Thus, a cover with granular materials and geosynthetic or topsoil is projected to cover the terraces of filtered tailings typically–this being done during the regular TSF buildup operations, constituting a progressive closure activity. This kind of project typically considers the placement of 0.5 m layers of a granular soil cover, both on the slopes as well as on the berms and terraces of the filtered tailings dry stack deposit. To promote the geochemical stability of the filtered TSF it is important to provide a cover, avoiding the acid rock drainage (ARD) of the tailings due to exposure to oxygen and water [14].

8. Environmental Management, Geochemical Stability and Seepage Monitoring System

8.1. Seepage Monitoring System

The TSF closure costs can be considerable; particularly if the tailings contain potentially acid-generating behavior, degrading the strength properties of cycloned tailings sand dams and borrow dams, or if extensive reshaping of the structure is required to create a landform that will need to be stable in the long term. Some governments are addressing the issue by requiring mining companies to establish some form of bonding arrangement to guarantee the safe and successful decommissioning of the facilities at no cost to the community, which is a closure plan which includes the direct, indirect, and post-closure operating, maintenance and surveillance costs for the TSFs. Filtered tailings with a high compaction degree can be placed in a relatively dense state, allowing an aggressive use of the TSF site on any type of topography, and reducing the TSF footprint [74].

The following Table 4 presents a closure cost estimate comparison between conventional TSF and filtered TSF.

Table 4. Conventional/Filtered TSF Closure Cost Estimate–Comparative Analysis [73,74].

TSF Characteristics(M: Millions)	Flat Topography Configuration		Valley Topography Configuration	
	Conventional TSF	Filtered TSF	Conventional TSF	Filtered TSF
Throughput/Mine lifetime (mtpd/years)	50,000/20	50,000/20	10,000/20	10,000/20
Tailings Dry Density (t/m^3)	1.3	2.0	1.3	2.0
TSF Capacity/Footprint (Mt/Mm^3/Km^2)	347/267/8	347/173/4	208/160/2.0	208/104/1.5
Number of TSF Sites (Impoundment/Dry Stack)	2/0	0/1	1/0	0/1
Max. TSF Height (Dam or Dry Stack) (m)	35	65	150	120
Total Closure Cost (MUS$)	122	55	47	20

The above table shows that the estimated closure cost of conventional TSFs is double compared with filtered TSFs. This has a clear impact on aspects of financial guarantee for decommissioning activities and sustainable mining development.

Filtered TSF offers progressive reclamation and mine closure activities during the TSF operation, controlling dust with cover material placement, TSF side slope reclamation, and revegetation (if required) such as the use of the Phytostabilization technique. The combination of these technologies decreases TSF seepage to very low levels, almost avoiding tailings leachate, and reducing the risk of geochemical contamination. Under dry climate conditions, no supernatant TSF ponds need to be monitored during operation, and no freeboard requirement of tailings dams, reducing water losses significantly and eliminating an overtopping event risk, respectively [1–3].

A filtered tailings TSF solution is very competitive today compared to conventional technologies because the cost of closure and post-closure are lower, the environmental and construction permits are easier to obtain, and the authorities and community are more prone to accept it [11].

9. New Trends

Filter suppliers are working to improve equipment reliability and capacity. Nowadays some equipment reaches an 8000 mtpd production rate, significant advancements have been achieved, and more filter plants with high production rates will be operating in the next years around the globe. For tailings production over 50,000 mtpd, it is possible to combine cycloning–high thickening–filtered plant with cyclones, thickeners, and filters, classifying the total tailings to filter the underflow coarse tailings and thicken the overflow fine tailings, providing flexibility and versatility against ore characteristic variations, such as the successful case of Mantos Blancos project [68].

An alternative process to obtain filtered tailings can be used, consisting of the recovery of the coarse fraction of tailings (cycloned tailings sand) through two cycloning stages followed by a drainage stage in dewatering vibratory screens to reduce tailings moisture and turn it into a paste to be easily transported to the adjoining dumping facility (Figure 78) [67].

New trends include the development of new agglomerating agents for filtered tailings. Regarding developments with conveyor systems, a new generation of "pipe" or "tubular" conveyor belts is a viable alternative to use with filtered tailings, allowing a confined transport that does not allow liberate dust emissions in very windy areas [11].

Figure 78. Mantos Blancos Cycloning Plant and Vibratory Screens for Tailings [68].

10. Discussions—Strengths and Limitations of Filtered Tailings from Practical Experiences Presented

The Table 5 presents the strengths and limitations of filtered tailings technology based on the practical experience recorded in Chile and Peru.

Table 5. Strengths and Limitations considering Practical Experience of Filtered Tailings in Chile and Peru.

Strengths	Limitations
May permit storage in areas with steep topography, space restrictions, or other circumstances where conventional tailings facility is impractical.	May not be applicable to all tailings types in all situations (tailings with high clay-sized particle content are more challenging to filter to required moisture contents).
Highest water recovery during processing is advantageous in regions where water is scarce.	Filters or filter plants require more operational attention and are subject to system "upsets" due to ore variability, gradation, or operator attention.
More flexibility for placement strategy and final configuration, which can allow for progressive reclamation.	Additional storage ponds are required in the case of system upset where filtering must stop.
More flexibility for placement strategy and final configuration, which can allow for progressive reclamation.	Trafficability of filtered tailings surfaces can be a challenge depending on tailings moisture contents from the filter plant and climate conditions.
Most amenable to dry closure and landform development.	More challenging placement in wet and cold climates. Covered sheds are typically required for periods of wet weather when placement must stop.
Failure of physical stability, if it occurs, would likely be local slumping and consequences would be restricted to the local area.	
Approval and good reputation by the communities neighboring the mining projects.	

11. Conclusions

The filtered dry stacked tailings cases presented in this paper show that there is important potential to achieve a sustainable tailings management solution in desert areas such as the north of Chile, and the south of Peru, among others. The important benefits offered by this technology are: (i) less total cost of a project over the mine lifetime (construction + operation + closure liability) mainly reducing the makeup water supply costs, (ii) failure risks reduction in seismic zones improving TSF physical stability, (iii) maximum water recovery, and (iv) an environmentally friendly solution allowing for smaller TSF footprints, a progressive reclamation, and effective dust control, allowing EIAs approvals.

These benefits are a good reason to promote new or existing large mining operations to shift from conventional slurry tailings disposal facilities to alternative solutions with highly dewatered tailings disposal facilities, which are day by day more accepted by the environmental authority and communities.

The application of this technology is an excellent alternative providing attributes to streamline and permitting instruments (EIAs and Closure Plans, and besides this, tailings management technology will have lower closure costs).

These are major challenges for engineers, scientists, vendors, and mining operators. To date, we have not solved all our difficulties and there are many interesting solutions waiting to be found.

Author Contributions: Conceptualization, C.C.V.: and G.P.C.; formal analysis, C.C.V.; investigation, C.C.V.; resources, G.P.C.; writing—original draft preparation, C.C.V.; writing—review and editing, C.C.V.; visualization, C.C.V.; supervision, G.P.C. All authors have read and agreed to the published version of the manuscript.

Funding: This research is funded by the Research Department of Catholic University of Temuco, Chile.

Data Availability Statement: The data presented in this study are available on request from the corresponding author.

Conflicts of Interest: The authors declare no conflict of interest.

Abbreviations

TSF	Tailings Storage Facility
ASTM	American Society for Testing and Materials
MSC	Mobile Stacker Conveyor
Cw	Slurry tailings solids content by weight
mtpd	Metric tonnes per day
QA/QC	QA (Quality Assurance) and QC (Quality Control)
CCD	Counter Current Decantation
TTD	Thickened Tailings Disposal
masl	Meters above sea level
PLC	Programmable Logic Controller
PSD	Particle Size Distribution
OMS	Operating, Maintenance, and Surveillance

References

1. Jewell, R.J.; Fourie, A.B. (Eds.) *Paste and Thickened Tailings—A Guide*, 2nd ed.; Australian Centre for Geomechanics: Crawley, Australia, 2006.
2. Smith, L. *Hydrogeology and Mineral Resource Development*; The Groundwater Project: Guelph, ON, Canada, 2021.
3. Davies, M. Filtered dry stacked tailings—The fundamentals. In Proceedings of the 15th International Conference on Tailings and Mine Waste, Vancouver, BC, Canada, 6–9 November 2011.
4. Consejo Minero. Plataforma de Relaves. 2022. Available online: https://consejominero.cl/comunicaciones/plataforma-de-relaves/ (accessed on 24 April 2022).
5. Cacciuttolo, C.; Valenzuela, F. Efficient Use of Water in Tailings Management: New Technologies and Environmental Strategies for the Future of Mining. *Water* **2022**, *14*, 1741. [CrossRef]

6. Rodari, G.; Oldecop, L. Secado de colas filtradas en contacto con la atmósfera. *Boletín Geológico Min.* **2021**, *132*, 551–571. [CrossRef]
7. Caldwell, J.; Crystal, C. Filter-Pressed Tailings Facility Design, Construction, and Operating Guidelines. October 2015. Available online: https://open.library.ubc.ca/soa/cIRcle/collections/59368/items/1.0320844 (accessed on 12 April 2022).
8. Copeland, A.M.; Lyell, K.A.; van Greunen, P. Disposal of belt filtered tailings—Skorpion zinc case study: Feasibility, design and early operation. In Proceedings of the Ninth International Seminar on Paste and Thickened Tailings, Limerick, Ireland, 3–7 April 2006.
9. Crystal, C.; Hore, C.; Ezama, I. Filter-pressed dry stacking: Design considerations based on practical experience. In Proceedings of the 24th International Conference on Tailings and Mine Waste, Fort Collins, CO, USA, 30 September–2 October 2018; pp. 209–219.
10. Gridley, N.; Salcedo, S. Cemented Paste for Ore Recovery. In Proceedings of the 16th International Seminar on Paste and Thickened Tailings, Belo Horizonte, Brazil, 17–20 June 2013.
11. Cacciuttolo, C.; Barrera, S.; Caldwell, J.; Vargas, W. Filtered Dry Stacked Tailings: Developments and New Trends. In Proceedings of the 2nd International Seminar on Tailings Management, Antofagasta, Chile, 20–22 August 2014; pp. 357–370.
12. Ulrich, B. Practical thoughts regarding filtered tailings. In Proceedings of the 22nd International Conference on Paste, Thickened and Filtered Tailings, Perth, Australia, 22 October 2019; pp. 71–79.
13. Minson, D.N.; Williams, C.E. Filtering systems for dry tailings deposition. In Proceedings of the 38th Annual Conference of Metallurgists, Montreal, QC, Canada, 22–26 August 1999.
14. Ulrich, B.; Coffin, J. Considerations for tailings facility design and operation using filtered tailings. In Proceedings of the 16th International Seminar on Paste and Thickened Tailings, Australian Centre for Geomechanics, Perth, Australia, 17–20 June 2013; pp. 201–210.
15. InfoMine Inc. Filter-Pressed, Dry-Stack Tailings. Available online: https://glacierrig.com/infomine/ (accessed on 10 April 2020).
16. Ulrich, B.; Rogers, J. Toward the Sensible Use of Tailings Filtering Technology. In Proceedings of the Tailings and Mine Waste, Keystone, CO, USA, 1–2 October 2018.
17. Caldwell, J.; Crystal, C.A. Dry stack tailings: Some ideas. In Proceedings of the 17th International Conference on Tailings and Mine Waste, Banff, AB, Canada, 14–17 October 2013.
18. Cacciuttolo, C.; Holgado, A. Dry Stacking of Filtered Tailings in the Andean Region of South America—An Engineered and Environmentally Friendly Solution. In Proceedings of the Congreso Panamericano de Mecánica de Suelos y Geotecnia, Buenos Aires, Argentina, 15–18 November 2015; 2015.
19. ASTM International. *ASTM D698; Standard methods for laboratory compaction characteristics of soil using standard effort; (12,400 ft-lbf/ft3 (600 kN-m/m^3)*; ASTM International: West Conshohocken, PA, USA, 2021.
20. Lara, J.L.; Pornillos, E.; Muñoz, H. Geotechnical-geochemical and operational considerations for the application of dry stacking tailings deposits—State-of-the-art. In Proceedings of the 16th International Seminar on Paste and Thickened Tailings, Australian Centre for Geomechanics, Perth, Australia, 23 July 2013; pp. 249–260.
21. Lara, J.L.; Pornillos, E.; Loayza, C. The application of highly dewatered tailings in the design of tailings storage facilities—Experience in Mining Projects in Peru. In Proceedings of the 16th International Conference on Tailings and Mine Waste, Keystone, CO, USA, 14–17 October 2012; pp. 29–41.
22. Lupo, J.; Hall, J. Dry stack tailings—Design considerations. In Proceedings of the 14th International Conference on Tailings and Mine Waste, Vail, CO, USA, 17–20 October 2010.
23. Trulsson, C. Mobile dry stack conveyor system for tailings management at the Karara iron ore project´. In Proceedings of the 10th Annual Bulk Materials Handling Conference, Brisbane, Australia, 23 June 2013.
24. Wisdom, T. *Filtration for Tailings Management*; FLSmidth Tailings Disposal Seminar: Santiago, Chile, 2012.
25. Lara, J.L. Experiencias de Operación de Depósitos de Relaves Espesados y Filtrados. In Proceedings of the Jueves Minero Instituto de Ingenieros de Mina del Peru IIMP, Lima, Peru, 6 October 2013.
26. Davies, M.P.; Veillette, M.F. Cold Regions Tailings and Water Management, Mining and the Environmental. In Proceedings of the 3rd International Conference of Tailings Management, Sudbury, ON, Canada, 12 October 2007.
27. AGORA SOL. Fotografias Deposito de Relaves Filtrados El Gato. Available online: https://www.agorasolspa.com/copia-de-testimonios/ (accessed on 23 April 2022).
28. East, D.; Fernandez, R. Managing water to minimize risk in tailings storage facility design, construction and operation. *Mine Water Environ.* **2021**, *40*, 36–41. Available online: https://doi.org/10.1007/s10230-020-00720-8 (accessed on 21 June 2022).
29. Condon, P.D.; Lear, K.G. Geochemical and geotechnical Characteristics of Filter—Pressed Tailings at Greenks Creek Mine, Admiralty Island, Alaska. In Proceedings of the 7th International Mine Water Association Symposium, St. Louis, MO, USA, 26–30 March 2006.
30. Kwan, K.; Dawson, R.; Boese, C.; Churcher, D. Design of the Eldorado Gold Efemcukuru Filtered Tailings Facility. In Proceedings of the 15th International Conference on Tailings and Mine Waste, Vancouver, BC, Canada, 6–9 November 2011.
31. Sikano, J.; Kalumba, B.; Mulenga, S.B.; Mwale, J. Recovery Improvement Strategies at Konkola Copper Mines PLC, Nchanga Tailings Leach Plant. In Proceedings of the 6th International Symposium Hydrometallurgy, Phoenix, AZ, USA, 17–20 August 2008.
32. Newman, L.; Arnold, K.; Wittwer, D. Dry stack tailings design for the Rosemont copper project. In Proceedings of the 14th International Conference on Tailings and Mine Waste, Vail, CO, USA, 17–20 October 2010.

33. Connelly, D. Lessons learned from thickening and filtering tailings for dry stacking. In Proceedings of the Paste 2021: 24th International Conference on Paste, Thickened and Filtered Tailings, Perth, Australia, 21–23 September 2021; pp. 31–42.
34. Oldecop, L.A.; Rodari, G.J. Unsaturated mine tailings disposal. *Soils Rocks* **2021**, *44*, e2021067421. [CrossRef]
35. Garino, L.; Rodari, G.; Oldecop, L.; Zabala, F.; Rodari, R. Almacenamiento de Colas Mineras Filtradas: Primera Experiencia en Argentina. In Proceedings of the Congreso Argentino de Mecánica de Suelos e Ingeniería Geotécnica CAMSIG XXI, Rosario, Argentina, 12–14 September 2012.
36. Poderosa—A. Gestion Minera en Poderosa. Presentación en el Instituto de Ingenieros de Mina del Perú. Available online: https://iimp.org.pe/pptjm/jm20141211_poderosamagistral.pdf (accessed on 18 April 2022).
37. Poderosa—B. Batolito Revista de Compañia Minera Poderosa. Revista N° 39. Available online: https://www.poderosa.com.pe/Content/descargas/batolito/Batolito_39.pdf (accessed on 15 April 2022).
38. Cutti, A. Tesis Titulada Tratamiento de los Relaves de Flotación para Optimizar el Proceso de Cianuración Orientado a Mejorar la Recuperación del Oro en la Compañía Minera Horizonte S.A. Available online: https://repositorio.uncp.edu.pe/bitstream/handle/20.500.12894/7408/T010_48147209_T.pdf?sequence=1&isAllowed=y (accessed on 11 April 2022).
39. Bancayan, P. Tesis Titulada Automatización de una Planta de Filtrado de Relaves Mineros. Available online: https://repositorio.utp.edu.pe/bitstream/handle/20.500.12867/1738/Pedro%20Bancayan_Tesis_Titulo%20Profesional_2019.pdf?sequence=1&isAllowed=y (accessed on 19 April 2022).
40. Almerco, D. Tesis Titulada Construcción de Dique con Tratamiento del Relave, en Mina Catalina Huanca—Región Ayacucho. Available online: https://repositorio.usmp.edu.pe/bitstream/handle/20.500.12727/1045/almerco_do.pdf?sequence=1&isAllowed=y (accessed on 10 April 2022).
41. Sarmiento, S. Trabajo de Investigación titulado Disposición de Relaves Filtrados Compactados en el Depósito de Ramahuaycco. Available online: https://app.ingemmet.gob.pe/biblioteca/pdf/Mine-21.pdf (accessed on 22 April 2022).
42. Morales, C. Tesis Titulada Evaluación de la Planta de Tratamiento de Aguas de Relaves (Ptare) de una Planta Polimetálica en el Distrito de Tapay—Provincia de Caylloma. Available online: repositorio.unsa.edu.pe/bitstream/handle/20.500.12773/12251/IMmoorce.pdf?sequence=1&isAllowed=y (accessed on 26 April 2022).
43. Samaniego, C. Tesis Titulada Instalación y Montaje del Tercer Filtro de Relaves del Área 620—Proyecto Tambomayo—Compañía Minera Buenaventura. Available online: https://repositorio.uncp.edu.pe/bitstream/handle/20.500.12894/4970/T010_19867699_T.pdf?sequence=1&isAllowed=y (accessed on 12 April 2022).
44. EIA Chungar. Segunda Modificación del EIA 4200 TMD, Unidad Minera Animón, Volcan Compañia Minera. 2020. Available online: https://www.volcan.com.pe/wp-content/uploads/2021/02/3.-Resumen-Ejecutivo_2da-MEIA-Animon.pdf (accessed on 21 June 2022).
45. Queirolo, C. Proyecto desarrollo fundición Potrerillos. In Proceedings of the 1st Seminar on Environmental Challenges for Smelters of Copper Concentrates, Santiago, Chile, 12 April 2013.
46. Sernageomin. Catastro de Depositos de Relaves en Chile. 2020. Available online: https://www.sernageomin.cl/datos-publicos-deposito-de-relaves/ (accessed on 24 April 2022).
47. DIA AK (Declaracion de Impacto Ambiental Atacama Kozan). Declaración de Impacto Ambiental Aumento de Vida Útil de Faena Minera Atacama Kozan, Incorporando Relaves Filtrados en Quebrada El Gato. Available online: https://seia.sea.gob.cl/documentos/documento.php?idDocumento=2138007308 (accessed on 23 April 2022).
48. DIA Tambillos (Declaracion de Impacto Ambiental Compañia Minera Florida) (2017). Declaración de Impacto Ambiental Depósito de Relaves Filtrados CM Florida. Available online: https://seia.sea.gob.cl/documentos/documento.php?idDocumento=2132106323 (accessed on 23 April 2022).
49. DIA Tambo de Oro (Declaracion de Impacto Ambiental Compañia Minera Tambo de Oro). Declaración de Impacto Ambiental Continuidad Operacional Faena Minera Tambo de Oro. Available online: https://infofirma.sea.gob.cl/DocumentosSEA/MostrarDocumento?docId=48/3c/6b758e1d25b40c96b3107bc21fca748adfef (accessed on 23 April 2022).
50. EIA Salares Norte. Estudio de Impacto Ambiental Proyecto Minero Salares Norte. Available online: https://infofirma.sea.gob.cl/DocumentosSEA/MostrarDocumento?docId=ab/c4/7adcf22236bb11c628d95e838cd6e4765505 (accessed on 23 April 2022).
51. Pucobre. Consejo Municipal de Illapel Sesión Ordinaria N° 69 Presentación del Proyecto Minero El Espino. Available online: https://municipalidadillapel.cl/trans_web/sitio/transparencia/full_actas/actas_2014/s69_21_10_14%20.pdf (accessed on 24 April 2022).
52. Conaf. Autoriza la Intervención o Alteración de Hábitat de Especie en Categoría de Conservación que se Indica, Presentada por el Espino S.A., Titular del Proyecto "el Espino". Available online: https://www.conaf.cl/LBN/RF/2014/Res_357-El-Espino.pdf (accessed on 24 April 2022).
53. Robledo, M.; Meyer, J. El Indio Mine Closure—A Significant Environmental Accomplishment. In Proceedings of the 2nd International Seminar on Environmental Issues in Mining Industry ENVIROMINE, Santiago, Chile, 23–25 November 2007.
54. EIA CAP. Estudio de Impacto Ambiental Compañia de Acero del Pacifico Deposito de Relaves Filtrados Planta de Pellets. Available online: https://infofirma.sea.gob.cl/DocumentosSEA/MostrarDocumento?docId=55/92/e4bcc15e918c4c15213794e4bae257c39c4c (accessed on 24 April 2022).
55. CAP. Proyecto Deposito de Relaves Filtrados Planta de Pellets Webpage. Available online: https://www.proyectorelavefiltrado.cl/ (accessed on 24 April 2022).

56. Williams, D. Presentation Dry Stacked Tailings and Seepage Management La Coipa Mine Chile. Available online: https://bc-mlard.ca/files/presentations/2012-9-WILLIAMS-ETAL-dry-stack-tailings-seepage-mgt-chile.pdf (accessed on 24 April 2022).
57. Tailings Info. La Coipa, Region III, Chile, South America. 2022. Available online: https://www.tailings.info/casestudies/lacoipa.htm (accessed on 24 April 2022).
58. Sernageomin, Guía tecnica de operación y control de depositos de relave, DSM/07/31 diciembre de 2007. Available online: https://www.sernageomin.cl/wp-content/uploads/2018/12/GuiaTecOperacionDepRelaves.pdf (accessed on 28 June 2022).
59. Saez, M. Experiencias de Movimientos de Tierra para Cierre de Faenas Mineras. In Proceedings of the 4th National Cierre de Faenas Mineras CIFAMIN, Santiago, Chile, 12 November 2011.
60. EIA Pucobre. Estudio de Impacto Ambiental Proyecto El Espino. Available online: https://seia.sea.gob.cl/documentos/documento.php?idDocumento=8063381 (accessed on 24 April 2022).
61. Atacama Kozan. Fotografias Proyecto Relaves Filtrados Quebrada El Gato Obtenidas de Webpage de Atacama Kozan. Available online: https://www.atacamakozan.cl/ (accessed on 23 April 2022).
62. Yamana Gold. NI 43-101 Technical Report El Peñón Gold-Silver Mine Antofagasta Region, Chile. 2020. Available online: https://s28.q4cdn.com/334653565/files/Yamana-El-Pe%C3%B1%C3%B3n-NI-43-101-March-25-2021.pdf (accessed on 24 April 2022).
63. DIA El Peñon. Expansion Deposito de Relaves Filtrados Faena El Peñón. 2010. Available online: https://seia.sea.gob.cl/documentos/documento.php?idDocumento=4601644 (accessed on 24 April 2022).
64. Goldfields. Salares Norte Project Presentation. Available online: https://www.goldfields.com/pdf/investors/presentation/2019/salare-norte-presentation.pdf (accessed on 23 April 2022).
65. Goldfields. The Gold Fields Mineral Resources and Mineral Reserves Supplement to the IAR. 2018. Available online: https://www.goldfields.com/reports/annual-report-2018/mrr/introduction.php (accessed on 23 April 2022).
66. AGORA SOL. Fotografias Deposito de Relaves Filtrados Tambillos. Available online: https://www.agorasolspa.com/ (accessed on 20 April 2022).
67. Bouso, J.L.; Renner, P. Application of dewatering screens in tailings filtration. In Proceedings of the 8th International Seminar on Paste and Thickened Tailings, Santiago, Chile, 20–22 April 2005.
68. Cacciuttolo, C.; Barrera, S.; Bouso, J.L.; Astete, A. Dewatering Tailings on Horizontal Vibration Screens: An Alternative to Filtration Systems to Obtain Tailings to be Dispose at Dry Stack Tailings Storage Facilities. In Proceedings of the 3rd International Seminar on Tailings Management, Santiago, Chile, 19–21 August 2015.
69. DIA Potrerillos Modificación RCA N° 227/2011 Flotación de Escorias Potrerillos. 2018. Available online: https://seia.sea.gob.cl/documentos/documento.php?idDocumento=2138993240 (accessed on 24 April 2022).
70. Buenaventura. Presentación Titulada Gestion Moderna de Depositos de Relaves en Buenaventura, Realizada en Jueves Minero del Instituto de Ingenieros de Minas del Perú. Available online: https://www.youtube.com/watch?v=c9OoMQtWF5A (accessed on 5 April 2022).
71. San Martin. San Martin Webpage, Fotografias Depositos de Relaves Filtrados Tambomayo. Available online: https://sanmartin.com/proyectos/construccion-de-deposito-de-relaves-%EF%AC%81ltrados-tambomayo/ (accessed on 9 April 2022).
72. San Martin. Presentacion de Proyectos, Fotografias Depositos de Relaves Filtrados Tambomayo. Available online: https://www.isem.org.pe/portal/intranet/files/docs/evento/2016-10-27%20norma%20silva%20update%201.pdf (accessed on 7 April 2022).
73. MEND. Study of Tailings Management Technologies by Mine Environment Neutral Drainage (MEND) Project. 2017. Available online: mend-nedem.org/wp-content/uploads/2.50.1Tailings_Management_TechnologiesL.pdf (accessed on 24 April 2022).
74. Robins, M. Closure of Tailings Storage Facilities, Current Practice Review & Guidelines for Success. Available online: http://wiredspace.wits.ac.za/bitstream/handle/10539/1850/Closurefinal.pdf (accessed on 2 June 2022).

minerals

MDPI

Article

An Alternative Technology to Obtain Dewatered Mine Tailings: Safe and Control Environmental Management of Filtered and Thickened Copper Mine Tailings in Chile

Carlos Cacciuttolo [1,*] and Edison Atencio [2,3]

1 Civil Works and Geology Department, Catholic University of Temuco, Temuco 4780000, Chile
2 School of Civil Engineering, Pontificia Universidad Católica de Valparaíso, Av. Brasil 2147, Valparaíso 2340000, Chile
3 Department of Management, Economics, and Industrial Engineering, Politecnico di Milano, Via Lambruschini 4b, Bovisa, 20156 Milan, Italy
* Correspondence: ccacciuttolo@uct.cl or carlos.cacciuttolo@gmail.com

Abstract: An alternative process to obtain a high degree of dewatering tailings that produces a high-density product is presented in this article. This technology involves the combination of tailings particle grain size classification by hydrocyclones (HC) and tailings dewatering by horizontal vibratory screens (HVS). It makes it possible to dewater tailings to a high grade of performance. This alternative technology (HC-HVS) involves the recovery of water from the coarse fraction of tailings (sands) through two hydrocycloning stages, followed by a dewatering stage of cycloned tailings sands on horizontal vibratory screens, to reduce moisture content and turn it into a "cake". The resulting coarse fraction tailings are easily transported to a dry stack tailings storage facility (TSF). The fine fraction of tailings (slimes) can be dewatered on thickener equipment to recover part of the process water. Finally, this article describes the main benefits of this alternative dewatered tailings technology with an emphasis on (i) dewatering technology evolution over the last 17 years; (ii) process stages features; (iii) pilot test results; (iv) tailings properties analysis (such as particle grain size distribution, fines content) and (v) lessons learned about the experience gained in the operation of Mantos Blancos case study with dry stack tailings storage facility.

Keywords: filtered tailings; vibratory dewatering screens; hydrocyclones; dry stack tailings storage facility; tailings dewatering; tailings cake

1. Introduction

1.1. *Efficient Use of Water and Safe Management in Mine Tailings*

Engineers, scientists, mine operators, and authorities are working to improve the design and operations of tailings storage facilities (TSF) in areas of safe storage and efficient water management [1]. Concerning this last issue, i.e., water management, the effort has been focused on the development of optimal solutions, which consider the following: (i) reliable performance of technologies; (ii) a dynamic water balance (keeping in mind the specific conditions of the site); and (iii) an efficient management of water with emphasis on the control of water losses (evaporation and seepage) [1–3]. If these key aspects are achieved successfully, such as (i) a decrease in the requirement for freshwater (make-up), (ii) a decrease in the negative environmental impacts (seepages, dust emission, groundwater pollution, among others), and (iii) an increase in the opportunity for others to use the natural water for other uses (urban zones, agricultural zones, environment, among others) sustainable development will be reached. Therefore, the resulting approach to water management will promote an advance in the development of sustainable technologies for the mining industry [1,2,4–6].

The application of tailings dewatering technologies to increase water reclaim is an important step in decreasing water losses (water coming from freshwater supplies or the sea in some cases) [2,4,7–9]. Water losses occur because of evaporation, infiltration, and water retention in the pores of the tailings in Tailings Storage Facilities (TSF) [1,2]. This challenge requires new designs for tailings management, and such designs must make tailings management less negative impacts on the environment and use water more efficiently [4,10]. Considering the Best Available Technologies (BATs) for tailings management incorporating tailings dewatering techniques is possible to mention four main categories: (i) Conventional Tailings, (ii) Thickened Tailings, (iii) Paste Tailings, and (iv) Filtered Tailings [2,4,11,12].

Conventional copper tailings typically range from 25%–40% solids weight concentrations (Cw), thickened copper tailings 40%–65% (Cw), paste copper tailings 65%–80% (Cw), and filtered copper tailings over 80% (Cw) (solid concentrations may vary with particle size and shape, clay content, mineralogy, electrostatic forces, and flocculant dosing). Conventional, thickened, paste and filtered tailings refer to a continuum of tailings with high solid concentrations and higher yield stress due to greater fluid removal from tailings before disposal [2,11].

In current Chilean large-scale mining in dry climate areas, most typical tailings disposal schemes consist of conventional or slightly thickened at modest levels of tailings solids weight concentration (Cw 25%–40%). Conventional TSFs have dams built of the coarse fraction of tailings (cycloned tailings sand) obtained by hydrocyclones or have slightly thickened tailings deposits with dams built of borrowed material. Conventional tailing dams may have water recoveries as high as 65%–75% in well-operated TSFs, which means they have appropriate tailings distribution, good control of the pond (volume and location), and adequate seepage recovery. In conventional dams, water at the settling pond is decanted by floating pumps or decant towers, and dam seepages are collected by drainage and cutoff trench systems. However, a high seasonal evaporation rate can substantially reduce water recovery from the pond area, and infiltration from the pond in contact with natural soil can produce water losses. Some mining operations with this technology in Chile are Chuquicamata, El Soldado, Los Pelambres, and Los Bronces [2,11].

Thickened Tailings Disposal (TTD) technology requires more background data than conventional tailings disposal. In the conventional approach, the properties of tailings are fixed by the concentrator plant. In contrast, in a TTD tailings storage facility, the properties of the tailings and their placement are "engineered" to suit the topography of the disposal area. The behavior of tailings in the two approaches is entirely different. In conventional disposal, tailings segregate as they flow and settle out to an essentially flat deposit, whereas in TTD technology, a sloping surface is obtained. The principal difference is that, in TTD technology, tailings are thickened before discharge to a homogeneous heavy consistency, resulting in a laminar non-segregating flow. In this way, TTD produces high water recovery (70% of tailings water recovery) and a self-supporting deposit with sloping sides, requiring small dams. Some mining operations with this technology in Chile are Centinela (Esperanza), Spence, and Sierra Gorda [2,11].

Paste Tailings Technology has been applied on a small production scale because equipment manufacturing ability is limited. This method permits obtaining a medium make-up water requirement (80% of tailings water recovery). However, in some cases, there are difficulties in tailings transportation requiring the use of positive displacement pumping (PD Pumps), resulting in the highest capital/operating costs. The main advantages of this method are that: large dams are not required; Only small dams are needed, and the emission of dust is negligible because the surface of the deposited tailings remains as a solid hard crust due to the bonding of the tailings particles by the action of the flocculant. Some mining operations with this technology in Chile are Planta Demo Collahuasi, Delta Plant Enami, Coemin, Las Cenizas, and Alhue [2,11,13].

In the last 20 years, many mining projects worldwide have applied a tailings disposal technology called dry stacking of filtered tailings. This technique produces an unsaturated cake that allows storing this material without the need to manage large slurry tailings ponds. The application of this technology has accomplished: (i) an increase in water recovery from tailings (90% of tailings water recovery), (ii) a reduction of TSF footprint (impacted areas), and (iii) a decrease in the risk of physical instability because TSFs are self-supporting structures under compaction (such as dry stacks), and (iv) a better community perception. Some mining operations with this technology are: Salares Norte, Potrerillos, San Jose Plant Pucobre, El Gato, Tambo de Oro, Huasco, La Coipa, El Peñon, and Mantos Blancos [2,12,14–20].

The risk associated with TSFs is very large, considering that a potential dam failure could easily damage the environment and people with pollution or other adverse impacts [21,22]. Accidents or failures at these facilities have always been associated with social, environmental, or public safety issues, some more catastrophic than others [21,23]. The above has contributed to a reduction in the rate of TSF failures in recent decades. In most countries around the world where responsible mining is practiced daily, regulatory frameworks have become more stringent, requiring a higher level of responsibility, with the intent of minimizing risks posed to society and the environment [4,10,24].

Filtered tailings are an environmentally friendly technology since it generates negligible seepages, significantly decreasing the risk of contaminants' transportation [12,25–29]. However, a relevant feature to consider is to avoid dust or fine particulate entrainments carried by local winds throughout the life of the TSF. Dust emission controls will play an important role as good design and proper implementation will provide the primary control mechanism for dust following regulatory air quality requirements [12,25–29]. Some dust control alternatives are soil cover (borrowed material), topsoil/revegetation cover, phytostabilization, binder material, or chemical agglomeration [12].

Filtered TSF offers progressive reclamation and mine closure activities during the TSF operation, controlling dust with cover material placement, TSF side slope reclamation, and revegetation (if required) such as using the phytostabilization technique. The combination of these technologies decreases TSF seepage to very low levels, almost avoiding tailings leachate and reducing the risk of geochemical contamination. Under dry climate conditions, no supernatant TSF ponds need to be monitored during operation, and no freeboard requirement of tailings dams, reducing water losses significantly and eliminating an overtopping or piping (internal dam erosion by seepage) event risk, respectively [12].

A filtered tailings TSF solution is very competitive today compared to conventional technologies because the cost of closure and post-closure are lower, the environmental and construction permits are easier to obtain, and the authorities and community are more prone to accept it [12].

From 1998 to date there have been major disasters worldwide due to tailings storage facility failures. Some examples are Los Frailes Aznalcóllar Spain (1998), Baia Mare Romania (2000), Kolontar Hungary (2010), Mount Polley Canada (2014), Fundao Samarco Brazil (2015), Corrego de Feijao Brumandinho Brazil (2019) and Jagersfontain South Africa (2022), which have caused the death of hundreds of people and irreparable environmental damage [21–24]. As a consequence of these disasters, the international community, institutions, and global groups such as the ICMM (International Council on Mining and Metals), the UN environment program, and the PRI (Principles for Responsible Investments), developed a "Global Tailings Management Standard for the Mining Industry" launched in August 2020, to regulate the operation throughout the entire life cycle of tailings storage facilities, including closure and post-closure (perpetuity), considering zero harm to people and the environment, and zero tolerance for human fatalities [30]. ICMM advocates that the application of appropriate design and management standards and good practices allow tailings storage facilities to be safe. ICMM members are committed to preventing catastrophic failures of tailings deposits with continuous improvement in these storage facilities' design, construction, and operation stages. This organization urges mining companies to improve their management by adopting the Global Industry Standard on Tailings Management, taking advantage of technological innovation and continuous improvement [30].

In this context, the current regulatory framework needs reassessment to consider the best available practices and technologies for tailings storage and eliminate tailings dam failures worldwide to establish safe tailings management by implementing environmentally friendly solutions [1,6,31].

1.2. Aim of the Article

This article presents a new alternative process to obtain a high degree of dewatering tailings. It is an emerging combination of dewatering tailings technologies that produce a high-density product. This technology combines tailings particle grain size classification by Hydrocuclones (HC) and tailings dewatering by Horizontal Vibrating Screens (HVS). The HC-HVS technology (combination of particle grain size classification tailings with the use of HC and dewatering of the tailings with the use of HVS) has been used in the last 17 years (2005–2022) in the dewatering of copper mine tailings. Dewatering vibratory screens were initially used to filter industrial fine-size minerals, such as silica, feldspar, kaolin, fluorite, salts, carbon, etc. This dewatering was done to allow for storage in silos and decrease the storage volumes [32,33]. In this manner, the optimal management of the products known as bulk solids was obtained, both in hoppers and conveyor belts.

In Chile, dewatering vibratory screens have been used mainly to treat carbon, silica sands for fusion and glass, rinse salts in the chemical industry, and the aggregates and fine minerals obtained from mining processing [33,34].

The experience developed in the North of Chile has enabled the dewatering of tailings to a high degree and the production of a copper tailings product with a low moisture content, which facilitates the disposal of tailings in the environment as a tailings storage facility in a dry stockpile.

This article describes the main benefits of this alternative dewatering tailings technology (HC-HVS) with an emphasis on (i) dewatering technology evolution over the last 17 years; (ii) process stages features; (iii) pilot test results; (iv) tailings properties analysis (particle grain size distribution, fines content, density, permeability, etc.); and (v) lessons learned about the experience gained in the operation of a case study in Chile called Mantos Blancos in a dry stack tailings storage facility.

1.3. Description of Mantos Blancos Dewatered Tailings Management-Study Case

The Mantos Blancos copper sulfide plant, located in the Atacama Desert, Chile, started operation in 1981 [35]. It has had a tailings treatment plant from its inception. It consists of a tailings classification circuit that thickens the fine fraction of the tailings and filters the coarse fraction. Since this plant is located in one of the world's driest areas, the coarse tailings were filtered to maximize the water recovery. Disc filters were used in the beginning. Later in 1984, during the expansion of the concentrator plants to a production of 8000 mtpd, vacuum belt filters were incorporated to dewater the coarse tailings, which were in use until 2005 (Figure 1). The tailings were finally deposited in a TSF with a dam to impound thickened slimes and in a dry stack with a dike to retain the coarse fraction in a 40/60% cut, respectively, and a solids concentration of 60% (Cw) and 80% (Cw), respectively [32].

The copper tailings from the concentrator plant were classified in a hydrocyclone station of 500 mm Eral Hydrocyclones, functioning by gravity, in a conical configuration, with four operating units and two on standby. This classification system produced a 50%–55% solids distribution and a cut size of d_{50} between 50–60 microns. The overflow of the hydrocyclones, solid concentration 24%–26% (Cw), was transported to three thickening units (Larox of 60 m diameter, Dorr Oliver of 45 m diameter, and Eimco of 45 m diameter) for the process of sedimentation, using a dosage of flocculent of 5 g/ton, that later feeds with a solids content of 55%–60% (Cw) the filtration units and the fine tailings dam in a ratio restricted by the capacity of absorption of fines in the filters. The thickeners recovered approximately 63 to 67% of water, which was returned to the concentrator plant (See Figure 2) [32].

Figure 1. Tailings vacuum Belt Filters in Mantos Blancos Project [32].

(a) (b)

Figure 2. Dewatering of Mantos Blancos Coarse Tailings with Vacuum Belt Filters and Slimes Thickener Overview (**a**) Filtered Coarse Tailings with Vacuum Belt Filters, (**b**) Slimes Thickening Overview.

The underflow of the hydrocyclones, with a solids content of 65%–67% (Cw), was sent to three filtration units (band filters of 100 m^2 each) and was previously mixed with parts of thickened fine tailings. Coarse tailings cakes were obtained with a 17% moisture content considering wet basis (equivalent to a 23% moisture content considering dry basis), which were later transported to the dry stack TSF of coarse material by conveyor belts. The filtered water, plus the wash water of the filtering cloth, was recirculated to the thickeners to sediment solids in suspension.

From the beginning, the objective of the classification and dewatering of the tailings was to decrease the amount of water in the tailings management. Later, in 2004 and 2005, the concept of replacing the vacuum belt filters with vibratory screens was analyzed, starting a pilot plant, driven mainly by the need to reduce operational and maintenance costs.

2. Materials and Methods

2.1. Process Description of Tailings Dewatering by HC-HVS Technology

An emerging technology (HC-HVS) is presented that makes it possible to dewater tailings to a high grade of performance, which involves the recovery of the coarse fraction of tailings (sands) through two hydrocycloning stages followed by a drainage stage or dewatering of cycloned tailings sands on horizontal vibratory screens, to reduce moisture content and turn it into filtered tailings or "cake" [36]. The resulting material is easy to transport to the adjoining dry stack TSF. The fine fraction of tailings (slimes) can be dewatered on high-density thickener equipment to recover part of the process water.

This technology's operational principle is forming a fine layer of tailings on the screen. This way, the filtering or dewatering process is carried out mainly via a stratified tailings material bed, which is in contact with the HVS. Although the opening of the screen in its lower parts may be larger than the smallest particles of the tailings being processed, the formation of a filtering bed on the upper part of the stratified tailings material achieves an efficient decrease of the moisture of the dewatered tailings. Also, it allows for a significant reduction in the loss of fine tailings [36]. The formation of this bed is done with hydrocyclones that recover the filter material on the HVS (water and fine tailings). These are the real pillars of the tailings dewatering with HVS; this alternative process consists of the recovery of solids > 45 microns content in the tailings (See Figure 3) [34,36].

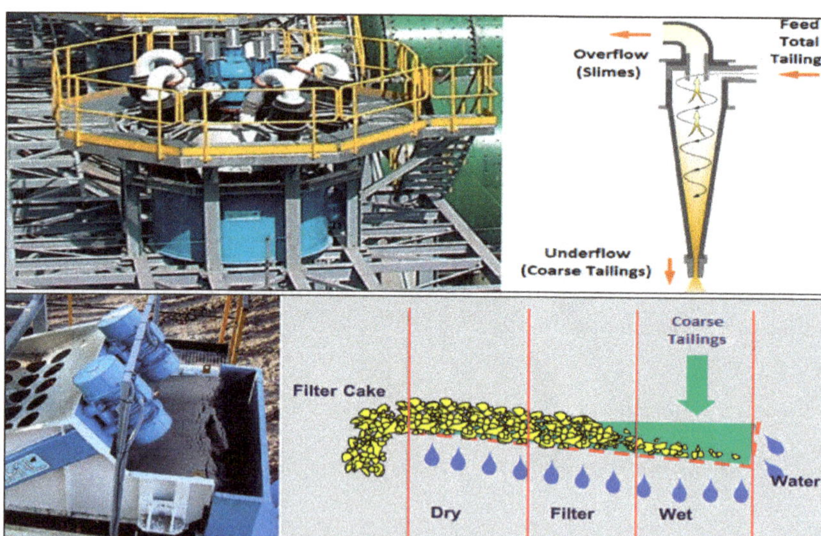

Figure 3. Schematic View of Dewatering Process on Hydrocyclone (HC) + Horizontal Vibratory Screen (HVS) (Adapted figure from [37]).

Figure 4 presents a typical process diagram of Mantos Blancos dewatering copper tailings with the HC-HVS technology. This diagram shows that the total tailings generated by the concentrator plant are classified in Hydrocyclone Station No. 1, to produce a fine fraction of the tailings (slimes) and a fraction of coarse tailings (sands). The coarse fraction of the tailings (underflow of the cyclones) is sent to the HVS [38].

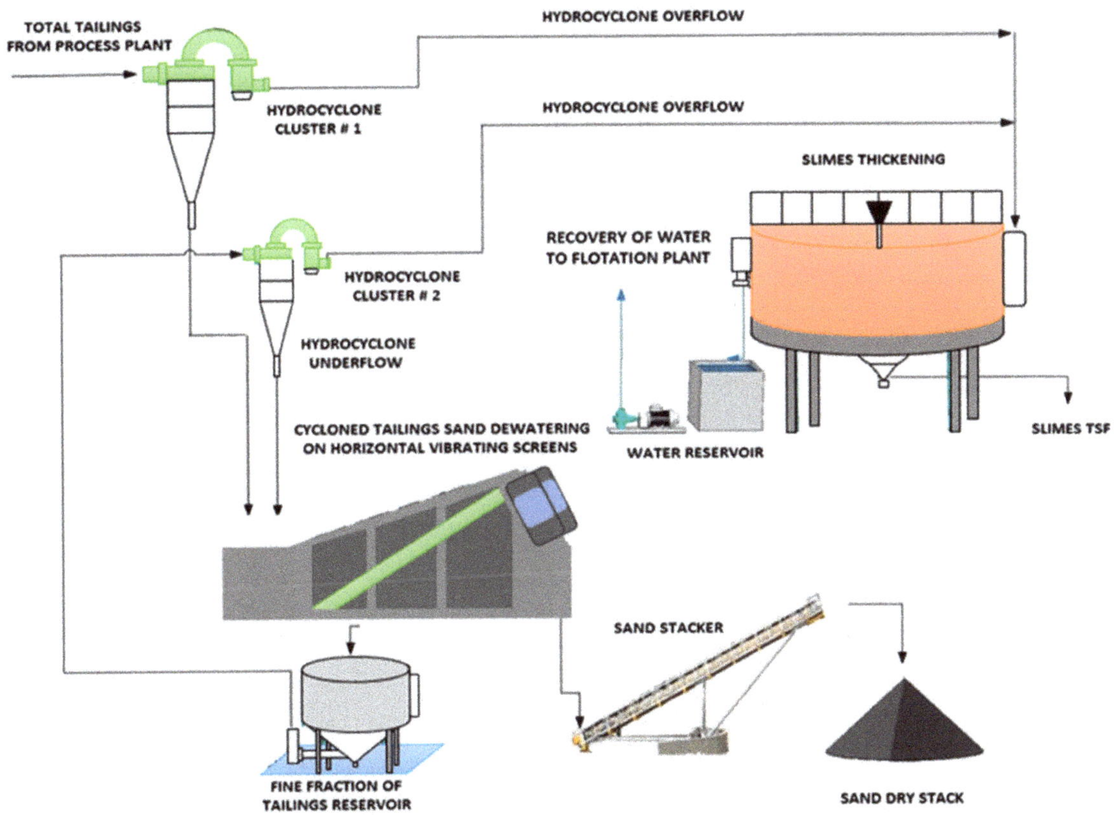

Figure 4. Copper tailings dewatering process with HC-HVS technology-Process Tailings Flowsheet (Adapted figure from [38]).

The coarse fraction of the tailings dewatered in the HVS is discharged to a conveyor belt to be transported to the TSF and deposited by a Mobile Conveyor Stacker and Tripper. The fine fraction of tailings obtained in the HVS, which has high water content, is sent to a box that feeds Hydrocyclone Station No. 2 located above the vibratory screen. Later, the coarse fraction of filtering obtained by Hydrocyclone Station No. 2 is discharged on the tailings "cake", which is located in the HVS. This "cake" is formed by the feed of the coarse fraction of the tailings resulting from the first classification stage using hydrocyclones. In this manner, the final tailings are filtered through this "cake", retaining the fines in the screen and preventing their filtration through the screen panels. The above is why this type of tailings dewatering plant can filter the solids to a size significantly finer than the opening of the screen installed in the HVS [32,38].

The overflow from Hydrocyclone Station No. 1 is sent to a fines thickener to recover water. This water is returned to the concentrator plant for reuse in the metallurgical process [32,38].

3. Results

3.1. Pilot Test Dewatering Application

From the beginning, the objective of the classification and dewatering of the tailings was to decrease the amount of water in the tailings management. Later, in 2004–2005, the concept of replacing the vacuum belt filters with a vibratory screen was analyzed, driven mainly by the need to reduce operational and maintenance costs. A dewatering pilot plant was assembled, installing a compact stage of HC and an HVS to dewater the coarse fraction of the tailings [39].

3.1.1. First Stage Process Description-Tailings Particle Grain Size Classification

The tailings from the flotation plant were classified into two fractions within an existing hydrocyclone station (Hydrocyclone Station No. 1). The fine fraction was sent to the tailings thickeners as was previously done. A fraction of coarse tailings was sent to the new dewatering pilot plant and a new hydrocyclone station (Hydrocyclone Station No. 2), instead of being sent to the vacuum belt filters [39].

In order to recover a maximum of coarse particles, the existing 500 mm hydrocyclones (Hydrocyclone Station No. 1) were maintained in operation. However, their configuration and geometry were modified to increase the solids content at the discharge of the underflow. To do this, it was necessary to decrease the separation size, d_{50} [34,39]. To achieve this objective, the future classification tailings with the 500 mm hydrocyclones were simulated with a 250 mm hydrocyclone, to obtain an appropriate amount of solids and thus accommodate the treatment capacity of the tailings dewatering pilot plant, installed as a second stage in the process [39].

3.1.2. Second Stage Process Description-Dewatering of Coarse Tailings

This dewatering stage consists of a compact hydrocyclone pilot plant, which includes a vibratory screen that directly receives the coarse fraction of the tailings obtained during the first stage of classification, and which produces a final "cake" of course tailings with sufficient concentration to be transported to the dry stack TSF with conveyor belts. The vibratory screen used in the pilot tests was located between the extremes of the vacuum belt filters and the conveyor belts [39].

The tailings slurry filtered by the vibratory screen is pumped to a hydrocyclone battery that performs a solid/liquid separation. Given that the objective of this classification stage (Hydrocyclone Station No. 2) is to recover almost all of the total solids in the dewatering operation, 100 mm hydrocyclones with a high shear capacity are used. The hydrocyclones produced overflow with a minimum solids content in weight, and this overflow is sent to the thickeners together with the fines obtained during the first stage of tailings classification.

The pilot plant used at Mantos Blancos (2004–2005), included the following equipment: (i) a compact hydrocyclone plant with two 100 mm hydrocyclones; (ii) a horizontal vibratory screen of 300 mm in width and 1600 mm in length; and (iii) a slurry centrifugal pump [39].

This equipment presents significant advantages in terms of a decrease of: (i) area; (ii) power consumption; (iii) wear elements (filter cloths); and (iv) reagents (flocculants). The result is a simple operating system for tailings dewatering [34,39]. Figure 5 presents the main equipment of the dewatering pilot plant used in the Mantos Blancos Project.

3.1.3. Pilot Test Dewatering Application Results

Preliminary tests were done to provide the circuit's necessary stability to achieve the pilot plant's maximum efficiency. Adjustments were performed to the plant to obtain a "cake", not overly dry but with the maximum moisture allowed in tailings for the transport in conveyor belts [34,39].

According to the results obtained in the pilot tests with vibratory screens, in addition to a fine selection of the configuration of the hydrocyclones, it was possible in this case to replace part of the existing belt filters with a compact plant with HVS together with HC, with a lower operating cost. The tailings solids content was 56%, with 82% (Cw) for test No. 1 and 68% with 78% (Cw) for test No. 2; this indicates a high degree of dewatering of the tailings [39].

The pilot tests performed during this period showed that a two-stage classification system with hydrocyclones and one stage with a vibratory screen obtained the same solids content in weight as the belt filters. Accordingly, it was decided to install two units with a capacity to treat 50% of the coarse tailings. Considering these results, Mantos Blancos decided to perform industrial-scale trials [34,39].

Figure 5. Dewatering Pilot Plant 600 mtpd: Hydrocyclone + Horizontal Vibratory Dewatering Screens Overview–Horizontal Vibratory Screens Operation–Cake Tailings Product [34].

3.2. *First Industrial Trial Phase Application*

The concept of using vibratory screens for tailings dewatering was undertaken to decrease the solids content in weight from 80% to 70% (Cw) of the coarse fraction of the tailings; the objective was to facilitate transport and disposal in the dry stack TSF. The coarse tailings with a solids content of 80% (Cw) transported to the TSF by a conveyor belt are stacked with steep slopes of 15%. By decreasing the solids content of the coarse tailings, coarse tailings flow further from the discharge point and occupy more surface area in the tailings storage facility [34,39].

3.2.1. First Stage Process Description-Tailings Particle Grain Size Classification

For the first stage of the hydrocyclones, the gravitational battery of 500 mm hydrocyclones (Hydrocyclone Station No. 1) was maintained. Only the configuration was changed to decrease the tailings split and to increase the solids fraction in the discharge, which is fed directly to the vibratory screen. The fine fraction of the tailings obtained from the overflow of the hydrocyclones was sent to the thickeners [34,39]. The first industrial unit used in Mantos Blancos (2007) included the following components described below, keeping in mind the first stage of the hydrocyclone:

Gravitational hydrocyclones battery, Model ERAL DEP-6-G4, with 02 hydrocyclones, Model PP050102 V, 500 mm in diameter, made of polyurethane.

3.2.2. Second Stage Process Description-Dewatering of Coarse Tailings

An HVS was installed to dewater the coarse fraction of the tailings. The fines that pass through the vibratory screen plus the filtered water are collected in a rectangular tank located underneath the vibratory screen, from which the suspension is pumped to the second stage of hydrocycloning (Hydrocyclone Station No. 2). Later, the fine tailings and filtered tailings are collected and recirculated to be placed on a pre-layer over the coarse tailings that are on the vibratory screen—this is done in order to dewater the resulting cake [34,39]. The first industrial unit used in Mantos Blancos (2007) included the following components described below:

Hydrocyclone battery, Model ERAL DEF-38-G4 with 38 hydrocyclones installed and 19 hydrocyclones in operation, Model PP010041 II, 100 mm in diameter, made of polyurethane.

A horizontal vibratory screen, Model EV-86, with a 38 kW installed power, with a functional area of filtering of 11.5 m^2.

Centrifugal pump group GB-86, with an 8/6 pump and 75 kW of installed power.

3.2.3. First Industrial Trial Phase Application Results

According to the results obtained from the first industrial phase, where part of the coarse tailings was dewatered with vibratory screens, in addition to performing an adequate selection of the configuration of the hydrocyclones, where 50% of the coarse tailings were dewatered with the use of vacuum belt filters and the rest is dewatered with the use of an HC-HVS, it is possible to conclude that in this case, it is feasible to replace all of the existing belt filters by a compact plant with vibratory screens and hydrocyclones, with a lower operating cost. It is relevant to mention that the solids content (Cw) at the discharge of the vibratory sieve was decreased from 80% to 70%, thus achieving the objective of facilitating their disposal in a dry stack TSF. With these results, Mantos Blancos decided to perform the second phase of industrial trials [34,38,39]. The following Table 1 and Figure 6 present the results of the first stage of the industrial trial and an overview of the dewatering plant:

Figure 6. 1st Industrial Trial Phase of Mantos Blancos Tailings Plant 4056 mtpd: Hydrocyclone Cluster + 01 Horizontal Vibratory Dewatering Screen [40,41].

Table 1. Process Performances and Results–1st Tailings Dewatering Industrial Plant (HC-HVS).

First Stage–Tailings Classification (Hydrocyclone Cluster #1)			**Second Stage–Tailings Dewatering (Vibratory Dewatering Screen + Hydrocyclone Cluster #2)**		
Tailings Flow at cluster #1 feed	11,592	mtpd	Tailings Flow at HVS feed	4056	mtpd
Solid content at cluster #1 feed	41	%	Solid content at cluster #2 feed	35	%
Solid content at cluster #1 overflow	21	%	Solid content at cluster #2 overflow	5	%
Solid content at cluster #1 underflow	70	%	Sand/slime split	92/08	-
Sand/slime split	70/30	-	Solid content at HVS discharge	70	%
Tailings Flow to underflow	8112	mtpd	Tailings Flow at HVS discharge	3744	mtpd

Note: Only 50% of coarse tailings are dewatered in the HC-HVS compact plant at this trial phase.

3.3. *Second Industrial Trial Phase Application and Results*

As a consequence of the good outcomes obtained with the first trials at an industrial scale, a second vibratory screen was installed in January 2010, with a capacity to dewater approximately 8112 mtpd of coarse tailings. The above replaced the tailings filtering activities performed by the vacuum belt filters. Conceptually, the tailings' primary grain size classification process in this second stage of trials at an industrial scale is the same as applied in the first stage of industrial trials, where only the number of hydrocyclones and HVSs used is changed [34,39].

3.3.1. First Stage Process Description-Tailings Particle Grain Size Classification

The quantity of hydrocyclones used is presented below:

Gravitational hydrocyclone cluster, Model ERAL DEP-6-G4, with 04 hydrocyclones, Model PP050102 V, 500 mm in diameter, made of polyurethane (PU).

3.3.2. Second Stage Process Description-Dewatering of Coarse Tailings

The number of hydrocyclones, HVSs, and pump equipment are presented below:

Hydrocyclone cluster, Model ERAL DEF-38-G4 with 38 hydrocyclones installed and all in operation, Model PP010041 II, of 100 mm in diameter, made of PU.

02 horizontal vibratory screens, Model EV-86, with a 38 kW installed power, with an area of filtering of 11.5 m^2.

Centrifugal pump group GB-86, with a 10/8 pump and 110 kW of installed power.

3.3.3. Second Industrial Trial Phase Application Results

The following Table 2 and Figure 7 present the results of the second stage of the industrial trials and an overview of the HC-HVS technology, respectively:

Table 2. Process Performances and Results–2nd Tailings Dewatering Industrial Plant (HC-HVS).

First Stage–Tailings Classification (Hydrocyclone Cluster #1)			**Second Stage–Tailings Dewatering (Vibratory Dewatering Screen + Hydrocyclone Cluster #2)**		
Tailings Flow at cluster #1 feed	11,592	mtpd	Tailings Flow at HVS feed	8112	mtpd
Solid content at cluster #1 feed	41	%	Solid content at cluster #2 feed	35	%
Solid content at cluster #1 overflow	21	%	Solid content at cluster #2 overflow	5	%
Solid content at cluster #1 underflow	70	%	Sand/slime split	92/08	-
Sand/slime split	70/30	-	Solid content at HVS discharge	70	%
Tailings Flow to underflow	8112	mtpd	Tailings Flow at HVS discharge	7488	mtpd

Note: The operational availability of this HC-HVS compact plant at the 2nd industrial trial has been of the order of ~90%.

Figure 7. 2nd Industrial Trial Phase of Mantos Blancos Tailings Plant 8000 mtpd: Hydrocyclone Cluster + 02 Horizontal Vibratory Dewatering Screens [40,41].

Finally, the tailings produced with the use of the HC-HVS technology present the following characteristics:

- P_{80}: 255 to 270 μm
- Fines content: 23 to 25%
- Solids content (Cw): 70%

The operational availability of the coarse tailings dewatering plant for its equipment is equivalent to 90%.

3.4. Tailings Storage Facilities Results

The thickened fine tailings (slimes) are transported by pumping pipelines to a tailings storage facility near the tailings dewatering plant. In this place, the tailings are deposited through a series of spigots, forming a tailings beach and a clear water lagoon. The process waters are recirculated to the metallurgical process plant through pumping and pipelines, as shown in Figure 8.

Figure 8. Mantos Blancos Disposal of Thickened Slimes in Tailings Storage Facility [39].

The filtered coarse tailings are transported to the filtered tailings deposit by a conveyor belt with Mobile Stacker Conveyor and Tripper. The Filtered Coarse Tailings Storage Facility is developed by the down valley placement method applied at Mobile Stacker Conveyor with Tripper forming a fan shape, where saturated filtered tailings flow along roughly repose angle slopes, buttressed at the toe by a containment dyke and downstream supported by a sedimentation collection pond. Dozers carried out spreading and compaction activities on filtered tailings lift of an average of 20–30 cm in thickness (Figures 9 and 10) [34,39].

Figure 9. Mantos Blancos Disposal of Filtered Coarse Tailings with Mobile Stacker Conveyor and Tripper [12].

Figure 10. Mantos Blancos Aerial View of Disposal of Filtered Coarse Tailings [12].

4. Discussion: HC-HVS Dewatering Technology Application

As this case history shows, appropriate management of the dewatering of tailings can lead to better environmental management from a seepage control and water reclaim point of view. With the increase of social pressure for the safety of the TSF at the surface and environmentally friendly disposal of tailings, the disposal of filtered tailings is an option to be considered to this day.

4.1. Advantages (Pros)

The following paragraphs present the main advantages of tailings dewatering with the use of HC-HVS technology:

4.1.1. Reduction of Operating Costs

The implementation of HC-HVS units for tailings dewatering the Mantos Blancos Mine has shown to be an alternative to belt filters given their low power demand, lower operational cost, and high efficiency. This application is based on pilot tests and industrial-scale trials performed directly in the mine, which present promising results. Table 3 presents Opex Comparison between the vacuum belt filter plant and the HC-HVS plant at Mantos Blancos Project [32]:

Table 3. Opex Comparison between vacuum belt filter plant and HC-HVS plant at Mantos Blancos Project [32].

Dewatering Technology Alternatives	Energy Cost kUS$/year	Operation, Maintenance, and Labour kUS$/year
3 Vacuum Belt Filters of 100 m^2 area	200	315
3 Hydrocyclone Compact Plants (HC-HVS) MUE 10/38-150.90-86	122	117

4.1.2. Performance at High Elevations

Filtering equipment suppliers have gained experience performing trials and projects during the last decade. The learning curve associated with each type of tailings differs given their particle size distribution (PSD) and the associated mineralogy. Each tailings type exhibits a unique filtering behavior for which solid/liquid separation units are required that are efficient and reliable. The performance of the dewatering of tailings by HC-HVSs is efficient at high elevations since they are not sensitive to the altitude, as is the case for vacuum filters (belt and disc filters). This characteristic is relevant depending on the project areas, such as the mountainous region of the Andes in South America.

The effect of the screens on the tailings is a consequence of the vibration and gravity. The first is a mechanical consequence of the energy applied and the second is a consequence of the mass through the force of gravity. Based on the above, the site's geographical elevation does not influence the screen's behavior since gravity is almost constant on earth, with only minor variations between sea levels and higher elevations. With this technology, atmospheric pressure, which does show important variations with elevation, does not affect the behavior of the equipment since it does not use a vacuum as part of the process.

4.1.3. HC-HVS Plant Location

The location of the HC-HVS plant must be determined with a trade-off study that includes factors such as (i) distance to the tailings storage facility; (ii) transport medium of the dewatered tailings; (iii) access; (iv) availability of power; and (v) requirements of maintenance and operational equipment. In general, factors (iv) and (v) indicate that this plant is located in the vicinity of the floatation plant, but this must be verified on a case-by-case basis.

4.2. Disadvantages (Con)

The main disadvantages of the HC-HVS dewatering technology are presented below.

4.2.1. Tailings Scale Production

The filtering technologies have been applied with success for production rates of up to 20,000 tmpd, as is the case of projects at La Coipa (Chile), Chingola (Zambia), and Karara (Australia) [12]; and have shown improvements in the performance with large scale productions. In this scenario, there is still a need for more reliable equipment for large-scale filtering processes. There is a need to focus on the tailings water reclaim to improve its reuse in the mining processes. Today, the HC-HVS technology is limited in expansion because it needs more experience with production rates in the order of 10,000 mtpd to develop its full potential and show safety in the operational process of obtaining tailings "cakes", thereby improving water recovery and decreasing tailings pond water seepages.

Another relevant aspect is the definition of the screen's geometry and the filtration bed's characteristics. In practice, this is done using the initial laboratory tests but mainly with the pilot scale trials. With regards to tailings, there still is not enough experience to have a solid numeric model that allows for the dimensioning of this technology's elements as a function of the grain size and solids content of the tailings.

4.2.2. Fine Tailings

The nature of the tailings material is important when filtration is considered a dewatering process. Not only is the grain size of the tailings important, but also the mineralogy has a relevant role. Specifically, the high percentages of fines < 74 microns and clayey minerals tend to make the use of filtration technology difficult [15]. A relevant part of the water reclaims process in applying the HC-HVS technology, as a whole, depends on the thickening of the slimes, which implies performing an efficient liquid-solid separation with a fine fraction of the tailings. If the total tailings contain a high content of clayey minerals, the global performance of water reclaim may decrease. Attention is required to control the sedimentation and slimes thickening process.

The vibratory screens have limitations in managing fine and plastic tailings. This characteristic would eliminate the application of this technology in the majority of the tailings with a grain size in which 30% passes the #200 ASTM sieve. This restriction includes the majority of the copper tailings and all the gold tailings. However, this limitation may be overcome by the decrease of the fine fraction with the use of hydrocyclones in order to generate a product that the screens may efficiently process. Consequently, the design process first involves determining the finest grain size that the screens may dewater. Once this grain size is determined, the classification (hydrocycloning) system and the ratio or split of the coarse fraction (screen material) and a fine fraction (material to thickeners or deposit) are defined. There is an optimization stage to improve this process's performance by incorporating a fine material bed above the screen.

Another relevant property of the tailings is permeability. A medium to high permeability facilitates water removal by vibration, making the vibratory process faster. For this reason, measuring the permeability of the tailings in a laboratory is recommended. The specific weight of the particle is also relevant since, in general, a greater specific weight means a greater permeability in non-plastic tailings.

4.3. Water Recovery Comparison with Other Mine Tailings Management Technologies

Water losses at TSFs come from water retained in deposited tailings and the evaporation from beaches formed at the TSF. To reduce these losses, new management technologies have been developed, which seek to maximize the reclamation of water before tailings are discharged to the TSF, by cycloning, thickening, and/or filtering tailings. Improvements in conventional, thickened, paste, and filtered tailings disposal technologies need to be managed to increase water recovery and decrease water makeup (freshwater) in mining operations. These challenges have been met during the past decade in copper mining.

Table 4 shows a comparative analysis of water recovery quantities obtained with different dewatering tailings technologies assuming a tailings production of 100,000 mtpd.

Table 4. Water Recovery Comparison between Different Mine Tailings Management Technologies.

Description	Unit	Conventional Tailings Management	Thickened Tailings Management	HC-HVS Tailings Management	Filtered Tailings Management
Tailings Production	mtpd	100,000	100,000	100,000	100,000
Cw before Dewatering	%	28	28	28	28
Water on Conventional Tailings	L/s	2976	2976	2976	2976
Cw after Dewatering	%	50	60	70 (*)	80
Water on Dewatered Tailings	L/s	1157	772	496	289
Water Recovery from Dewatering Devices	L/s	1819	2205	2480	2687
Water Recovery from TSF	L/s	382	255	164	95
Total Water Recovery	L/s	2201	2459	2644	2782
Water Recovery Efficiency	%	74	83	89	93

Note: The following terms mean: Cw: Tailings solid content by weight (%). (*): 70% signifies a mean target Cw value, considering dewatering tailings technologies applied.

Table 4 shows that HC-HVS technology is competitive to recover water from mine tailings, outperforming conventional tailings and thickened tailings technologies, and ranking second behind filtered tailings technology, with an efficiency of 89%. This indicates that the HC-HVS technology is an attractive option to be implemented to carry out an environmentally controlled management of mining tailings and maximize

5. Conclusions

Filtered tailings TSF solution today has become very competitive compared to conventional technologies because the cost of closure and post-closure are lower, the reliability of technology has increased, and the authorities and community are more prone to accept this approach considering the safety management of tailings. Under dry climate conditions, applying the HC-HVS technology, no supernatant TSF ponds need to be monitored during operation. There is no freeboard requirement for tailings dams, which significantly reduces water losses and eliminates overtopping or piping (internal dam erosion by seepage) event risks [42,43].

The filtered dry stacked tailings case presented in this article shows that there is important potential to achieve a sustainable and cost-effective tailings management solution in desert areas such as the north of Chile and south of Peru, among others. The important benefits offered by this technology are: (i) less operational cost of the project over the mine lifetime; (ii) failure risks reduction in seismic zones improving TSF physical stability; (iii) maximum water recovery; and (iv) environmentally friendly solution are allowing smaller TSF footprints, a decrease of pond seepages, and allowing a progressive TSF reclamation.

Both the information obtained from the pilot and industrial tests from field and theory indicate that the HC-HVS technology has a promising future in dewatering tailings that are not too fine and of low plasticity. This technology presents economic and operational advantages concerning vacuum filters, particularly in sites located at high topographical elevations (masl). An important part of the mining activity in Chile is located in the Andes Mountains, at high elevations, and in an environment with scarce water availability.

To date, this technology has been applied in operations of limited production (below 6000 mtpd). Moreover, the units used have a processing capacity of 1500 mtpd. These capacities are very small to be applied in large production projects (25,000 mtpd and above) since they would require many units that would make their operation difficult, particularly in maintenance and flow conveyance. Although theory does not identify any obstacles to designing equipment of larger dimensions, it is evident that the use of larger equipment requires a thorough investigation (as was the case for press filters) or the performance of trials at an industrial scale (demonstrative plant).

In addition to the study associated with larger productions, it is particularly necessary to advance the investigation of the efficiency of this equipment and technology with fine tailings and/or with larger clay content. The above would allow for the application of this technology to iron and gold tailings, among others, marking the limits of this technology.

Over about 17 years of operation with this interesting technology, it has been established that the horizontal vibrators, together with hydrocyclones, are a viable alternative for dewatering of copper tailings in concentrator plants of midsize such 10,000 mtpd; such plants can treat tailings with a fraction of about 30–45 microns, which in many concentrators can represent the order of 60%–70% of the total mass of tailings. The true architects of the success of tailings dewater are: creating a hydrocyclone of a bed to drain the tailings; collecting the filtrate from the vibratory dewatering screens; creating a closed circuit, and the mechanical devices described here.

These benefits are a good reason for new or existing large mining operations to shift from conventional slurry tailings disposal facilities with extensive water ponds to alternative solutions with highly dewatered tailings disposal facilities. This tailings management approach is more accepted by the environmental authority and communities daily.

Author Contributions: Conceptualization, C.C. and E.A.; formal analysis, C.C.; investigation, C.C.; resources, E.A.; writing—original draft preparation, C.C.; writing—review and editing, C.C.; visualization, C.C.; supervision, E.A. All authors have read and agreed to the published version of the manuscript.

Funding: The research is funded by the Research Department of the Catholic University of Temuco, Chile.

Data Availability Statement: The data presented in this study are available on request from the corresponding author.

Conflicts of Interest: The authors declare no conflict of interest.

Abbreviations

TSF	Tailings Storage Facility
BATs	Best Available Technologies
TTD	Thickened Tailings Disposal
Cw	Slurry tailings solids content by weight
mtpd	Metric tonnes per day
HC	Hydrocyclones
HVS	Horizontal Vibratory Screens
HC-HVS	Combination of Hydrocyclones and Horizontal Vibratory Screens
PD Pumps	Positive Displacement Pumps
masl	Meters above sea level
PU	Polyurethane

References

1. Barrera, S.; Cacciuttolo, C.; Caldwell, J. Reassessment of Best Available Tailings Management Practices. 2015. Available online: https://open.library.ubc.ca/cIRcle/collections/59368/items/1.0319135 (accessed on 14 September 2022).
2. Cacciuttolo, C.; Valenzuela, F. Efficient Use of Water in Tailings Management: New Technologies and Environmental Strategies for the Future of Mining. *Water* **2022**, *14*, 1741. [CrossRef]
3. Tuomela, A.; Ronkanen, A.K.; Rossi, P.M.; Rauhala, A.; Haapasalo, H.; Kujala, K. Using geomembrane liners to reduce seepage through the base of tailings ponds—A review and a framework for design guidelines. *Geoscience* **2021**, *11*, 93. [CrossRef]
4. Edraki, M.; Baumgartl, T.; Manlapig, E.; Bradshaw, D.; Franks, D.M.; Moran, C.J. Designing mine tailings for better environmental, social and economic outcomes: A review of alternative approaches. *J. Clean. Prod.* **2014**, *84*, 411–420. [CrossRef]
5. Cacciutolo, C. Review of Tailings Management in Chile: Lessons Learned, Incipient Progress of a Holistic Management, and Future Challenges in Reducing Socio-Environmental Impacts. In Proceedings of the 3rd International Seminar on Environmental Issues in Mining, Lima, Peru, 2–4 December 2015.
6. Dold, B. Sustainability in metal mining: From exploration, over processing to mine waste management. *Rev. Environ. Sci. Biotechnol.* **2008**, *7*, 275–285. [CrossRef]
7. East, D.; Fernandez, R. Managing Water to Minimize Risk in Tailings Storage Facility Design, Construction, and Operation. *Mine Water Environ.* **2021**, *40*, 36–41. [CrossRef]
8. Watson, A.H.; Corser, P.G.; Pardo, E.E.G.; Christian, T.E.L.; Vandekeybus, J. A comparison of alternative tailings disposal methods—the promises and realities. In *Mine Waste 2010, Proceedings of the First International Seminar on the Reduction of Risk in the*

Management of Tailings and Mine Waste, Perth, Australia, 29 September–1 October 2010; Jewell, R., Fourie, A., Eds.; Australian Centre for Geomechanics: Perth, Australia, 2010; pp. 499–514.

9. Jewell, R.J.; Fourie, A.B. *Paste and Thickened Tailings: A Guide/Editors R.J. Jewell and A. Fourie*, 3rd ed.; Australian Centre for Geomechanics: Perth, Australia, 2015; ISBN 9780992481001.
10. Franks, D.M.; Boger, D.V.; Côte, C.M.; Mulligan, D.R. Sustainable development principles for the disposal of mining and mineral processing wastes. *Resour. Policy* **2011**, *36*, 114–122. [CrossRef]
11. Cacciuttolo Vargas, C.; Marinovic Pulido, A. Sustainable Management of Thickened Tailings in Chile and Peru: A Review of Practical Experience and Socio-Environmental Acceptance. *Sustainability* **2022**, *14*, 10901. [CrossRef]
12. Cacciuttolo Vargas, C.; Perez Campomanes, G. Practical Experience of Filtered Tailings Technology in Chile and Peru: An Environmentally Friendly Solution. *Minerals* **2022**, *12*, 889. [CrossRef]
13. Cacciuttolo, C.; Holgado, A. Management of Paste Tailings in Chile: A Review of Practical Experience and Environmental Acceptance. In Proceedings of the 19th International Seminar on Paste and Thickened Tailings Paste, Santiago, Chile, 5–8 July 2016.
14. Cacciuttolo, C.; Barrera, S.; Caldwell, J.; Vargas, W. Filtered Dry Stacked Tailings: Developments and New Trends. In Proceedings of the 2nd International Seminar on Tailings Management, Antofagasta, Chile, 12 July 2014; pp. 357–370.
15. Davies, M. Filtered dry stacked tailings: The fundamentals. In Proceedings of the Proceedings Tailings and Mine Waste 2011, Vancouver, BC, Canada, 6–9 November 2011; p. 9.
16. Cacciuttolo, C.; Holgado, A. Dry Stacking of Filtered Tailings in the Andean Region of South America—An Engineered and Environmentally Friendly Solution. In Proceedings of the Congreso Panamericano de Mecánica de Suelos y Geotecnia; Sociedad Argentina de Ingeniería Geotécnica, International Society for Soil Mechanics and Geotechnical Engineering, Buenos Aires, Argentina, 30 November 2015.
17. Lara, J.L. Presentacion de Relaves Espesados y Filtrados en Perú, Instituto de Ingenieros de Minas. Available online: https://www.youtube.com/watch?v=0vcYR2gSPjk (accessed on 16 September 2022).
18. Lara, J.L.; Pornillos, E.; Loayza, C. The application of highly dewatered tailings in the design of tailings storage facilities—Experience in Mining Projects in Peru. In Proceedings of the 16th International Conference on Tailings and Mine Waste, Vali, CO, USA, 17–20 October 2010.
19. Lara, J.L.; León, E. Design and Operational Experience of the Cerro Lindo Filtered Tailings Deposit. In Proceedings of the Proceedings of the Paste and Thickened Tailings 2011 Seminar, Pert, Australia, 5–7 April 2011.
20. Lara, J.; Pornillos, E.; Muñoz, H. Geotechnical-Geochemical and operational considerations for the application ofdry stacking tailings deposits—State ofthe art. In Proceedings of the 16th International Seminar on Paste and Thickened Tailings, Belo Horizonte, Brazil, 17–20 June 2013; pp. 249–260.
21. Schoenberger, E. Environmentally sustainable mining: The case of tailings storage facilities. *Resour. Policy* **2016**, *49*, 119–128. [CrossRef]
22. Kossoff, D.; Dubbin, W.; Alfredsson, M.; Edwards, S.; Macklin, M.; Hudson-Edwards, K.A. Mine tailings dams: Characteristics, failure, environmental impacts, and remediation. *J. Appl. Geochem.* **2014**, *51*, 229–245.
23. Franks, D.M.; Stringer, M.; Torres-Cruz, L.A.; Baker, E.; Valenta, R.; Thygesen, K.; Matthews, A.; Howchin, J.; Barrie, S. Tailings facility disclosures reveal stability risks. *Sci. Rep.* **2021**, *11*, 5353. [CrossRef] [PubMed]
24. Williams, D.J. Lessons from tailings dam failures—Where to go from here? *Minerals* **2021**, *11*, 853. [CrossRef]
25. Davies, M.P.; Rice, S. An Alternative to Conventional Tailing Management—"Dry stack" Filtered Tailings. In Proceedings of the 8th International Conference on Tailings and Mine Waste'01, Fort Collins, CO, USA, 16–19 January 2001; Taylor & Francis: London, UK; pp. 411–420.
26. Ulrich, B.; Coffin, J. Considerations for Tailings Facility Design and Operation Using Filtered Tailings Summary of Sig-nificant Operating Filtered Tailings Operations. In Proceedings of the 16th International Seminar on Paste and Thickened Tailings, Belo Horizonte, Brazil, 17–20 June 2013; pp. 201–210.
27. Ulrich, B. Practical Thoughts Regarding Filtered Tailings. In Proceedings of the the 22th International Conference on Tailings and Mine Waste, Cape Town, South Africa, 8–10 May 2019; pp. 71–79.
28. Crystal, C.; Hore, C.; Ezama, I. Filter-Pressed Dry Stacking: Design Considerations Based on Practical Experience. In Proceedings of the 22th International Conference on Tailings and Mine Waste, Keystone, CO, USA, 17–20 October 2010; pp. 209–219.
29. Hore, C.; Luppnow, D. Karara Iron Ore TSF—Design Considerations for a Unique Large Scale Dry Stack Facility. In Proceedings of the 18th International Conference on Tailings and Mine Waste, Keystone, CO, USA, 5–8 October 2014; pp. 293–304.
30. Oberle, B.; Brereton, D.; Mihaylova, A. *Towards Zero Harm: A Compendium of Papers*; Global Tailings Review: St. Gallen, Switzerland, 2020.
31. Araya, N.; Quiñonez, O.M.; Cisternas, L.A.; Kraslawski, A. Sustainable Development Goals in Mine Tailings Management: Targets and Indicators. *Mater. Proc.* **2021**, *5*, 82. [CrossRef]
32. Bouso, J.L. Vibrating Squeegees in Sterile Filtering. *Chem. Eng.* **2006**, *435*, 144–149.
33. Bouso, J.L. La Filtración de Lodos y el Impacto Ambiental. *Rev. Min. Chil.* **2001**, *248*, 24.
34. Bouso, J.L.; Berry, R. Application of dewatering screens in tailings filtration. In Proceedings of the International Seminar on Paste and Thickened Tailings, Australian Centre for Geomechanics, Santiago, Chile, 20–22 April 2005; Australian Centre for Geomechanics: Perth, Australia, 2005; pp. 147–162.

35. Bouso, J.L. Filtración Con Escurridores Vibrantes. In *Revista Técnica-Cientifica Rocas y Minerales*; Tecnicas y procesos de minas y canteras: Madrid, Spain, 1987; ISBN 0378-3316.
36. Keller, K.; Stahl, W. Vibration screens for dewatering—Theory and Practice. *Miner. Metall. Process. J.* **1997**, *14*, 27–34. [CrossRef]
37. Doerffer, M.; Heinrich, R. Efficient Dewatering Solutions on Vibrating Screens. In Proceedings of the 12th International Conference on Tailings and Mine Waste, Vail, CO, USA, 19–22 October 2008.
38. Astete, A. Tratamiento de Relaves en Mantos Blancos, Mediante Hidro-Escurridores. Available online: http://eralchile.com/equipo/plantas-de-hidro-escurrido (accessed on 16 September 2022).
39. Cacciuttolo, C.; Barrera, S.; Bouso, J.L.; Astete, A. Dewatering Tailings on Horizontal Vibration Screens: An Alternative to Filtration Systems to Obtain Tailings to be Dispose at Dry Stack Tailings Storage Facilities. In Proceedings of the Seminario Internacional en Gestión de Relaves TAILINGS, Gecamin, Chile, 12 October 2015; University of British Columbia: Vancouver, BC, Canada, 2015.
40. ERAL Horizontal Vibratory Dewatering Screens-Equipment Catalog. Available online: http://eralchile.com/equipo/secadoresrotativos (accessed on 16 September 2022).
41. ERAL Compact Hydrocyclone Plants-Equipment Catalog. Available online: http://eralchile.comequipo/hidrociclones (accessed on 16 September 2022).
42. Cacciuttolo, C.; Cano, D. Environmental Impact Assessment of Mine Tailings Spill Considering Metallurgical Processes of Gold and Copper Mining: Case Studies in the Andean Countries of Chile and Peru. *Water* **2022**, *14*, 3057. [CrossRef]
43. Cacciuttolo, C.; Atencio, E. Past, Present, and Future of Copper Mine Tailings Governance in Chile (1905–2022): A Review in One of the Leading Mining Countries in the World. *Int. J. Environ. Res. Public Health* **2022**, *19*, 13060. [CrossRef]

minerals

MDPI

Article

The Use of Lightweight Penetrometer PANDA for the Compaction Control of Classified Sand Tailings Dams

Gabriel Villavicencio [1,*], Claude Bacconnet [2,†], Pamela Valenzuela [1], Juan Palma [1], Alex Carpanetti [3], Gonzalo Suazo [4], Matías Silva [5] and José García [1]

1 Grupo de Geotecnia, Escuela de Ingeniería de Construcción y Transporte, Pontificia Universidad Católica de Valparaíso, Avenida Brasil 2147, Valparaíso 2362804, Chile
2 Institut Pascal, Université Clermont-Auvergne, BP 10448, F-63000 Clermont-Ferrand, France, CNRS, UMR 6602, F-63171 Aubière, France
3 Escuela de Ingeniería Química, Pontificia Universidad Católica de Valparaíso, Avenida Brasil 2180, Valparaíso 2362804, Chile
4 School of Civil Engineering, Universidad Técnica Federico Santa María, Avenida España 1680, Valparaíso 2390123, Chile
5 Golder Associates, Las Condes, Magdalena 181, Santiago 7550055, Chile
* Correspondence: gabriel.villavicencio@pucv.cl
† In memory of Claude Bacconnet.

Abstract: Sand tailings dams have historically been the most commonly used technology for tailings storage in Chile. Although engineering advances have resulted in the construction of approximately 250-m-high facilities, some operational challenges still remain, including compaction control. Control is currently performed at a few control points in a dam embankment, without considering a series of factors that affect its mechanical behavior (e.g.,layer thickness and material variability). Within this context, geostatistics can be applied in combination with low-cost geotechnical tools as an alternative to improve compaction control in tailings storage facilities. In this study, an extensive field investigation was carried out. A total of 91 PANDA penetrometer tests were conducted to monitor the degree of compaction in an experimental classified sand tailings dam. The results were analyzed using stochastic interpolation for ordinary kriging and considering the spatial distribution of the cone resistance and the degree of compaction determined for the dam. The results showed that spatial variability was associated with the material variability of sand tailings and the compaction method used, and deviations from design requirements. The article shows the value of the use of geostatistics in decision-making in the case of classified sand tailings dams. This is mainly due to the fact that it allows optimization of the compaction process used in these tailings dams. Additionally, a useful database is generated to continue deepening studies of physical stability during the useful life of the tailings storage facilities.

Keywords: sand tailings dams; compaction control; variability; Lightweight Penetrometer PANDA

1. Introduction

The Chilean Copper Commission (Comisión Chilena del Cobre—COCHILCO) has projected that copper production in Chile will increase over the next 9 years to reach 7.06 million tons in 2029 and peak at 7.25 million tons in 2025 [1]. However, an increase in copper production in deposits with an average grade of 1% or lower translates into a significant increase in tailings production, which is currently up to approximately 1.5 million t/day. A total of 757 tailings storage facility were identified in the last cadaster of mining operations in Chile [2]. Of all these deposits, 81.7% (606) correspond to classified sand tailings dams, of which 388 are inactive, 167 are abandoned and 51 are active. A total of 38 cases of physical instability have been reported and documented among the failures that have occurred in classified sand tailings dams between 1915 and 2010 in Chile. These cases have been attributed to seismic liquefaction or flow failure (50%), slope instability with seismically induced deformations (32%) and overflow (18%). Some of these failures, such as those that occurred at the Barahona (1928), El Cobre (1965) and Las Palmas (2010) dams, have released tailings with catastrophic consequences in terms of the loss of human lives and environmental damage.

The main causes of physical instabilities in classified sand tailings dams in Chile were identified in [3], these causes corresponded to, the construction method, a poor cycloning process (fines content, FC > 20%), inefficient degree of compaction, a high degree of saturation and inappropriate embankment geometry (downstream slopes, the grade, freeboard and crest width). These causes have been directly related to poor construction/operation, inadequate deposit monitoring or a combination of these factors, where compaction control is among the key parameters affecting the physical stability of these types of deposits. The compaction control in classified tailings dam is commonly undertaken by measuring soil's in-situ unit weight (γ_t) in a few points within the dam (technical specifications indicate minimum number of tests per volume or weight of compacted material). Depending on the operation, descriptive statistics is used to analyse the gathered data while spatial analysis of datasets is uncommon.

In [4] the relevance of carrying out compaction control considering both the variability in the physical characteristics of the sand tailings (e.g., mineralogy and particle size) and the variability resulting from the deposition, construction, and compaction of the embankment of classified sand tailings dam. A standard method was developed for the in-situ control of the degree of compaction with depth and the layer thickness of the sand tailings deposits using a PANDA penetrometer as a measuring instrument [5,6].

The method allows controlling economically, quickly, and efficiently, both the degree of compaction and the thickness of the deposited layer, considering both the material variability of the tailings sands, as well as the variability generated as a result of the construction process used. This tool generates in situ data more quickly over a larger depth range and in higher quantities than the sand-cone method, and nuclear densimeter, currently used in Chile to perform the compaction control of classified sand tailings dams [7]. Thus, the penetrometer is an effective tool for compaction control in sand tailings dams since it can cover a large surface area and deeper exploration depths in comparison to conventional superficial techniques (e.g., sand-cone method and nuclear densimeter). As part of the operational control of this type of facility, the use of these surface techniques implies carrying out a very small number of tests on the surface and reaching depths of less than 0.3 m, leaving questions about the state of compaction for more important depths. This information is not subsequently processed or spatialized, due to the compaction methodology applied for the construction of classified sand tailings dams, it is empirically validated by the same operators, without considering the optimization of the process.

The analysis of the generated compaction datasets can be performed using simple geostatistical techniques for spatial interpolation of a wide study area, considering a preferential distance and direction. This spatial estimation technique covers a large control surface for unsampled locations based on the location of field tests [8]. Thus, a geostatistical analysis can be used to model the spatial distributions of a study parameter over large areas, and the control volumes can be changed by switching to additive variables [9]. These methods have been used in geotechnics to develop spatial stratigraphic models and determine the index properties and geotechnical parameters, among other properties, of natural soil deposits, primarily using databases of surveys, standard penetration test (SPT), and cone penetration tests (CPT), e.g., [10–15].

Geostatistical techniques were used in the present study to develop spatial models of the γ_t and compaction degree for periodic compaction control in embankments of sand tailings dams based on the PANDA cone resistance (q_d). First, variations in the physical characteristics, index properties and q_d were analyzed for classified sand tailings dams, which were considered representative of medium and large Chilean mining operations. The percentage of homogeneity or heterogeneity of each parameter was determined. Second, geostatistics was applied to compaction control in a experimental classified sand tailings dam. A test area of the embankment was selected and compacted by the method typically used for the deposit. A geometric grid of survey points was defined for the study area within which 91 PANDA tests were performed down to a 1.5 m depth. The representativeness of the study area was assessed by performing a statistical analysis to comparatively analyze the means and variances of data from compaction control conducted over 3 years in the classified sand tailings dam and from field tests in the study area. The results of the Student's t-test and from Bartlett's test indicated the same means and variances for both datasets. The results for each control point were used to construct experimental vertical and horizontal semivariograms in different directions of analysis for q_d, the in situ dry unit weight (γ_d) and the degree of compaction Standard Proctor (SP = $100x\frac{\gamma_d}{\gamma_{max}}$, where γ_{max} corresponds to the maximum dry unit weight Standard Proctor test). These three parameters were best fit by a theoretical spherical semivariogram. After determining the spatial distribution of the three parameters, a stochastic interpolation method known as ordinary kriging (OK) was applied. The real and estimated q_d values were cross validated to compare the modeling results with the measured results for the survey grid and to validate the quality of the selected theoretical semivariogram model. From the results obtained, it is concluded that from the use of the Lightweight Penetrometer PANDA together with simple geostatistical techniques of spatial interpolation, it is feasible to perform quality control (QC) or quality assurance (QA) in a faster and more effective way compared to other techniques currently used (sand cone method [7], or nuclear density meter [16]). This speed and effectiveness were evaluated for classified sand tailings dams during construction. In this way, it is possible to more effectively identify SECTORS that do not comply with technical specifications defined as part of the facility design. Deviations from the design that can be identified using the geostatistical techniques in sand dams are: (i) zones of weak resistance and consequently with less degree of physical stability (e.g., greater potential for seismic liquefaction, a key aspect in countries with significant seismic activity, such as Chile [3,4]); (ii) over compacted areas, where the applied compaction process could be improved, optimizing the economic resources allocated for this purpose. (e.g., fewer machinery hours used for the compaction process).

2. Background

2.1. Compaction Control in a Classified Sand Tailings Dam

2.1.1. Criteria

In Chile, a seismically active country, sand tailing dams must be mechanically compacted. The above is to ensure the dilative behavior of the sandy material upon shearing. Compaction results in material above the critical state line avoiding liquefaction risk and improving slope stability under seismic and monotoning loading [3,4,17,18].

Chilean engineering criteria (e.g., [3,17,19]) include the FC (<75 μm) of classified sand tailings used to build a embankment and the compaction degree among the key requirements for adequate physical stability (under static and seismic conditions) of tailings deposits during operational and subsequent closure stages. A FC range from 10 to 25% reconciles considerations of the material geotechnical properties (the densities, permeabilities and shear strength, among others) with the costs of the cycloning process. Compaction degree higher than 95% SP are deemed acceptable. Sand tailings with a FC between 10 and 25% and 95% SP have a relative density (Dr) of 60 to 65% [18]. This Dr or SP is adopted in practice as a preliminary threshold, between contract (loose) and dilatant (dense) behavior of the sand tailings [20,21].

During the operational stage of the sand tailings dams, it must be checked periodically that the degree of compaction resulting from the applied compaction process complies with the value of SP or Dr established in the design project. This way, it ensures dilative behavior in the entire embankment and, therefore, a stable condition from the physical point of view [22–24].

Concerning the periodic monitoring of operational parameters in the sand tailings dams, the compaction control is carried out considering as a basis the requirements established by the National Service of Geology and Mining (SERNAGEOMIN, Ministry of Mining of Chile), together with the control criteria established internally by each mining company. SERNAGEOMIN requires mining companies to carry out a quarterly control through the official document called "Form E700" (https://www.sernageomin.cl/formularios-seguridad-minera/accessed on 5 November 2022). However, regarding compaction control, this control document has the following drawbacks:

1. It requests limited and punctual information (matrix with six control points for each quarter).
2. The E700 do not follow common control measures as to clearly characterise the in situ density and corresponding SP or Dr.
3. Reported Dr or SP values have some degree of uncertainty concerning the reference values used for compaction control.
4. The thickness of the compacted layer is not controlled.

For its part, current engineering practices use a limited number of reference samples (e.g., 3 to 5 control points are used for every 3000 m^2 or 50 m/L of the crest) and assume material homogeneity, albeit disregarding the thickness of the compacted layer.

Regarding the variability of deposited sand tailings, no standard methods are currently available in the country to control the degree of compaction efficiently, quickly, and economically, both on the surface and in-depth. Consequently, these factors and their effects on the actual mechanical behavior of this type of deposit remain unknown for noncontrolled areas due to limited checkpoints with limited space/time representativeness. The previous translates into uncertainties regarding the behavior of the tailings sands (contractive or dilative) and, consequently, the potential for liquefaction and slope instability of embankment.

2.1.2. Compaction Control Tests

The sand-cone method [7] and/or nuclear densimeter [16] are typically used to perform periodic compaction control of the sand layers deposited during the construction of an embankment of a classified sand tailings dam. However, the devices employed in these methods cannot be used to control the layer thickness. The test execution time of the sand-cone method limits the surface area that can be controlled. Ref. [4] addressed this issue by developing a method for compaction control in classified tailings dams in which the measuring instrument is a lightweight variable-energy dynamic penetrometer PANDA (Pénétromètre Autonome Numérique Dynamique Assisté par Ordinateur). The PANDA device is a 14-mm-wide rod with a metal cone at one end and a hammering head at the other end. The head is impacted by a standard 2-kg sledgehammer to drive the instrument into the soil (Figure 1). The cone area is 2 cm^2 in quality control (QC) mode and 4 cm^2 in quality assurance (QA) mode. The maximum penetration depths in QC and QA modes are 1.5 m and 6 m, respectively, in soils with a maximum size below 50 mm.

The impact of the hammer generates a stress wave in the hammering head that propagates towards the tip to drive the rod into the soil. The penetration (e, mm) and the cone resistance (q_d, MPa) for each blow are continuously recorded using the Dutch formula [25–28], assuming a perfectly inelastic impact between the hammer and the hammering head and neglecting the penetrometer and soil deformation.

The assumptions of the Dutch formula can be met during a PANDA test by controlling the driving energy to ensure that the penetration from a hammer blow ranges from 2 to 20 mm [29,30]. An almost continuous record of q_d v/s e is obtained. Therefore, the layer thickness can be estimated, and compaction problems can be identified.

During the PANDA test, a microprocessor receives a signal generated by the sensor in the cone and automatically records q_d and e for each hammer blow. The in-situ background noise of the penetrogram is eliminated by applying a signal filter for signal smoothing or regularization with a fixed 25 mm wide moving window. The penetrometer signal is compared with a preset control curve (e.g., [27,29,30]) in compaction control mode based on prior soil classification, by the American Association of State Highway and Transportation Officials (AASHTO) or the United States Geological Survey (USGS) systems, and given the compaction level and the moisture content (dry, medium, and wet) for the compaction process. As a reconnaissance tool in tailings deposits, the results obtained from the PANDA test have the following applications: estimation of geotechnical parameters in-depth, operational control of thickened tailings deposits, estimation of liquefaction potential (e.g., [27,29,30]).

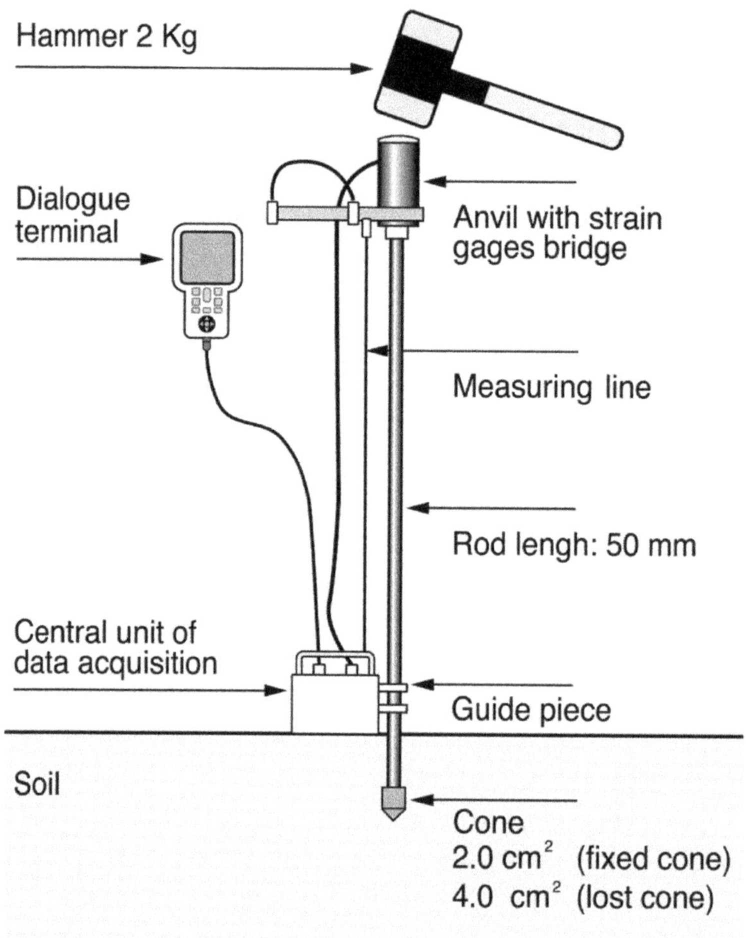

Figure 1. PANDA lightweight penetrometer scheme [18].

The existing correlation between q_d and γ_d of the soil depends on the nature and the water content (w_{nat}) of the material to be controlled. It is based on hypotheses derived from the micromechanics of granular media, which postulate that the characterization of the mechanical behavior of a material is carried out from the description of its microstructure [31], considering three families of data: nature of the grain, the interaction between two grains and fluids, set of grains. Based on this hypothesis, according to a micromechanical logic and within a geotechnical context [32,33] proposed the following classification: natural parameters (mineralogy, particle size distribution, and grain shape), natural parameters of the set of particles (γ_d or void ratio) and interaction parameters (w and Atterberg limits). This translates into a particular mechanical response for a type of soil. In this way, if the parameter to be measured in-situ is q_d, under the assumption of reversibility (unbreakable grains, resistance to wear, and suction reversibility) and knowing the nature and interaction parameters, it is possible to determine in-situ γ_d of the material.

The works carried out in [34,35], experimentally and analytically demonstrated (cavity expansion theory), that in granular soils it is possible to estimate γ_d from the point resistance (e.g., q_d) obtained from penetration tests. Although this function would not meet the limit conditions, regarding the state of minimum and maximum compactness, it is valid for the range of density that would have to be controlled in-situ (e.g., 95% SP in classified sand tailings).

2.2. Variability in the Geotechnical Properties of Sand Tailings

An extensive study on the variation in geotechnical properties of sand tailings is not available in the literature. A few studies have been conducted to establish the ranges of values selected index or shear strength properties (e.g., [36–39]). In [40] it was identified potential sources of variability in the physical and mechanical characteristics of the sand tailings. These sources were classified into 4 levels: These sources were classified into four levels: (i) level 1: Production size (mining scale): (ii) level 2: tailing manufacturing: crushing, pressing and flotation; (iii) level 3: deposition: cyclone, construction and compaction method (e.g., simple hydraulic deposition and mechanical compaction layer-by-layer); (iv) level 4: deposit age: aging effects, under progressive drying conditions over time. Two types of variability were identified: material, which is associated with the intrinsic properties of the tailings (e.g., mineralogy, particle size distribution and plasticity of the fraction of fines smaller than 75 μm), and structural, which is associated with the construction process used (e.g., γ_d and layer thickness). Figure 2 shows the variability of the particle size distribution (PSD) obtained as part of the operational control carried out in the embankment of sand tailings dams of medium copper mining in Chile. The embankment of a classified sand tailings dams exhibits both temporal and spatial variability. Temporal variability depends on material aging induced by physicochemical bonding between particles, gradual drying and changes in stress states, among others [20,41]. Table 1 presents a group of factors that generate material and structural variability in sand tailings on both spatial and temporal scales. Another factor to consider, which contributes to the variability of the q_d values, corresponds to the prospecting depth reached in the embankment, particularly if it is required to carry out a quality assurance using penetrometers (e.g., SPT, CPTu, Lightweight Penetrometer PANDA, among others), for depths greater than 1.5 (m) For shallower depths, the influence of depth is not considered on the values of q_d, within the context of compaction control [29] . Within this context [28], analyzed the effects of soil conditions (dry unit weight, overburden stress and water content of the surrounding soil, and fine content) on the cone resistance (q_d) of lightweight penetrometers (PANDA), from a series of tests in a calibration chamber. An overburden correction factor (C_N) for sand tailings was obtained as part of the results. This factor allows considering the influence of depth on the measurements of q_d.

Table 1. Types of variability, associated factors, effects, and variation in geotechnical properties sand tailings.

Level	Factor	Variability		Effects	Variation in Geothecnical Properties
		Type	Scale		
1	Production size (mining scale)	Material	Spacial	Degree of homogeneity or heterogeneity of whole tailings in terms of constituent residual minerals (mineralogical species and grade of the field of origin)	Particle size distribution (PSD) and specific gravity of solid particles (G)
2	Crushing, pressing and flotation	Material		Physical characteristics and incorporation of chemical reagents into whole tailings	
3	Deposition: cycloning, construction and operation method	Material and structural		Physical characteristics and geotechnical properties of sand tailings	PSD of sand tailings, in-situ dry unit weight and water content, shear strength parameters
4	Depostit age	Material	Temporal	Aging of sand tailings: generation of cohesion by case-hardening	Gradual increase in geotechnical properties and rigidity

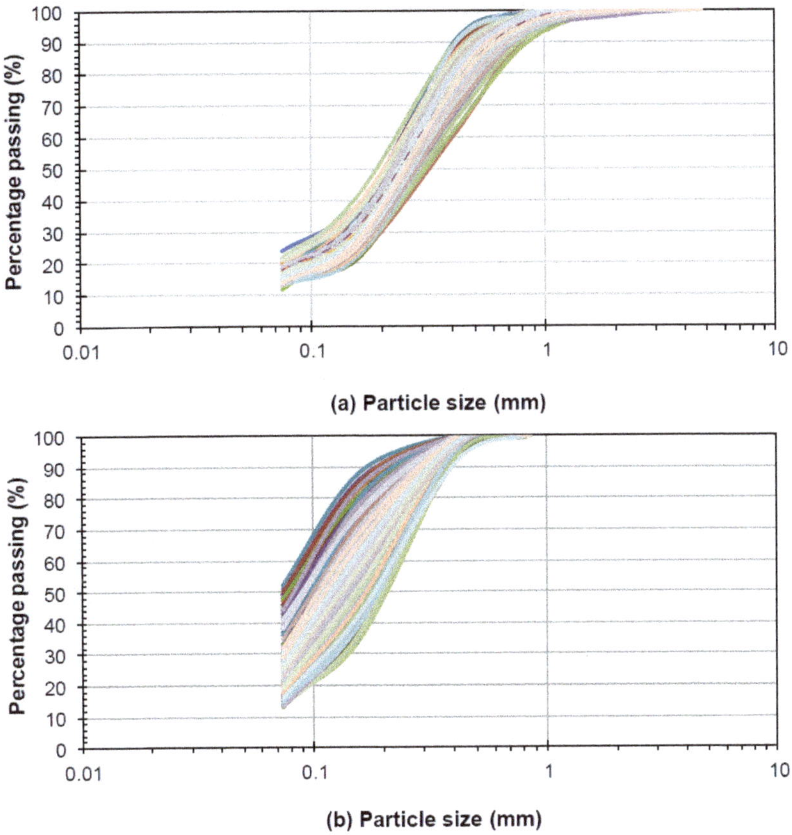

Figure 2. Variation of the particle size distribution (PSD). Operational control of two sand tailings dams of medium copper mining in Chile. (**a**) Sand tailings T1. (**b**) Sand tailings T2. (see Table 2).

Table 2. The construction method and technical specifications defined in the design project of different tailings dams of medium and large Chilean copper mining companies.

Sand Tailings Dam	Type	DS Slope V/H	Height m	Production of Sand Tailings	Construction of the DS Slope		
					Deposition and Distribution of Sand Tailings	Compaction	Specified Degree of Compactation
T1	DS and CL	1:3.5	30	Cycloned sand tailings derived from various deposits	From cyclones (underflow) in the crest of the deposit and levelling with a bulldozer	DS slope and crest in 50–60-cm layers in the direction of the slope by a 20-t bulldozer	95% SP
T2	CL	1:2.7	20			Crest in 20–40-cm layers in the horizontal direction by a 300-kg vibratory plate compactor	
T3	CL	1:3.5	60	Cycloned sand tailings derived from a single deposit		DS slope and crest in 30–40-cm layers in the slope direction by a 20-t vibratory roller	
T4	DS	1:2.5	40			DS slope and crest in 30–40-cm layers in the slope direction by an 18-t bulldozer	
T5	DS	1:2.5	40		From dump trucks and distribution with a bulldozer	DS slope and crest in 30–40-cm layers in the horizontal direction by a 20-t vibratory roller	95% Modified Proctor (MP)

Table 2. *Cont.*

Sand Tailings Dam	Type	DS Slope V/H	Height m	Production of Sand Tailings	Construction of the DS Slope		
					Deposition and Distribution of Sand Tailings	Compaction	Specified Degree of Compactation
T6	DS	1:3.0	150		From cyclones (underflow) in the crest of the deposit and leveling with a bulldozer	DS slope and crest in 50-cm layers in the slope direction by a 10-t vibratory roller	95% SP

3. Materials and Methods

3.1. Variation in q_d for Sand Tailings Dams

As an illustration, data from various PANDA experiments were collected and analyzed to show the variation in q_d for various classed sand tailings dams (Figure 3).The PANDA tests were performed for compaction control during the operation phase of six deposits located in Chile. The range of embankment heights from 20 to 150 m is representative of medium and large Chilean mining operations (Table 2).

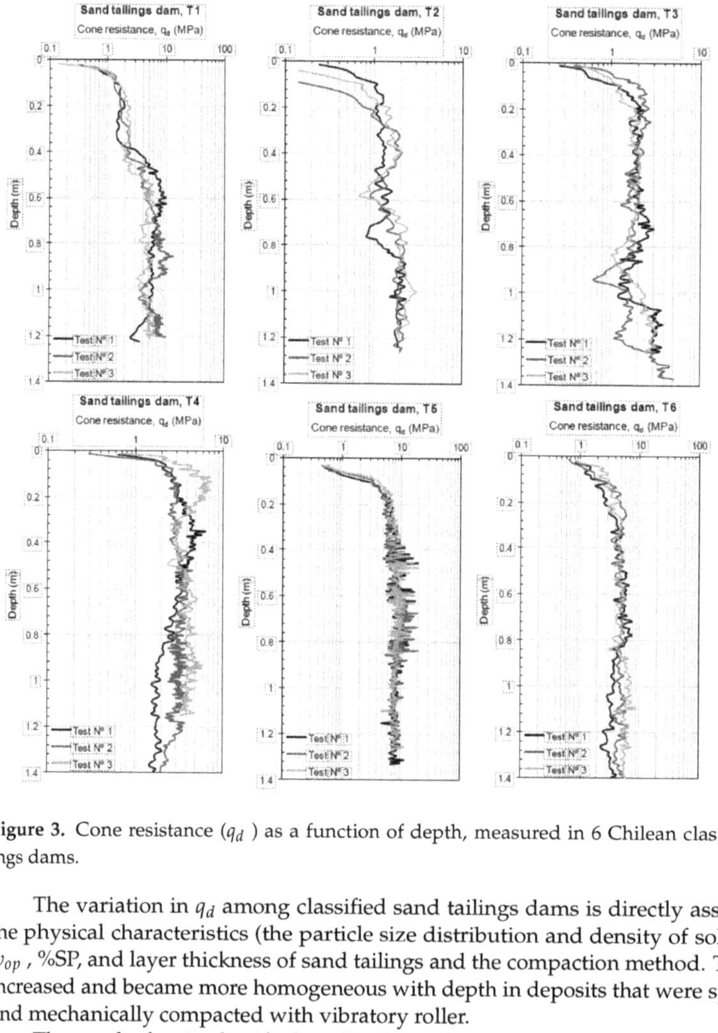

Figure 3. Cone resistance (q_d) as a function of depth, measured in 6 Chilean clasified sand tailings dams.

The variation in q_d among classified sand tailings dams is directly associated with the physical characteristics (the particle size distribution and density of solid particles), w_{op} , %SP, and layer thickness of sand tailings and the compaction method. The q_d values increased and became more homogeneous with depth in deposits that were systematically and mechanically compacted with vibratory roller.

The results for six classified sand tailings dams (Table 3) were statistically analyzed using the criteria of homogeneity or global heterogeneity defined by [42] as a function of the coefficient of variation (CV%). The embankment materials were consequently classified as nonplastic silty sands (SM). The specific gravity (G) of the sand tailings was homogeneous across different deposits. The FC varies among sand tailings from different fields, and the resulting mean heterogeneity is associated with changes in the production process that affect the mineralogical species and the grade of the extracted mineral.

Table 3. Geotechnical properties of mine tailings based on statistical analyses of experimental data from Control compaction or quality control (QC) for six Chilean sand tailings dams.

Geotech. Prop.	Dataset: Operational Control in Sand Tailings Dams																							
	T1 [a]				T2 [a]				T3 [a]				T4				T5				T6			
	Data	Av.	CV %	Var. Class	Data	Av.	CV %	Var. Class	Data	Av.	CV %	Var. Class	Data	Av.	CV %	Var. Class	Data	Var.	CV %	Var. Class	Data	Var.	CV %	Var. Class
G	108	3.09	4.6	Homo.	45	3.36	8	Homo.	40	3.07	2.2	Homo.		3.07	2.2		161	2.7	1.5		19	2.69	0.9	Homo.
D_{50} [mm]	3266	0.13	19.0	Homo.	262 [b]	0.11	15.2	Homo.	2958	0.25	8.7	Homo.	40	0.17	17.3		1206	0.3	7.8		44	0.32	32.2	Medium hetero.
FC [%]		28	28.7	Medium hetero.	262 [b]	33	26.3	Hetero. media		17	10			27	19.4			13	18.3			16	8.7	
γ_{dmax} [kN/m³]	392 [b]	18.2	6.2		262 [b]	20.8	8		495 [b]	18.5	2.3		8	18.2	5.1	Homo	17 [b]	18.2	5.1	Homo	23 [c]	16.7	1.0	
ω_{op} [%]		15.2	9.4		262 [b]	14.4	10.3	Homo		14.3	6.2	Homo		15.6	5.5			11.1	14.5			17.8	3.9	Homo.
γ_d [kN/m³]	3266	17.5	6.6	Homo	275 [b]	20.1	8.2		2958	18.1	2.9			19.1	2.3			17.4	3.0			16.0	1.1	
ω_{nat} [%]		11	22.3		275 [b]	3.3	43.1	Hetero. media		7.5	27.3	Medium hetero.	79	6.1	18.7		79	8.6	15.5		356	12.7	15.2	
γ_t [kN/m³]		19.4	6.7		275 [b]	20.7	8.2	Homo.		19.5	3.5	Homo.		20.3	2.6			18.8	2.5			18.1	2.2	
q_d [MPa]	275	4.8	50.6	Highly hetero.	75	2.87	45.9	Highly hetero.	100	1.95	52.8	Highly hetero.	75	2.71	42.4	Highly hetero	20	6.48	48.6	Highly hetero.	15	7.04	31.4	Medium hetero.

Note: Av: Average, CV %: coefficient of variation, D_{50}: median grain size, [a] Database from [30], [b] Standar Proctor test, [c] Modified Proctor Test.

The γ_{dmax}, and w_{op} values are case-dependent and vary with the mineralogical species, grade of the field of origin and production process. However, these parameters are homogeneous for the same deposit. In turn, the γ_t and γ_d can be controlled for the individual compacted layers from the surface and are classified as homogeneous for a single deposit. The w_{nat} can vary from medium heterogeneous to homogeneous, primarily depending on the time and climatic conditions (temperature and relative humidity) under which the test is performed. All classified sand tailings dams exhibit a highly heterogeneous variation in q_d with depth that is mainly generated by the multilayer structure created by the embankment construction process (deposition, distribution and compaction). Ref. [40] have reported that the index properties of the sand tailings can be described by a normal distribution, whereas q_d follows a log-normal distribution. Thus, the variability of sand tailings can be considered well defined and determined given the nature of the deposited materials and the structure resulting from the construction process.

The spatial structural variability in the embankment of a sand tailings dam is reflected in the variation in q_d at the surface and with depth. This variation reflects two key factors: the state of the structure between strata and the variability inherent to the compaction process (the machinery efficiency and compaction energy homogeneity). The variation in q_d also reflects changes in the w_{nat}, which is associated with the temporal variability of these materials.

The simplicity, speed and accuracy of the PANDA test facilitates the reconstruction of the multilayer structure along the depth of classified sand tailings dams during the construction phase. Correlating q_d with other geotechnical parameters, (e.g., the friction angle, compactness state and cyclical resistance) can be used to identify areas of low resistance and facilitate the assessment of the actual physical stability at different points of the structure and at any time during operation. Thus, the utility of geostatistics tools for the characterization and modeling of the spatial structure generated during operation of a sand tailings dam emerges within the context of routine compaction control.

3.2. Geostatistical Analysis for the Characterization of Structural Variability in a Classified Sand Tailings Dam

Geostatistics is generally applied to structural geotechnical data in the following stages [43]: (i) an exploratory database analysis is performed to determine basic statistical parameters (the mean, standard deviation, coefficient of variation and maximum and minimum values), identify outliers and calculate the data distribution; (ii) a structural data analysis is performed to determine the vertical and horizontal correlation distances that can be fitted by a theoretical semivariogram (spherical, exponential, Gaussian, cubic and linear, among others); (iii) estimation and simulation based on spatial correlation data and a theoretical model are performed to estimate (virtual) profiles of the geotechnical properties of interest and (iv) a 2-or 3-dimensional map is constructed to facilitate the interpretation of the numerical estimation and simulation results.

An experimental semivariogram (1) is defined as the semiaverage of the square differences between of all pairs of observations separated by a distance h (e.g., [44–48]).

$$\gamma(h) = \frac{1}{2N(h)} \Sigma_{i=1}^{N} [Z(x_i) - Z(x_i + h)]^2 \tag{1}$$

In the (1), $\gamma(h)$ is the semivariance estimated for a distance h; $N(h)$ is the number of pairs of observations separated by a distance h; $Z(x_i)$ is the value of the random variable at a point and x_i and $Z(x_i + h)$ is the value of the random variable at a distance h from point x_i.

A semivariogram model is only stationary if a Random Function is stationary, the variogram model must be of stationary type [8]. In [15] the authors have proposed that most soil science or geological engineering data are homogeneous and meet at least the criterion of weak stationarity.

After identifying and modeling the spatial structure of the observations, the Kriging stochastic interpolation procedure is applied [46]. The Kriging estimates a regionalized variable (q_d and the % SP) at a site without information through a linear combination of the information available from sampled sites. Each observed value is weighted based on the structure of the regionalized phenomenon (given by the variogram) and the geometric characteristics of the sampling (values and location), subject to a universal quantification procedure, as indicated in (2) and (3).

$$Z_v^* = \sum_{i=1}^{n} \lambda_i Z(x_i) \quad (2)$$

$$\sum_{i=1}^{n} \lambda_i = 1 \quad (3)$$

This methodology has been widely used to interpolate cartographic data. The main advantage of this methodology against classical interpolation methods (e.g., inverse distance weighting (IWD) is the determination of the minimum estimation error, given by the Kriging variance. The estimation variance is minimized to ensure the optimal use of the available information (e.g., [10,44,49,50]). The Stanford Geostatistical Modeling Software (SGeMS) was used as a calculation tool in the present study.

3.3. *Classified Sand Tailings Dam under Study*

The classified sand tailings dam investigated in this study is located at northern of Chile and stores waste generated from the flotation processing of copper sulfides from a single field. Therefore, the physical characteristics and mineralogical characteristics of the generated tailings can be considered homogeneous.

This deposit called T3 in the present study began operating at the beginning of the 21st century, storing 50 million tons of tailings. The dam was constructed using the CL method from a starter embankment built with loan material. The tailings are classified by placing hydrocyclones on the crest of the embankment. The classified sand tailings have deposition w_{nat} levels ranging from 28 to 30%. Subsequently, these tailings sands are distributed and compacted using the following methodology: (i) distribution and profiling of the tailings sands cycloned by a bulldozer after having obtained the optimum moisture content w_{op} Standard Proctor test until reaching the necessary layer thickness considering the percentage of swelling that it experiences the material. (ii) once the optimum humidity for compaction has been reached, compaction is carried out in layers of thickness equal to 0.4 m. on average, by round-trip cycles of a vibrating smooth roller of 20 t of static weight, carried out in the longitudinal direction of the slope. (iii) finally, a bulldozer pass is applied in order to roughen the surface of the deposited layer and thus mitigate the effects of wind erosion. The compaction control parameters defined in the design project are the following: FC = 12%, γ_{dmax} = 19.87 kN/m^3, w_{op} = 16.3% and % SP = 95%.

Field Study and Survey Grid

A field study of the classified sand tailings dam was performed in a 120-m-wide by 45-m-long study area of the DS of the recently compacted embankment, which was overlaid with a survey grid (Figure 4). The regionalized variable analyzed in this study was q_d, which was assessed as a function at different depths in each observation point.Note - the variable is a regionalised function of the field coordinates $q_d(x, y, z)$.

The grid was divided into 4 slabs on which 91 PANDA tests were performed in control compaction mode or quality control (QC) up to a depth of approximately 1.5 m. The grid points were arranged 7 m apart in the DS direction and 10 m apart in the direction parallel to the embankment crest. The regionalized variable analyzed in this study was q_d, which was assessed as a function of the 3D field coordinates point.

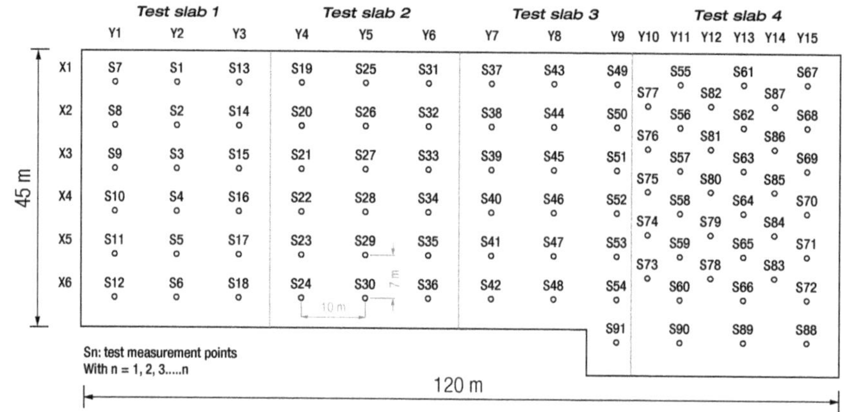

Figure 4. Geometry of the survey grid and PANDA test points. DS slope embankment of classified sand tailings dam T3.

4. Results and Discussion

This section has objective describes and analyzes the results. As a starting point, q_d is statistically analyzed. Subsequently, semivariograms are used to finally use ordinary kriging in order to model q_d, and estimate γ_d and % SP.

4.1. Physical Characteristics and Compaction Parameters of Classified Sand Tailings in the Study Area

The representativeness of the study area selected in the embankment classified sand tailings dams were assessed in terms of the physical characteristics and compaction parameters of the deposited sand tailings. Thus, a descriptive statistical analysis was carried out on data from compaction controls performed over the preceding 3 years and data collected in the area of the embankment with the survey grid (Table 4).

A preliminary analysis of the results shows that the study area is representative of the entire embankment in terms of the particle size distribution or PSD (FC and D_{50}), γ_{dmax}, ω_{op}, and γ_d of the classified sand tailings used to build the embankment. These results reflect material and structural (layer-by-layer) spatial homogeneity, which is primarily related to an invariant and well-controlled production process (pressing, flotation and cycloning) and a suitable construction and compaction method. The w_{nat} levels exhibit medium heterogeneity because this parameter depends on the time of and ambient temperatures during in situ sampling.

The representativeness of the study area with the survey grid was assessed using the Student's t-test and Bartlett's test to statistically analyze differences between means and variances at global and local scales. The same means and variances are found for each parameter (Tables 5 and 6).

Table 4. Summary of statistics for the total sample and the study area for the embankment and classified sand tailings dam T3.

Statistical Parameter	FC [%]		D_{50} [mm]		γ_{dmax} [kN/m^3]		ω_{op} [%]		γ_d [kN/m^3]		ω_{nat} [%]	
Sample	M A	M B	M A	M B	M A	M B	M A	M B	M A	M B	M A	M B
Data	3266	178	3266	178	495	35	495	35	3266	202	3266	202
Minimum Value	11	14	0.25	0.2	17.4	18.1	12.0	13.5	16.4	16.9	2.3	4.0
Maximum Value	24	21	0.46	0.28	19.8	19.5	16.3	16	19.8	19.8	19	14.3
Media	17	17	0.25	0.24	18.5	18.8	14.3	18.7	18.1	18.5	7.5	8.7
σ	1.7	1.5	0.022	0.016	0.43	0.36	0.89	0.98	0.52	0.61	2.05	2.68
CV%	10	8.9	8.7	6.6	2.3	1.9	6.2	5.2	2.9	3.3	27.3	30.8

Note: M A: total sample, M B: grid sector sample, σ: standard deviation.

Table 5. Results from the test of equality of means at global and local scales. Confidence level: 99.9%.

Parameter	Level	Data	Average	Student's *t*-Test Value	Confidence Interval Δ Means	Equality of Means
FC [%]	Global	3266	17	−2.8984	[−0.784, 0.049]	Yes
	Local	178	17			
D_{50} [mm]	Global	3266	0.25	6.507	[0, 0.018]	Yes
	Local	495	0.24			
γ_{dmax} [kN/m^3]	Global	495	18.5	−3.7449	[−0.053, 0.013]	Yes
	Local	35	18.8			
γ_d [kN/m^3]	Global	3266	18.1	−9.4948	[−0.532, 0.012]	Yes
	Local	178	18.5			

Table 6. Results from the test of equality of variances at global and local scales. Confidence level: 99.9%.

Parameter	Level	Data	Variance σ^2	Coefficient Between σ^2	σ^2 Variation	Equality of Variances
FC [%]	Global	3266	2.721	1.01	[0.755, 1.526]	Yes
	Local	178	2.474			
D_{50} [mm]	Global	3266	0.00048	1.92	[1.286, 2.661]	Yes
	Local	495	0.00025			
γ_{dmax} [kN/m^3]	Global	495	0.190	1.52	[0.579, 1.037]	Yes
	Local	35	0.125			
γ_d [kN/m^3]	Global	3266	0.281	0.76	[0.536, 1.037]	Yes
	Local	178	0.368			

4.2. Statistical Analysis of q_d

The geostatistical evaluation was performed in the following stages: (i) a descriptive statistical analysis is performed of the PANDA penetrograms recorded collected in the study area and at the local level for each test slab; (ii) experimental variograms are calculated in different directions (at global and local levels); (iii) a theoretical variogram model is fitted to the experimental variogramand iv) the OK technique is used to estimate the study variables (q_d and the % SP) [46]. An analysis of the frequency histograms shows that the q_d values assessed at the global level of the survey grid and for each test plan follow the typical distribution of the penetrograms recorded for the classified sand tailings dams (Figure 5). The assumption that the q_d values follow a log-normal distribution is thus verified (e.g., [51,52]). The q_d v/s e penetrograms were smoothed using a 2.5 cm moving average to remove background noise from the signal recorded in-situ from each trial PANDA performed on the survey grid. The statistical analysis of the q_d values is presented in Table 7.

A maximum value of 8 MPa was identified from the statistical analysis results for the smoothed variable q_d. The q_d values assessed above this threshold (0.88% sample) for the case under study would correspond to overcompaction near a 100% SP and some case-hardening resulting from the progressive drying that sand tailings are subjected to, particularly in northern Chilean deposits.

Table 7. Statistical summary of in-situ q_d measurements for the full survey grid and test slabs.

Sector	PANDA Tests	No. of Pairs [q_d(MPa), z(cm)]	Minimum and Maximum Values	Average	σ	CV%
Full grid	91	18,806	0.1 and 29.5	2.79	1.74	62.4
Slab N° 1	18	3850	0.14 and 22.7	2.96	1.88	63.5
Slab N° 2	18	4218	0.15 and 9.4	3.17	1.55	48.9
Slab N° 3	19	3728	0.19 and 29.5	3.24	1.96	60.5
Slab N° 4	36	7010	0.1 and 9.34	2.22	1.48	66.7

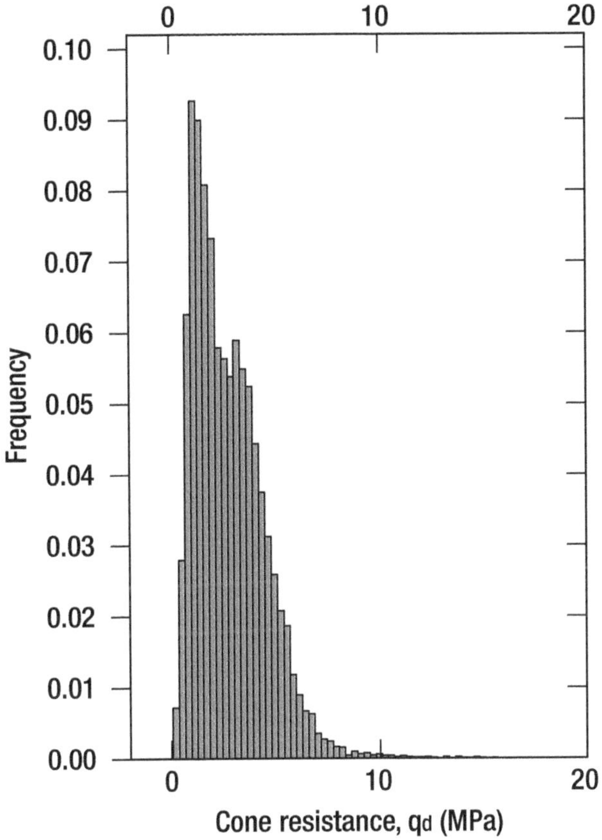

Figure 5. Histogram of q_d values. Full survey grid.

4.3. Experimental and Theoretical Semivariograms

4.3.1. Calculation Parameters

Calculation parameters were defined to obtain the experimental semivariograms of the survey grid both globally and for the test slab (Table 8). The anisotropy of the study area was analyzed considering different azimuth values in the xy-plane and in vertical direction (the z-axis). "Lag separation" values were defined considering the geometry of the survey grid and the PANDA test results, smoothed every 2.5 cm.

Table 8. Parameters of the experimental semivariograms of q_d at a global level using the survey grid and the local level using test slabs.

Scale	Direction of Analysis	Semivariogram Calculation Parameters					
		Lag Separation [m]	Lag Tolerance [m]	Azimuth [°]	Tolerance [°]	Dip [°]	Bandwith [°]
Survey grid	Horizontal x-y	7	3.5	0, 45, 90, 135	22.5, 90	0	5
	Vertical z	0.025	0.0125	0	22.5	90	3
Test slabs	Horizontal x-y	7	3.5	0, 45, 90, 135	22.5, 90	0	5
	Vertical z	0.0125	0.0125	0	22.5	90	3

4.3.2. Semivariogram Modeling

- Results obtained in the vertical direction

In the vertical direction, the 2.5 cm "lag separation" defined for the q_d values produces a virtually continuous experimental semivariogram that facilitates fitting the theoretical model for the full survey grid and each test slab. A Different spherical models are found to fit all the experimental results (Figures 6 and 7). This type of model reasonably accommodates strata or layered structures and is therefore widely used in several fields of geoscience (e.g., [13,46,53,54]).

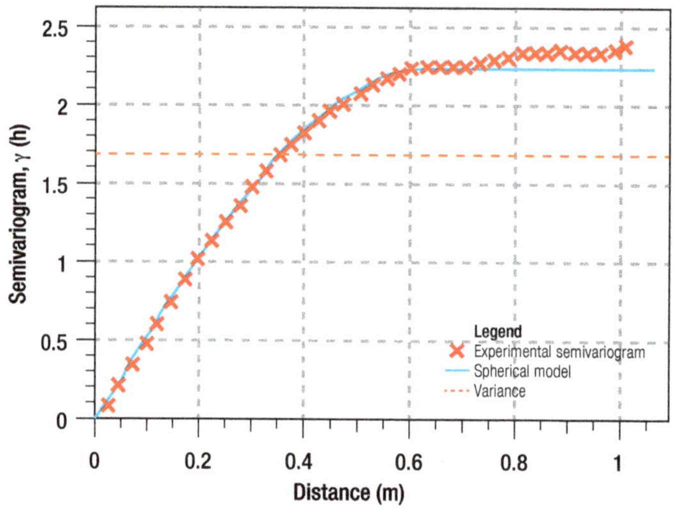

Figure 6. Experimental semivariogram and omnidirectional theoretical model for q_d in the vertical direction. Full survey grid.

Well-defined sills and ranges (Table 9) reflect the multilayer structure produced by the construction and compaction process. The range is approximately between 0.4 and 0.7 m and represents the global variability in the layer thickness. In turn, the increase in the experimental semivariogram values, put in evidence a different behaviour of the local variograms (in the different test slabs): quasi-stationary up to 0.4–0.6 m and not stationary or periodic for larger distances. A periodic behaviour of the experimental variogram reflects similar q_d values at distances approximately twice the theoretical range, especially for test slabs 1 and 2 [46]. The higher experimental semivariogram values for test slab 3 reflect a layered structure in which q_d gradually increases with depth. Note that test slab 4 does not have a well-defined sill over a range of approximately 1.2 m, which is near the maximum survey depth of 1.5 m reached in the PANDA tests. Consequently, the multilayer structure cannot be clearly identified for this specific area. This result may be associated with deviations in the method usually used to build the embankment, especially because the deposited layers were considerably thicker than those specified in the project (theoretical thickness: 0.5 m).

- Results obtained in the horizontal direction

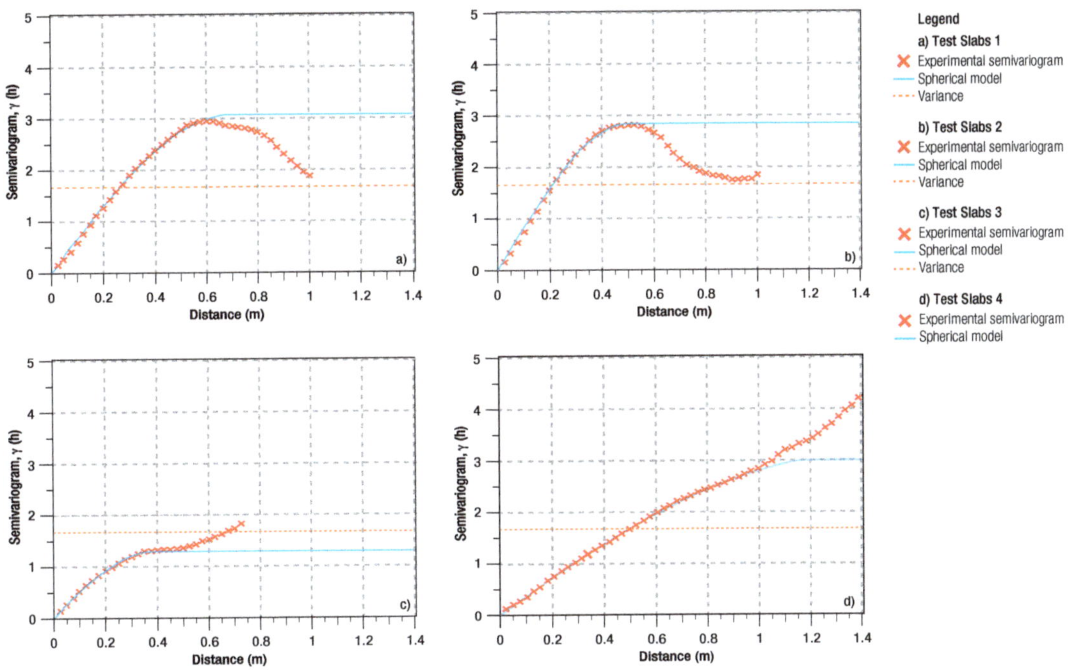

Figure 7. Experimental semivariograms and theoretical models for q_d. Vertical analysis plane.

Table 9. Experimental results for vertical and horizontal semivariograms of q_d based on the full survey grid and test slabs.

Scale	Plane of Analysis	Estimate Semivariogram					
		Azimuth/Inc [°]	Structure	Model	Nugget, Co[MPa]^2	Sill Co + C [MPa]^2	Range [m]
Survey grid	Vertical z	0°/−90°	Single	Spherical	0	2.23	0.6
	Horizontal x-y	0°/0°	Double	Spherical 1	0	1.67	8.9
				Spherical 2	0	0.38	39.9
		45°/0°		Spherical 1	0	1.62	8.6
				Spherical 2	0	0.40	36.1
		90°/0°		Spherical 1	0	1.61	8.2
				Spherical 2	0	0.40	35.1
		135°/0°		Spherical 1	0	1.62	7.9
				Spherical 2	0	0.42	33.7
Test Slab 1	Vertical z	0°/−90°	Single	Spherical	0	3.08	0.7
Test Slab 2					0	2.85	0.5
Test Slab 3					0	1.21	0.4
Test Slab 4					0	2.98	1.2

A theoretical semivariogram model in the horizontal plane of the embankment was designed at the level of the full survey grid by analyzing the experimental semivariograms in different directions or azimuths (Table 9). Experimental semivariograms in the horizontal direction could not be generated for the test slab because of an insufficient number of points in the xy-plane. The semivariograms for the full survey grid exhibited some anisotropy in the xy-plane to the downstream talus of the retaining dyke (Figure 8). The best fit in the horizontal direction was obtained using two nested spherical models [55] with a null nugget justified by the small scale structure shown by the vertical variogram. The ranges of the first and second structures were 7.9 to 9 m and 34 to 40 m, respectively. A comparative analysis of the experimental variograms results in the different directions on xy-plane show some of anisotropy, because globally his behavior they do not differ from each other in the analyzed directions. Considering the ranges variation between vertical and horizontal, a geometric anisotropy can be considered in the study area and therefore representative of the entire T3 embankment sand tailings dam, because the construction process is globally repetitive.

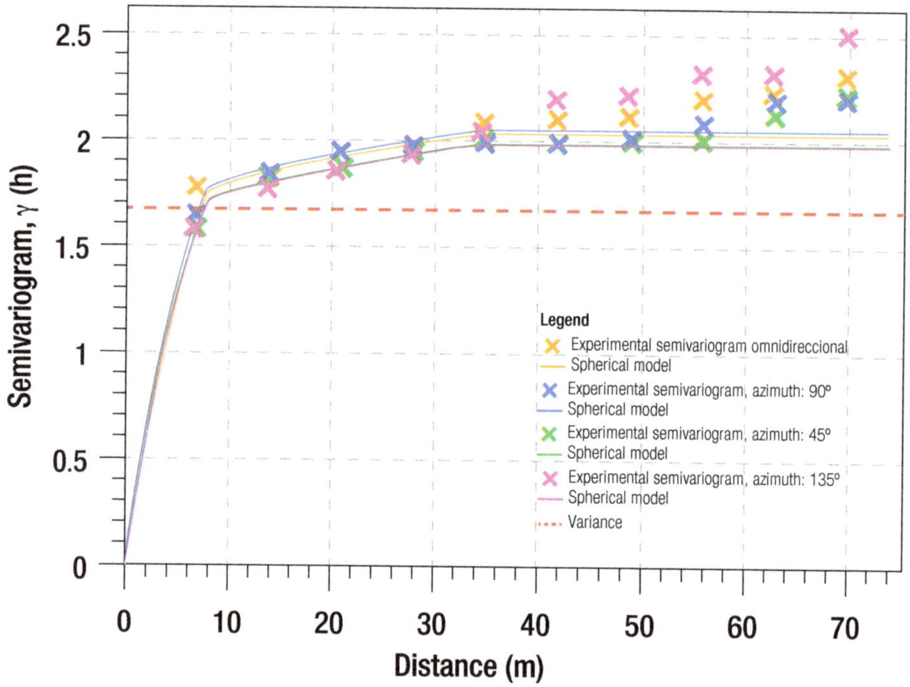

Figure 8. Experimental semivariograms and theoretical models for q_d. Horizontal analysis plane.

From the analysis of the results presented in (Table 9), it is possible to consider in a general way that the range in the vertical direction is equal to 0.6 m, while in the horizontal direction two structures at about 8 m and 33–40 m. This allows defining a global spherical variogram model for the survey grid. This variogram model (Equation (4)) is made up of three spherical structures with sills of 0.21, 0.40, and 1.62, horizontal ranges of 8.4, 36, and 40 m, and a vertical range of 0.6 m. In this case, there is no nugget effect.

$$\gamma(h) = 1.62sph(8.4, 8.4, 0.6) + 0.4sph(36, 36, 0.6) + 0.21sph(40, 40, 0.6) \tag{4}$$

4.4. Using the Ordinary Kriging (OK) Method to Estimate q_d

The Ordinary kriging (OK) method was performed on a parallelepiped equals to the extension of study volume, subdivided in 50 cm × 50 cm × 2.5 cm blocks or cells along the x, y and z directions, respectively, using a search ellipsoid based on the dimensions of the survey grid (120 m × 45 m). A the OK with a spherical variogram model has been utilised to estimate the q_d variable according to Equation (4). The results indicate that a 95% of the estimated q_d values are equal to or lower than 4.5 MPa. This value has been used as the upper limit to facilitate graphical representation (Figure 9). Figures 10 and 11 show longitudinal profiles (xz-plane) for the first and second third of the survey grid and medium cross-sections (yz-plane) for each test slab in terms of q_d. The following conclusions are drawn.

- The OK estimates of q_d generally reflect an anisotropic internal structure for the embankment corresponding to layers with variable thicknesses. The q_d values decrease towards the crest and increase in the downstream direction towards the foot of the embankment. These results reflect the spatial structural variability produced by the compaction process.
- All the longitudinal and cross-sectional profiles show a clear decrease in q_d at the beginning of test slab 4, which is associated with an increase in the layer thickness of the material with low q_d deposited from 0.6 to 1.2 m and matches the range assessed using the spherical model for the experimental semivariogram. This is justified by the different behaviour of the experimental vertical variogram of slab 4.

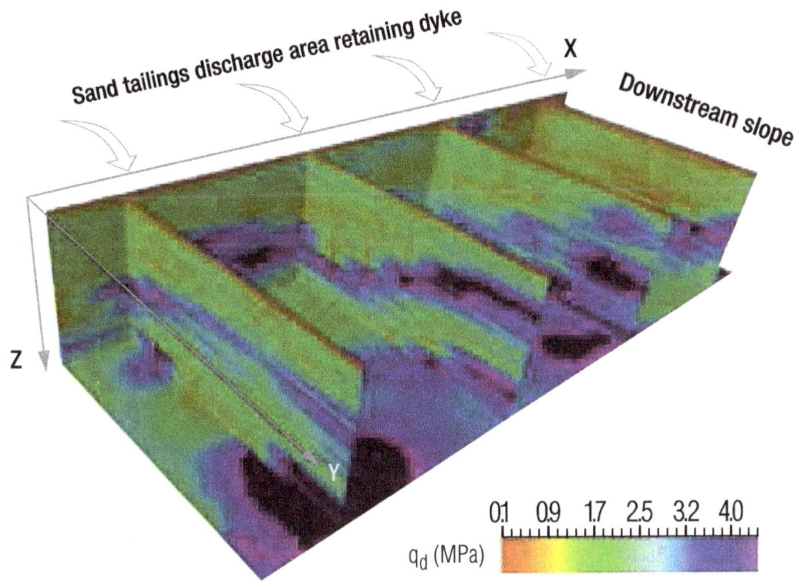

Figure 9. OK estimation of q_d. Full survey grid.

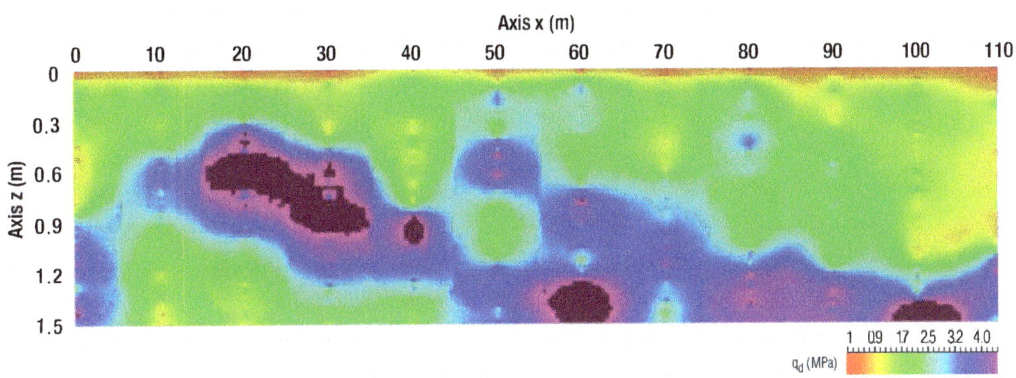

Figure 10. OK estimation of q_d. xz-plane longitudinal profile. Survey grid: First third, near the embankment dam crest.

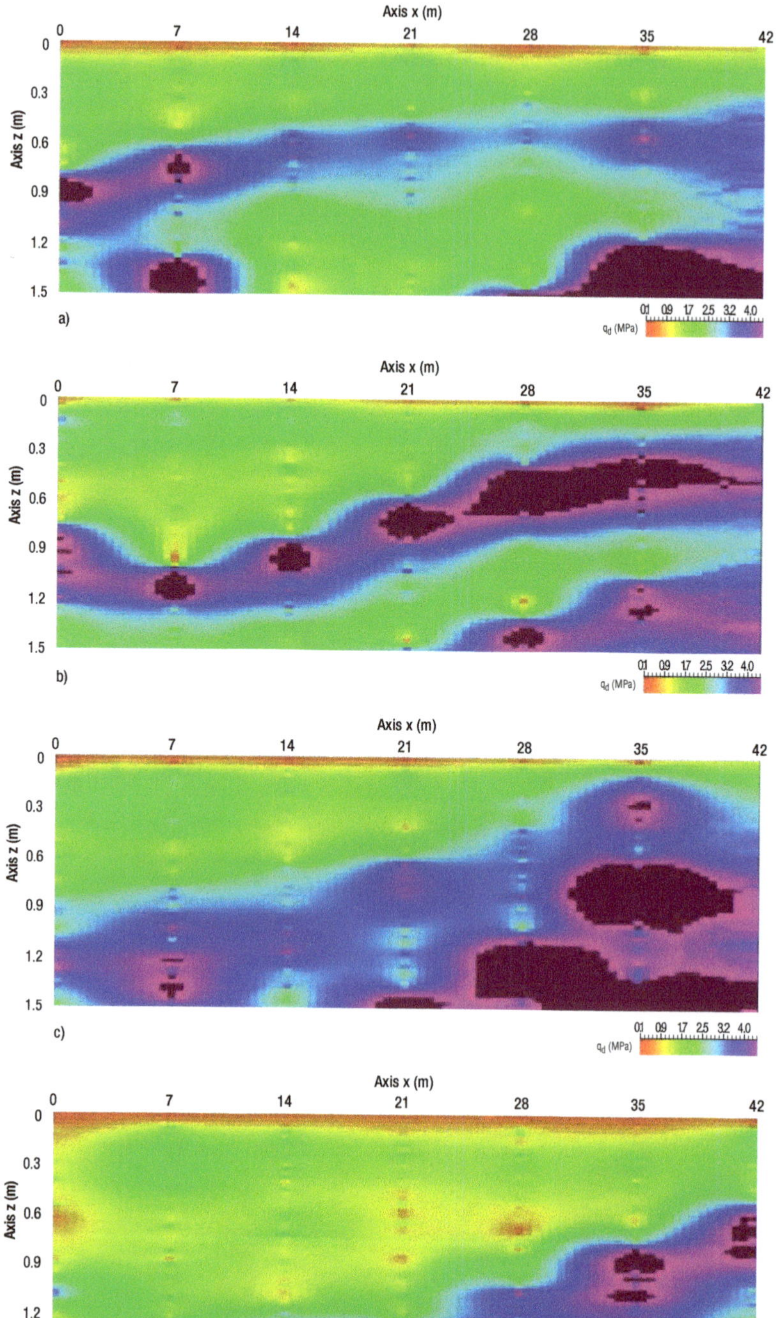

Figure 11. OK estimation of q_d. yz-plane cross-sectional profile. (**a**) Test slab 1. (**b**) Test slab 2. (**c**) Test slab 3. (**d**) Test slab 4.

To compare the estimated q_d results with the results for the grid measured points and to validate the quality of the selected theoretical semivariogram model, a cross validation was performed. For this purpose, test slab 4 which has the highest number of PANDA tests points surveyed (35), was selected. Of the total, 21 PANDA tests were used, with the aim of calculating the following theoretical semivariogram parameters (Spherical model): Nugget = 0 [MPa]^2; Sill = 2.86 [MPa]^2; Range = 1 m. These parameters were used to perform the estimation by OK, considering the following search parameters: (i) ellipsoid dimension: x = 50 m, y = 50 m, z = 1 m; (ii) points used in the estimation: 35 to 50 and 100 to 200. The OK results indicate a good fit between the interpolated and real values, with coefficients of determination (R^2) of 0.82 and 0.86 for each penetrogram (Figure 12).

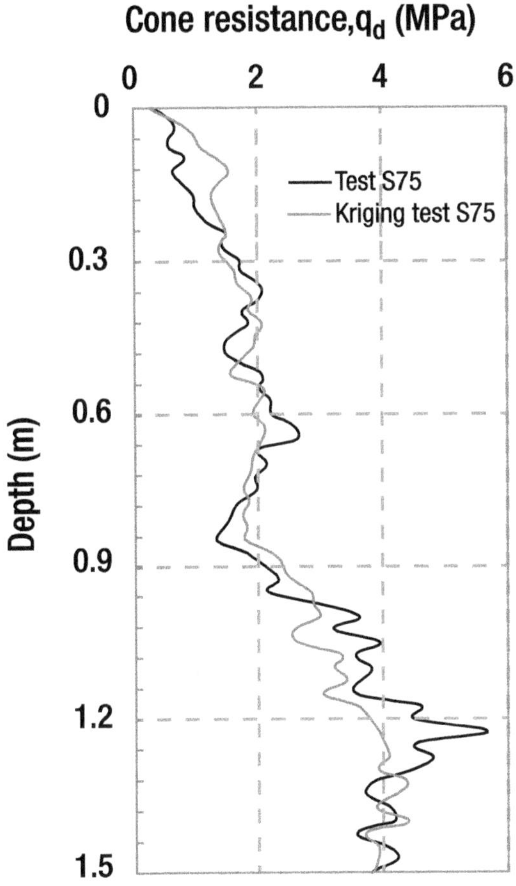

Figure 12. OK estimation of qd. yz-plane cross-sectional profile. PANDA test 75. Test slab 4.

4.5. Estimate of γ_d and % SP

Ref. [30] conducted experimental tests on sand tailings dams derived from copper extraction and used the results to correlate γ_d and q_d. Thus, the experimental procedure proposed by [27] was used to determine the q_d/γ_d ratio for the case study case, as shown in Equation (4). Where q_d is expressed as Mpa.

$$\gamma_d = 1.11ln(q_d) + 17.25(\text{kN/m}^3) \quad (5)$$

A good fit is obtained between γ_d estimated from q_d and γ_d determined using the sand-cone method shows ($R^2 = 0.92$), which validates the correlation found in the present study. Vertical and horizontal theoretical semivariograms of γ_d estimated from q_d and the % SP compaction were derived for the full survey grid with a null nugget effect (Co = 0) and identical ranges (Table 10). The following conclusions are drawn from the results for γ_d and the % SP compaction estimated using the OK method (Figure 13).

Table 10. Results for γ_d and %SP semivariogram modeling in vertical and horizontal directions for full survey grid.

Parameter	Plane of Analysis	Theoretical Model	Sill, Co + C	Range, [m]
γ_d	Vertical z	Spherical	0.12 $[kN/m^3]^2$	0.7
	Horizontal x–y	Spherical 1	0.11 $[kN/m^3]^2$	5.2
		Spherical 2	0.02 $[kN/m^3]^2$	40.7
% SP	Vertical z	Spherical	35.82 $[\%]^2$	0.7
		Spherical 1	31.18 $[\%]^2$	5.2
		Spherical 2	5.54 $[\%]^2$	40.7

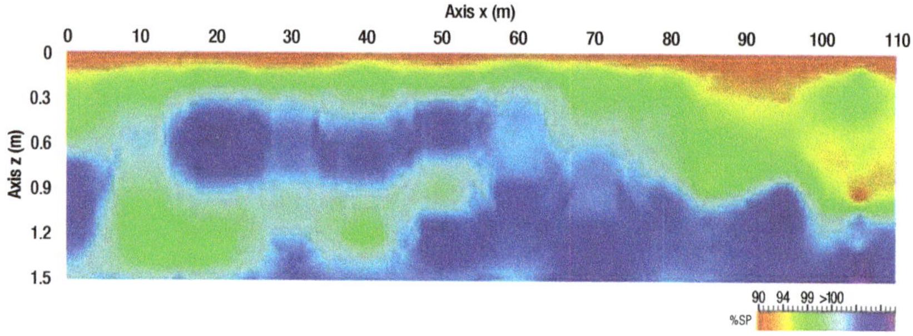

Figure 13. % SP compaction percentages estimated by OK. x-z plane longitudinal profile. Survey grid: second third, near the foot of the embankment.

- The sand tailings mechanically compacted into layers exhibit in situ structural variability on a spatial scale directly associated with the construction and compaction process.
- The estimated γ_d values range from 17 to 21.5 (kN/m^3), and the % SP compaction ranges from 90 % to 100 %. These ranges are similar to the statistical analysis results for the data obtained from routine layer-by-layer compaction control.
- The models have direct practical applications for performing post-construction control of an embankment of classified sand talings dams to identify areas with weak resistance or compaction percentages below the project design values (SP $\geq$ 95%) and optimize the compaction process.

5. Conclusions

The application of simple geostatistical techniques to the compaction control of classified sand tailings dams by the PANDA test provides an efficient and a rapid alternative to estimate spatially the variation in the degree of compaction with depth and layer thickness, considering the material (production sand tailings), and structural variability (deposition, distribution, and compaction) of an embankment, a relevant aspect not considered in current practice, for technical and economic reasons. In this way, it is possible to perform QC checks layers recently compacted, and QA of deep layers.

The incorporation of the material and structural variability in the compaction control together with the incorporation of geostatistical models as a tool for geotechnical monitoring, constitutes an important aid for the decision-making of the operators in charge of the construction of classified sand tailings dams, periodically optimizing the compaction process in terms of the thickness of the layer a compacting and the compaction energy to be applied (type of machinery to be used and a number of operating hours). The combination of compaction control using the Lightweight PANDA Penetrometer and a simple geostatistical model allow for quick and efficient verification of the degree compaction (Dr or SP), both on the surface and in sand depth. In this way, it is possible to identify the zones that present a dilatant behavior (physically stable) in the embankment and those that show a contractive behavior with low compaction whose risk of liquefaction and slope instability in seismic conditions must be evaluated. The structural analysis has allowed having an idea of the anisotropic structure of the classified sand tailings dams, and to estimate the thickness of the internal layers, a fundamental variable for controlling the compaction of this type of deposit not measured with the commonly used control techniques.

A structural analysis of the variable q_d shows that the experimental semivariograms analyzed in different directions of the deposit can be fit using a model of three nested spherical structures. The assessed range of 0.4 to 0.7 m in the vertical direction (z-plane) is directly associated with the theoretical thickness of the compacted layers in the project (0.4 m). This range highlights the spatial structural variability of the layer thickness of the dam. This range is a key parameter for physical stability, which is typically inadequately controlled during the compaction process. The horizontal semivariograms (xy-plane) obtained in different directions (azimuth: 0°, 45°, 90°, and 135°) do not differ from each other globally and therefore do not exhibit anisotropy in the xy-plane. The second spherical model ranges from 34 to 40 m and is related to the dimensions of the test slabs composing the survey grid defined in the study area of the deposit. Confirmation of the xy-plane ranges for other classified sand tailings dams could provide an order of magnitude estimate to significantly optimize routine compaction controls that are currently performed based on only empirical criteria.

The OK estimates of the q_d, γ_d and % SP variables enable the location of both weak areas with a high potential of mechanical instability and overcompacted areas, thereby facilitating evidence-based decision-making for improving the construction and compaction process.

In turn, the values determined by OK constitute an important database. Under properly recording and management procedures, these data could be processed using artificial intelligence methods (e.g., artificial neural networks) to continuously monitoring the variables that define the physical stability of these deposits over time and detect early deviations from the project design. The estimation by OK of the variable q_d, and the incorporation of the relation q_d/γ_d allows studying the degree of compaction of a certain area, thus being able to better identify the existence of weak areas in the embankment

Classified sand tailings dams typically have large volumetric and geometric dimensions. Thus, the size of the surface over which a given geotechnical variable of interest is estimated by OK could be increased by integrating the results from different survey methods or in situ tests (e.g., geotechnical surveys with SPT and CPTu penetration tests and near-surface geophysical methods, such as multichannel analysis of surface waves (MASW) and Microtremor Array Measurements (MAM). A common indicator for assessing failure mechanisms (e.g., a safety factor to assess the liquefaction potential) could be developed as a tool to periodically monitor the physical stability of sand tailings dams over time during operation. Although the results obtained represent the internal structure of the embankment of a classified sand tailings dams and constitute a great help for the decision-making of the operators responsible for the operation and construction, with the objective of evaluating possible improvements in the models obtained, such as future works are proposed: (i) analyze the multi-layer structure considering a zonal anisotropy;

(ii) to perform a more robust cross-validation that validates the OK models, exploit the variance of the point-by-point estimate.

Author Contributions: Conceptualization, G.V. and C.B.; methodology, C.B., G.V. and A.C.; software, A.C. and M.S.; validation, G.S. and J.G.; formal analysis, A.C., G.V. and P.V. and J.P.; investigation, C.B., P.V., J.P. and J.G.; writing—original draft preparation, G.V. C.B., J.G. and G.S.; writing—review and editing, G.V., C.B., J.P., J.G. and P.V.; visualization, G.V. and J.G. All authors have read and agreed to the published version of the manuscript.

Funding: This research was funded by DI Interdisciplinaria Pontificia Universidad Católica de Valparaíso (PUCV), grant number 14ENI2-26905: "Inteligencia Artificial para el Monitoreo de la estabilidad de Depósitos de Relaves".

Data Availability Statement: Not applicable.

Acknowledgments: Our sincere thanks and forever appreciation to Claude Bacconnet †, for his relevant contributions in the discipline of geotechnics, in particular to the mechanics of granular media and spatial variability of geotechnical parameters.

Conflicts of Interest: The authors declare no conflict of interest.

References

1. Cochilco. *Proyección de la Producción de Cobre en Chile 2018–2029*; Cochilco: Santiago, Chile, 2019.
2. SERNAGEOMIN. Datos Públicos Depósito de Relaves. Catastro de Depósitos de Relaves en Chile. 2020. Available online: https://www.sernageomin.cl/datos-publicos-deposito-de-relaves/ (accessed on 5 November 2022).
3. Villavicencio, G.; Espinace, R.; Palma, J.; Fourie, A.; Valenzuela, P. Failures of sand tailings dams in a highly seismic country. *Can. Geotech. J.* **2014**, *51*, 449–464. 10.1139/cgj-2013-0142. [CrossRef]
4. Villavicencio, G.; Breul, P.; Bacconnet, C.; Fourie, A.; Espinace, R.A. Liquefaction potential of sand tailings dams evaluated using a probabilistic interpretation of estimated in-situ relative density. *Rev. Constr.* **2016**, *15*, 9–18. [CrossRef]
5. *NF P94-105*; Grounds: Investigation and Testing Measuring Compaction Quality. Method Using Variable Energy Dynamic Penetrometer. Penetrometer Calibration Principle and Method. Processing Results. Interpretation. AFNOR.M: La Plaine Saint-Denis, France, 2012.
6. *Nch 3261-12*; Tailings Deposits-Control of Compaction with Light Dynamic Penetrometer; INN: Santiago, Chile, 2012.
7. *ASTM D1556/D1556M-15e1*; Standard Test Method for Density and Unit Weight of Soil in Place by Sand-Cone Method. ASTM: West Conshohocken, PA, USA, 2015.
8. Vennapusa, P.K.R.; White, D.J.; Morris, M.D. Geostatistical analysis for spatially referenced roller-integrated compaction measurements. *J. Geotech. Geoenviron. Eng.* **2010**, *136*, 813–822. [CrossRef]
9. Soulié, M.; Montes, P.; Silvestri, V. Modelling spatial variability of soil parameters. *Can. Geotech. J.* **1990**, *27*, 617–630. [CrossRef]
10. Sitharam, T.G.; Samui, P. Geostatistical modelling of spatial and depth variability of SPT data for Bangalore. *Geomech. Geoengin.* **2007**, *2*, 307–316. [CrossRef]
11. Altun, S.; Göktepe, A.; Sezer, A. Geostatistical interpolation for modelling SPT data in northern Izmir. *Sadhana* **2013**, *38*, 1451–1468. [CrossRef]
12. Costa, Y.; Cunha, E.; Costa, C.; Pereira, A. Correlations between SPT and CPT data for a sedimentary tropical silty sand deposit in Brazil. In *Geotechnical and Geophysical Site Characterisation 5*; Lehane, B.M., Acosta-Martínez, H.E., Kelly, R., Eds.; Australian Geomechanics Society; Sydney, Australia, 2016; pp. 407–412.
13. Fisonga, M.; Wang, F.; Mutambo, V. The estimation of sampling density in improving geostatistical prediction for geotechnical characterization. *Int. J. Geotech. Eng.* **2018**, *15*, 724-731. [CrossRef]
14. Kim, H.-S.; Kim, H.-K. Optimizing site-specific geostatistics to improve geotechnical spatial information in Seoul, Korea. *Arab. J. Geosci.* **2019**, *12*, 104. [CrossRef]
15. Kim, M.; Kim, H.-S.; Chung, C.-K. A Three-Dimensional geotechnical spatial modeling method for borehole dataset using optimization of geostatistical approaches. *KSCE J. Civ. Eng.* **2020**, *24*, 778–793. [CrossRef]
16. *ASTM D2922-05*; Standard Test Methods for Density of Soil and Soil-Aggregate in Place by Nuclear Methods. ASTM: West Conshohocken, PA, USA, 2005.
17. Valenzuela, L. Design, construction, operation and the effect of fines content and permeability on the seismic performance of tailings sand dams in Chile. *Obras Proy.* **2016**, *19*, 6–22. [CrossRef]
18. Verdugo, R. Compactación de relaves. In Proceedings of the IV Chilean Congress of Geotechnical Engineering, Valparaíso, Chile, 1997.
19. Barrera, S.; Valenzuela, L.; Campaña, J. Sand tailings dams: Design,construction and operation. In *Tailings and Mine Waste 2011*; Academic Press: Vancouver, BC, Canada, 2011; p. 13.
20. Troncoso, J. *Envejecimiento y Estabilidad síSmica de un Depósito de Residuos Minerales en Condición de Abandono*; Apuntes de Ingeniería: Santiago, Chile, 1986; pp.147–158.

21. Bhanbhro, R. Mechanical Behavior of Tailings: Laboratory Tests from a Swedish Tailings Dam. Ph.D. Thesis, Luleå University of Technology, Luleå, Sweden, 2017.
22. Martin, T.E.; McRoberts, E.C. Some considerations in the stability analysis of upstream tailings dams. In Proceedings of the Sixth International Conference on Tailings And Mine Waste, Fort Collins, CO, USA, 24–27 January 1999; AA Balkema: Rotterdam, The Netherlands, 1999; Volume 99, pp. 287–302.
23. Verdugo, R. Seismic performance of slopes and earth and tailings dams (2010 Maule Earthquake). In Proceedings of the Fifth International Conference on Geotechnical Earthquake Engineering (5-ICEGE), Santiago, Chile, 10–13 January 2011.
24. Sernageomin. Guía Metodológica para Evaluación de la Estabilidad Física de Instalaciones Mineras Remanentes. 2018. Available online: https://www.sernageomin.cl/wp-content/uploads/2019/06/GUIA-METODOLOGICA.pdf (accessed on 3 June 2022).
25. Cassan, M. *Les Essais In Situ en Mécanique des Sols. Volume 1: Réalisation et Interpretation*, 2nd ed.; Eyrolles: Paris, France, 1988.
26. Gourvès, R.; Barjot, R. The Panda ultralight dynamic penetrometer. In Proceedings of the 11th European Conference on Soil Mechanics Foundation, San Francisco, CA, USA, 12–16 August 1985 ; Danish Geotechnical Society: Copenhagen, Denmark, 1995; pp. 83–88.
27. Benz, M.A. *Mesures Dynamiques lors du Battage du Pénétromètre Panda 2*; University Blaise Pascal: Clermont Ferrand, France, 2009.
28. Villavicencio, G.; Suazo, G.; Zúñiga, R.; Valenzuela, P. Effects of Soil Conditions on the Cone Resistance of Lightweight Penetrometers. *J. Geotech. Geoenviron. Eng.* **2021**, *147*, 04021049. [CrossRef]
29. Chaigneau, L. *Caractérisation des Mileux Granulaires de Surface à l'Aide d'un Pénéromètre*; Université Blaise Pascal: Clermont Ferrand, France, 2001.
30. Villavicencio, G.; Breul, P.; Espinace, R.; Valenzuela, P. Control de compactación con penetrómetros ligeros en tranques de relaves, considerando su variabilidad material y estructural. *Rev. Constr.* **2012**, *11*, 119–133. [CrossRef]
31. Cambou, B. *Mécanique des Milieux Granulaires: L'Approche Microstructurale. Rhéologie des Géomatériaux*; Darve, F., Ed.; Presse E.N.P.C: Paris, France 1987; pp. 261–278.
32. Biarez, J.; Favre J.L. *Statistical Estimation and Extrapolation from Observations. Reports of Organisers*; IX ICSMFE: Tokyo, Japan, 1977; Volume 3, pp. 505–509.
33. Favre, J.L. Milieu Continu et Milieu Discontinu: Mesure Statistique Indirecte des Paramètres rhéOlogiques et Approche Probabiliste de la Sécurité. Ph.D. Thesis, Université Pierre et Maire Curie, Paris, France, 1980.
34. Salgado, R.; Mitchell, J.K.; Jamiolkowski, M. Cavity expansion and penetration resistance in sand. *J. Geotech. Geoenviron. Eng.* **1997**, *123*, 878–888. [CrossRef]
35. Rahim, A.; Prasad, S.Y, George, K. Dynamic Cone Penetration Resistance of Soils-Theory and Evaluation. In Proceedings of the Geo-Trans 2004 Conference, Los Angeles, CA, USA, 27–31 July 2004.
36. Arnaouti, S.; Angelides, D.; Chatzigogos, T.; Pytel, W. Variability of Soil Strength Parameters and its Effect on the Slope Stability of the Želazny Most Tailing Dam. *Int. J. Geol. Environ. Eng.* **2012**, *6*, 415–421. [CrossRef]
37. Hamade, T.; Mitri, H. Reliability-based approach to the geotechnical design of tailings dams. *Int. J. Mining Reclam. Environ.* **2013**, *27*, 377–392. [CrossRef]
38. Bhanbhro, R. *Mechanical Properties of Tailings: Basic Description of a Tailings Material from Sweden*; Luleå University of Technology: Luleå, Sweden, 2014.
39. Hu, L.; Wu, H.; Zhang, L.; Zhang, P.; Wen, Q. Geotechnical properties of mine tailings. *J. Mater. Civ. Eng.* **2017**, *29*, 04016220. [CrossRef]
40. Villavicencio, A.G.; Breul, P.; Bacconnet, C.; Boissier, D.; Espinace, A.R. Estimation of the Variability of Tailings Dams Properties in Order to Perform Probabilistic Assessment. *Geotech. Geol. Eng.* **2011**, *29*, 1073–1084. [CrossRef]
41. Troncoso, J.; Garcés, E. Ageing effects in the shear modulus of soils. *Soil. Dyn. Earthq.* **2000**, *19*, 595–601. [CrossRef]
42. Paikowsky, S.; Birgisson, B.; McVay, M.; Nguyen, T.; Kuo, C.; Baecher, G.; Ayyub, B.; Stenersen, K.; OíMalley, K.; Chernauskas, L.; et al. *Load and Resistance Factor Design (LRFD) for Deep Foundations*; Transportation Research Board: Washington, DC, USA, 2004.
43. Juárez-Camarena, M.; Auvinet-Guichard, G.; Méndez-Sánchez, E. Geotechnical zoning of Mexico Valley subsoil. *Ing. Investig. Tecnol.* **2016**, *17*, 297–308. [CrossRef]
44. Journel, A.; Huijbregts, C. *Mining Geostatistics*; Academic Press: London, UK; New York, NY, USA, 1978.
45. Matheron, G. *Traité de Geostatisque Appliquée, Volumen 1. Mémories du BRGM, 14*; Technip: Paris, France, 1962.
46. Matheron, G. *The Theory of Regionalized Variables and its Applications. Volumen 5. Les Cahiers du Centre de Morphologie Mathématique de Fontainebleau*; École Nationale Supérieure des Mines: Paris, France, 1971.
47. Baczkowski, A.J.; Clark, I. Practical Geostatistics. *J. R. Stat. Soc. Ser.* **1981**, *144*, 537. [CrossRef]
48. Hohn, M.E. An Introduction to Applied Geostatistics. *Comput. Geosci.* **1991**, *17*, 471–473. [CrossRef]
49. Burgess, T.M.; Webster, R. Optimal interpolation and isarithmic mapping of soil properties. I. The semi-variogram and punctual kriging. *J. Soil Sci.* **1980**, *31*, 315–331. [CrossRef]
50. De Rubeis, V. Application of kriging technique to seismic intensity data. *Bull. Seismol. Soc. Am.* **2007**, *95*, 540–548. [CrossRef]
51. Deplagne, F.; Bacconnet, C. Analyse structural d'une digue en argile. *Cah. Géostat.* **1993**, *3*, 188–191.
52. Villavicencio, G. Méthodologie pour Évaluer la Stabilité Mécanique des Barrages de Résidus Miniers. Ph.D. Thesis, University Blaise Pascal, Clermont Ferrand, France, 2009

53. Raspa, G.; Innocenti, C.; Marconi, F.; Mumelter, E.; Salmeri, A. Evaluation of an Automatic Procedure Based on Geostatistical Methods for the Characterization of Contaminated Sediments. In *GeoENV VI-Geostatistics for Environmental Applications*; Soares, A., Pereira, M., Dimitrakopoulos, R., Eds.; Springer: New York, NY, USA, 2008; pp. 421–441.
54. Vessia, G.; Di Curzio, D.; Castrignanò, A. Modeling 3D soil lithotypes variability through geostatistical data fusion of CPT parameters. *Sci. Total Environ.* **2020**, *698*, 134340. [CrossRef] [PubMed]
55. Alfaro, M. Estimación de Recursos Mineros. Available online: cg.ensmp.fr/bibliotheque/public/ALFARO_Cours_00606.pdf (accessed on 2 November 2020).

minerals MDPI

Article

Commingling of Waste Rock and Tailings to Improve "Dry Stack" Performance: Design and Evaluation of Mixtures

Ralph Burden [1,*] and G. Ward Wilson [2]

1 Frontera Geotechnical, Squamish, BC V8B 0K8, Canada
2 Department of Civil & Environmental Engineering, University of Alberta, Edmonton, AB T6G 2W2, Canada
* Correspondence: ralph@fronterageo.ca

Abstract: Mine tailings are typically deposited as slurry and stored in impoundment dams. These structures pose a serious geotechnical risk, and are difficult to successfully reclaim at the end of mining. An increasingly common alternative to traditional tailings disposal is "dry stacking": the placement of tailings dewatered using filtration in a self-supporting stack. It has been demonstrated that the addition of rock (termed "commingling") to a filtered tailings stack has the potential to improve the geotechnical performance of the stack and may make large-scale dry stacking more economically viable. This paper discusses the application of commingling to tailings dry stacking, specifically relating to the design and evaluation of commingled blends of waste rock and tailings. The authors present a review of existing mix design theory, and present an extended theoretical model to predict the structure and behavior of blends of waste rock and tailings, based upon mix ratio and density. This paper is based upon established theory, but is extended to consider the case of loosely placed materials and the effect of volume change on structural configuration. The extended model may be used to describe the geotechnical behavior of commingled filtered tailings and waste rock. It is postulated that the geotechnical behavior of blends, and the primary mechanism of volume change, is governed by particle configuration. A brief discussion of experimental methods to evaluate the structure and configuration of commingled mine wastes is also presented.

Keywords: filtered tailings; commingling; mix design; dry stacking

1. Introduction

Conventionally, mine tailings are discharged and transported as a slurry and stored behind dams, which are often constructed from the tailings themselves [1]. Tailings dams can present a challenge for reclamation and closure, and are a long-term geotechnical risk and liability. Several large, high-profile and sometimes tragic failures have occurred in recent years, such as Mount Polley [2], Fundão [3] and Feijão [4], the latter of which resulted in the loss of over 250 lives. In response, traditional tailings dams are increasingly becoming seen as unacceptable by many stakeholders, and tailings management practices are now coming under increasing scrutiny by major investors.

The global appetite to reduce risk associated with tailings management, combined with advances in filtration technology, has led to a resurgence of interest in recent years in "dry stacking", the deposition of tailings in a self-supporting pile that has the potential to eliminate or reduce the need for a dam. Dry stacking is typically achieved by using pressure or vacuum filtration to rapidly dewater tailings (usually to around 80% solids), which are then transported by truck or conveyor and then spread and compacted. While dry stacking is now considered for most major projects, it has thus far only been applied at small to medium-sized mines; the largest currently operating mines have throughputs of around 30,000 tons per day [5,6]. A potential barrier to the successful scale-up of filtered tailings is the cost and operation complexity involved in the placement and compaction of filtered tailings using conventional construction equipment.

Commingling of waste rock and tailings may offer a solution to this problem. The Merriam-Webster dictionary defines the word "commingle" as to "blend thoroughly into a harmonious whole". The word commingled has been in the English vocabulary perhaps since Shakespeare used it in his play *Hamlet*. Commingled, or blended, waste rock and tailings have been the subject of numerous laboratory [7–13] and field-scale [14] trials over the years that have demonstrated that this material can be engineered to have preferential geotechnical or geochemical properties compared to tailings or waste rock alone. The addition of waste rock to a filtered tailings stack during transit, using the same conveyor system used to transport the material to create a blended material, has the potential to offer significant advantages over established dry stacking methods, which may make large-scale dry stacking more economically viable.

This paper discusses the application of commingling to tailings dry stacking, specifically relating to the design and evaluation of commingled blends of waste rock and tailings. A brief, high-level review is presented of the current state of practice in tailings dry stacking and commingling. A review of existing mix design theory is presented, and its applicability to dry stacking applications is discussed. An extended theoretical model is presented for the case of commingled filtered tailings and waste rock. In addition, the paper includes a brief discussion of experimental methods that may be used to evaluate the structure and configuration of commingled mine wastes.

2. Background

2.1. Filtered Tailings "Dry Stacking"

Densified tailings technologies are broadly divided into three categories based on the mechanical properties and method of dewatering: thickened tailings (typically ~50%–65% solids by mass), paste tailings (typically 70%–80% solids) and filtered tailings (>80% solids). Figure 1 summarizes the properties, and pros and cons, of different densified tailings technologies.

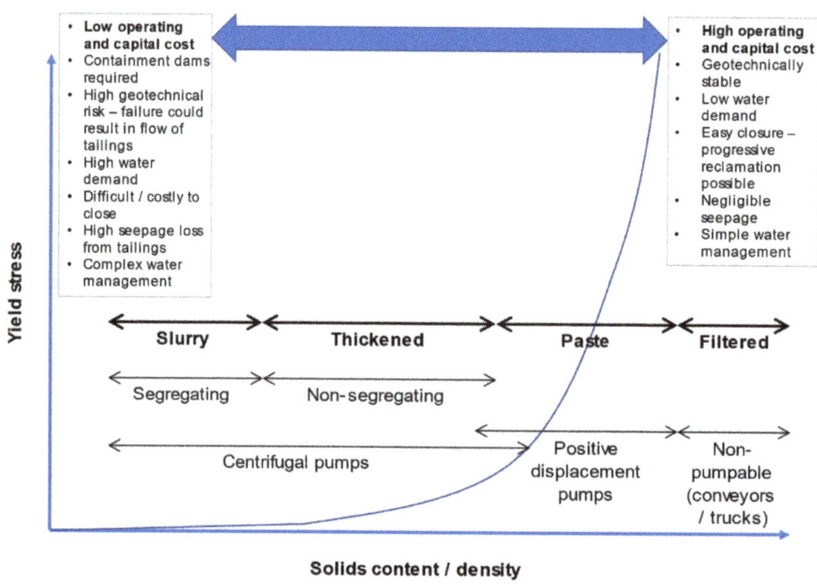

Figure 1. The tailings continuum, after Davies and Rice (2001) [15] and Jewell and Fourie (2006) [16].

Thickened and paste tailings technologies have many potential benefits, namely reduced water consumption, reduced waste volumes and improved geotechnical stability, whilst maintaining relatively low operating costs and good geochemical performance [16,17]. However, some form of containment structure is still required for the deposition of thickened tailings or paste tailings alone, as well as uncompacted filtered tailings. Construction of a stable, self-supporting "dry stack" requires both filtration and compaction or co-disposal technologies.

Filtration is widely used in many fields to separate solids and liquids. The material is placed on a porous medium (such as filter paper or cloth), and a pressure gradient is applied across the medium, causing flow of filtrate through the medium and leaving behind a densified "cake". In tailings applications, the pressure gradient may be applied by vacuum, mechanical squeezing action or via the feed pump. The most commonly used filter plant configurations are drums, stacked plates in either horizontal or vertical alignment, or horizontal belts. The tailings are typically run though a conventional thickener prior to filtration. The filter cake is often left to dry to the optimum moisture content, or slightly dry of optimum, before compaction into the stack. The efficiency of the process is very much dependent upon the properties of the tailings feed. Low fines content, low clay content and high specific gravity are all preferable for filtration [18].

The first published use of filtered tailings technology is in the coal mining industry [18]. Coal tailings, produced in the washing of coal, have a high fines content and low permeability, which makes dewatering by more conventional means difficult. As of 2008, there were reported to be 37 operating mines worldwide employing filtered tailings technologies, with production rates of up to 24,000 t/day [19]. The number of operating facilities has undoubtedly increased in recent years, although no comprehensive list exists in the literature. Crystal and Hore [5] prepared an inventory of 19 currently operating dry stacks, although almost all of them are small operations below 10,000 t/day. The largest currently operational mine employing filtered tailings that has published a case study is Karara [6] in Western Australia, a 30,000 t/day magnetite mine. Tailings at this facility are pressure filtered using vertical plates to around 85% solids, whereupon they are transported using an overland conveyor system and stacked in a radial pattern using conveyors and stackers. Recent technological advances led to the development of very high volume filter plates with lower maintenance requirements, which may assist in making filtration viable for large-scale mining [20].

The published literature on the geotechnical properties of filtered tailings is sparse and generally restricted to case histories [21–24]. Generally speaking, filtered metal mine tailings have been shown to have similar behavior to unsaturated silty sand, which is amenable to compaction and stacking.

2.2. Co-Disposal of Mine Wastes

The term "co-disposal" generally refers to the disposal of different waste streams in the same facility. Co-disposal can take on many forms with varying degrees of mixing of the wastes. Figure 2 shows common forms of co-disposal, ordered by degree of mixing.

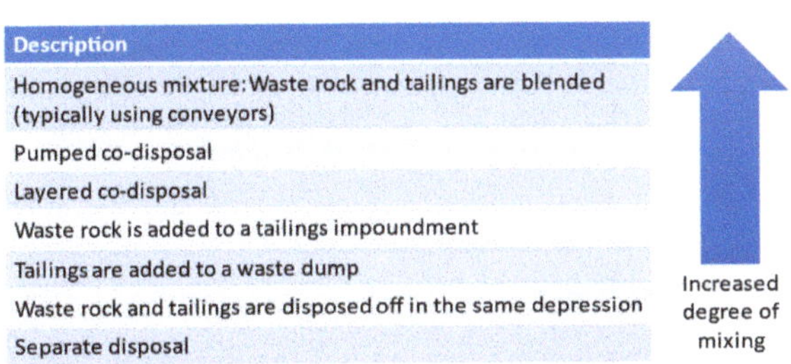

Figure 2. Forms of co-disposal, after Wickland (2006) [7].

The earliest use of co-disposal is probably in underground backfills, often combined with a binder such as Portland cement [25]. Pumped co-disposal has also been implemented successfully at several coal mines around the world, and has been shown to reduce waste volumes and the need for containment [26]. Other approaches that have been demonstrated include layered co-disposal, or "commingling", which typically involves the placement of fine tailings layers inside a waste rock pile to act as a barrier to seepage or oxygen flux, and waste rock "inclusions" which are waste rock dykes placed inside tailings ponds as the ponds are raised to promote drainage and improve overall stability and seismic resistance. A summary of these techniques is given by Bussiere [27]. The focus of this paper is on blended, or "commingled", co-disposal.

2.3. Commingling of Tailings and Waste Rock

Numerous laboratory [7–13] and field-scale [14] trials focusing on blended waste rock and tailings have demonstrated that these materials can have favorable geotechnical and geochemical properties when compared to tailings and waste rock alone, combining the high shear strength and comparatively low compressibility of the waste rock with the permeability and water retention properties of the tailings, and potentially allowing tailings stacks to remain unsaturated. Field trials have demonstrated that waste rock and filtered tailings can successfully be blended using conveyors [28]. Field trials in the Canadian oil sands have demonstrated the successful stacking of very-clay-rich oil sands tailings, which are widely regarded as being challenging to dewater by conventional methods, by commingling with clay shale overburden using conveyors [29,30].

3. Mixture Design Theory for Commingled Waste Rock and Tailings

3.1. Introduction

Wickland et al. [7] presented a theoretical basis for the design of "paste rock" blends, based on particle structure, as well as methods for predicting blend structure and properties from the mix ratio. The optimum mix ratio was found to be the "just-filled" point, where the dewatered tailings fill the voids between the waste rock particles. Figure 3 shows structural configurations of "paste rock"-style blends.

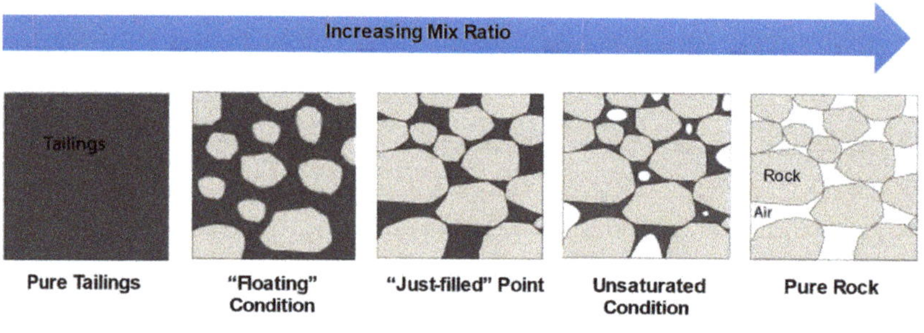

Figure 3. Blend configurations after Wickland et al. (2006) [7].

Figure 4 shows the general relationship between the density and mix ratio for commingled tailings and rock. The "optimum" mix ratio, which forms the densest possible blend, is a function of the tailings density, the waste rock skeleton porosity and the waste rock moisture content. Increasing rock porosity and tailings density results in an optimum mix ratio with lower rock content.

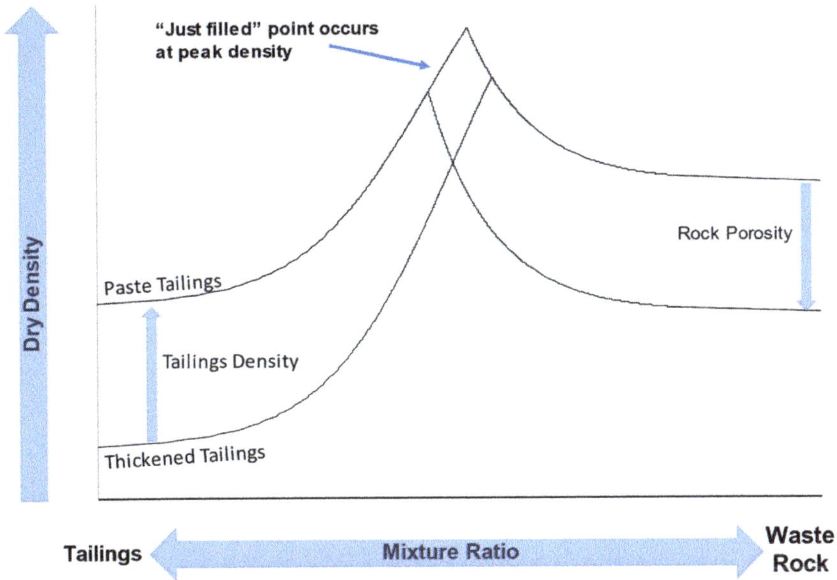

Figure 4. Mix design theory after Wickland et al. (2006) [7].

3.2. Ternary Diagrams

Ternary diagrams constructed using the approach developed by Charles and Charles [31] to characterize three-phase mixtures are widely used in the oil sands industry [32] to describe the structure and behavior of composite tailings, which contain both a fines and a sand component. The same approach provides a useful framework to describe the structure and behavior of waste rock–tailings blends, with respect to mass relations. Any theoretical combination of rock solids, tailings solids and water can be represented by a point inside the triangle. The ternary diagram approach provides a simple and effective tool for quickly estimating the bulk properties of a blend of a given tailings solids content and mix ratio, which may assist in high-level tailings planning and conceptual design.

Figure 5 shows the construction of the ternary diagram. The apices of the triangle represent 100% by mass water (top), rock solids (bottom left) and tailings solids (bottom right). Points inside the triangle represent any theoretical blend of tailings solids, rock solids and water. Points along the axis connecting 100% water and 100% tailings apices represent tailings only. Straight lines drawn from this line to the 100% rock apex (shown in magenta on Figure 5) represent blends with constant tailings solids content. Straight lines drawn from the bottom axis to the 100% water apex represent constant mix ratios. It should be noted that unlike ternary diagrams used in the oil sands industry, which consider coarse solids and fine solids as defined by a specific particle size, the diagrams discussed in this paper consider rock solids and tailings solids as defined by origin.

Figure 6 shows the ternary diagram annotated with general material property and blend configuration boundaries. The figure was generated using a waste rock moisture content of zero and specific gravity of tailings and waste rock solids of 2.7. The solid black lines show the "just-filled" point, or optimum mix ratio described by Wickland et al. [7], for waste rock skeleton porosities of 40% and 50%, calculated based upon the assumption that the tailings are fully saturated at the "just-filled" point. These lines define the division between blends whose behavior is dominated by fines (blends in the "floating" condition) and blends whose behavior is rock-dominated (blends with a continuous phase of rock particles in contact). The dashed lines show assumed tailings property boundaries based on solids content. For consistency with other researchers, the mix ratio R is defined as dry mass of waste rock solids/dry mass of tailings solids. Figure 7 shows the ternary diagram annotated with typical solids contents of different tailings dewatering technologies.

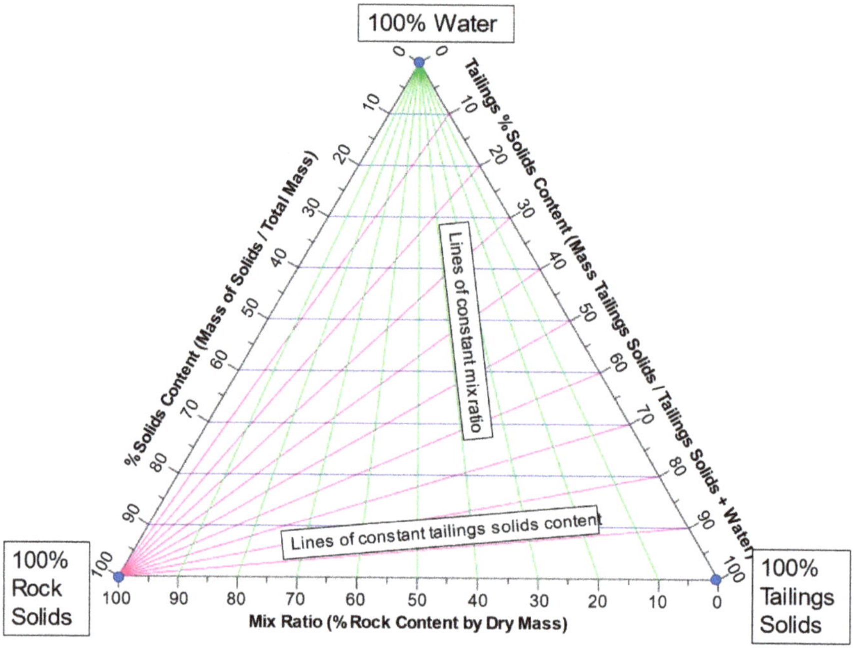

Figure 5. Ternary diagram for waste rock–fine tailings blends.

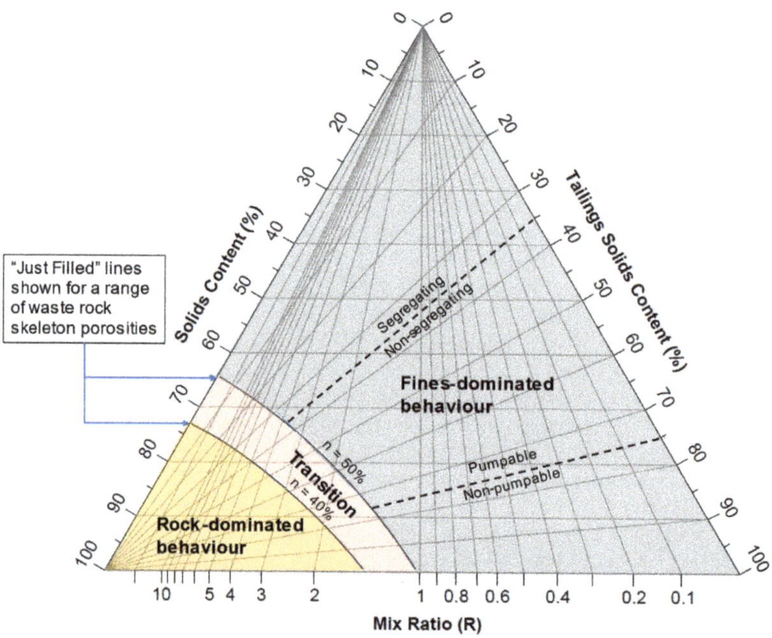

Figure 6. Ternary diagram showing blend configuration vs. mix ratio for fine tailings–waste rock blends.

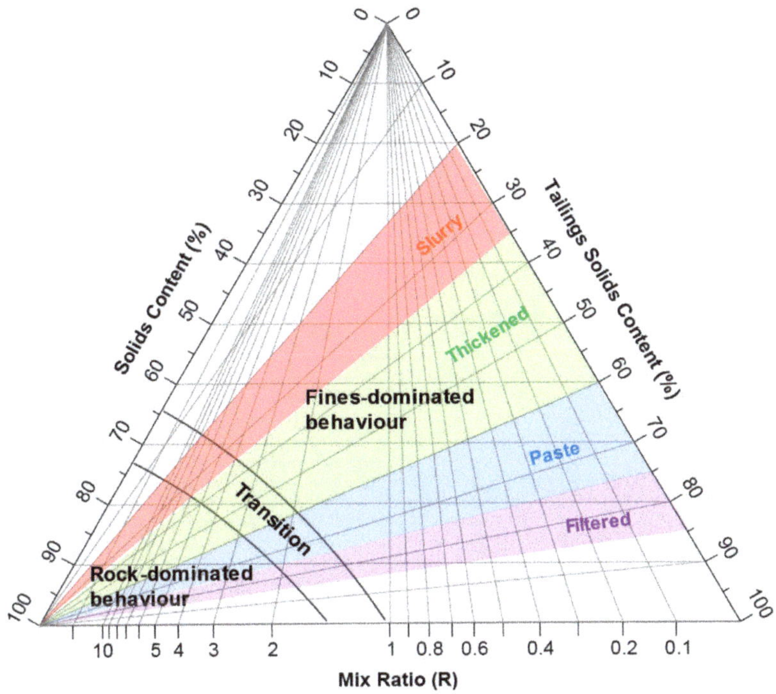

Figure 7. Ternary diagram showing blend configuration vs. mix ratio for different tailings dewatering technologies.

3.3. Commingled Filtered Tailings and Waste Rock

Established mix design theory has shown that the geotechnical behavior of a blend is governed by particle packing configuration, and that the configuration can be predicted from the mix ratio. However, the established theory is not applicable in the case of filtered tailings and waste rock blends that are placed loose and compacted under self-weight. This section presents an extended theoretical model to predict the structure and behavior of blends of waste rock and tailings, based upon mix ratio and density. It is based upon the co-disposal mix design theory developed by Wickland et al. [7], but is extended to consider the case of loosely placed materials and the effect of volume change on structural configuration.

When in an uncompacted state, filtered tailings and waste rock blends consist of discrete "lumps" of filter cake, rock particles and macro-scale (inter-lump) air voids. The lumps typically consist of compact-to-dense fines which are generally saturated or close to saturated at the time of deposition. In general terms, the geotechnical behavior of the blend is controlled by the structural configuration of the blend. The structural configuration is dependent upon the mix ratio and the packing density. Figure 8 presents a theoretical framework for describing the behavior of commingled filtered tailings and rock. The structural configurations and key processes that govern volume change under compression are shown in Figure 8.

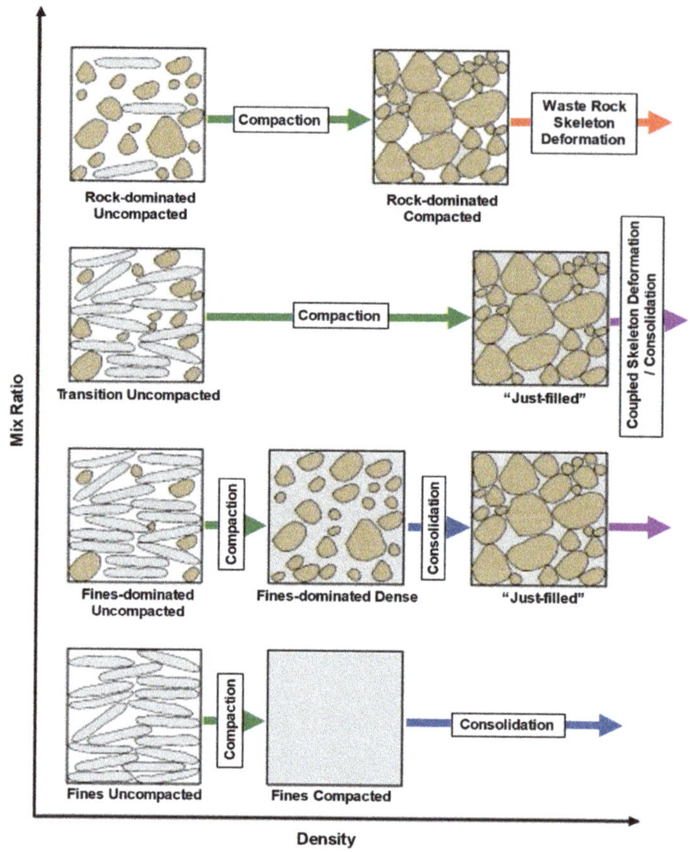

Figure 8. Structure and compression behavior of filtered tailings–waste rock blends.

In summary, five key structural configurations are proposed:

- Rock-dominated uncompacted;
- Fines-dominated uncompacted;
- Rock-dominated compacted;
- Fines-dominated compacted;
- "Just-filled".

"Just-filled" refers to a condition where a continuous, load-bearing waste rock "skeleton" exists, with the void spaces between the rock particles fully occupied with tailings. Figure 9 shows just-filled commingled tailings and waste rock at Porgera Mine in Papua New Guinea. "Fines-dominated compacted" refers to a condition where a discontinuous waste rock phase is "floating" in a continuous tailings matrix. "Rock-dominated compacted" refers to a condition with a continuous rock "skeleton", where tailings and air voids take up the void space between rock particles. Uncompacted blends (rock or fines dominated) consist of lumps of filter cake, rock particles and air voids.

Four distinct processes that govern volume change under compression are proposed:

- Compaction (shown as green arrows in Figure 8);
- Waste rock skeleton deformation, or deformation of a continuous waste rock phase (shown as red arrows in Figure 8);
- Tailings consolidation, or deformation of a continuous tailings phase (shown as blue arrows in Figure 8);

- Coupled consolidation and skeleton deformation (shown as purple arrows in Figure 8).

Figure 9. Example of "just-filled" commingled tailings and waste rock at Porgera Mine.

Compaction refers to the process of removing large, inter-lump air voids by the deformation and breakdown of the filter cake lumps. Skeleton deformation refers to the deformation of the load-bearing waste rock skeleton after filter cakes are de-structured. Waste rock deformation occurs due to the rearrangement of the waste rock particle structure and particle breakage. Consolidation refers to the gradual reduction of the volume within the tailings fraction of the blend, due to drainage of pore water caused by an increase in pore pressure due to applied stress. Once the blend is in the "just-filled" condition, i.e., a continuous waste rock skeleton with the voids fully occupied by fine tailings, the compression behavior is governed by both consolidation and skeleton deformation.

Material property boundaries can be defined based on the initial tailings solids content and minimum void ratio of the waste rock skeleton, shown below in Figure 10.

Figure 10 shows the full range of possible conditions for a typical blend of waste rock and tailings with $0 \leq R \leq 5$. The chart defines the structural configuration and dominant volume change mechanism of a blend of a given mix ratio and density. A chart can quickly be generated for blends of specific materials based on easily obtainable properties. Inputs to the chart are the specific gravity of tailings solids and rock solids, the initial moisture contents of tailings and waste rock and maximum waste rock skeleton porosity.

A blend of a given mix ratio will trace a path horizontally across the chart from its initial dry density as it compresses. The blue line (tailings consolidation line) represents the point at which the blend becomes fully saturated, positive pore pressures are generated and the tailings start to consolidate. The position of this line is controlled primarily by the tailings solids content at deposition. The red line (waste rock skeleton deformation line) represents the point at which a continuous, load-bearing skeleton of waste rock is formed and deformation is governed by the waste rock skeleton. The position of this line is controlled primarily by the maximum void ratio of the waste rock skeleton. Once the blend has crossed both lines, deformation is governed by a coupled process involving both skeleton deformation and consolidation.

The model proposed here is intended for mixtures where the waste rock has a low fines content and low moisture content, and the tailings are saturated and predominately fine. This is typical for run-of-mine waste rock blended with tailings in a hard rock, metal mining operation. The model presented here is not considered to be applicable for cases where the waste rock has a high fines content, high moisture content, high clay content or may be considered a very weak, soil-like rock that is prone to weathering. Considerations relating to these cases, for example in an oil sands operation, are discussed in the following section.

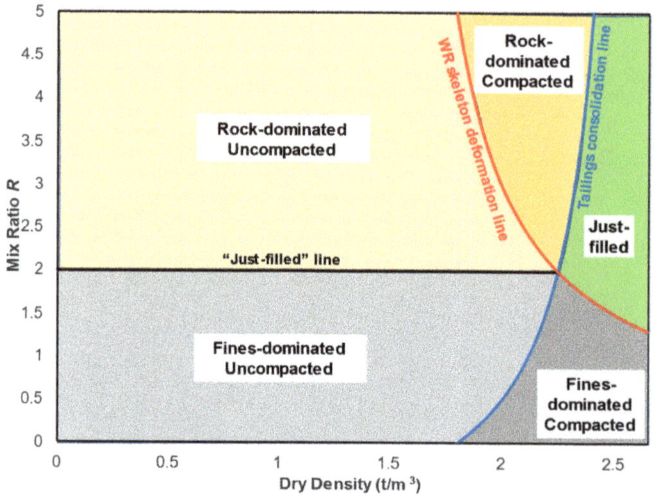

Figure 10. Material property boundary chart for a typical filtered tailings–waste rock blend.

3.4. Considerations Relating to Shales and Other Weak Rock Types

In general, the mix design theories described in this paper are not applicable to blends composed of clay shale overburden or other highly weatherable, or soil-like, rock types. In contrast to waste rock particles, shale lumps are highly compressible due to the deformation and breakdown of the lumps. Consequently, it is difficult to predict the maximum void ratio of the shale lump skeleton. Furthermore, moisture is transferred from the tailings to the unsaturated shale lumps, causing them to swell. The internal porosity of the shale lumps must be considered.

At a high level, the blends can be characterized by final blend solids content, as the relationship between solids content and mix ratio can be easily calculated. A recommended method to predict blend structure and density from mix ratio is to do so by using experimental compaction tests, as described in the following section.

3.5. Experimental Methods

This section provides a brief discussion of experimental methods that may be used for the evaluation of commingled mine wastes.

The Proctor test is widely used in geotechnical practice to evaluate the compaction properties of soils, and provides a quick and effective method to evaluate the density of commingled materials at a range of mix ratios. The Proctor test may be effective in assessing the properties of blends of tailings and soil-like overburden materials. Figure 11 shows an example of a series of standard and modified Proctor tests carried out on oil sands tailings and clay shale overburden, prepared to a range of mix ratios. The plot is annotated with a line representing fully saturated conditions and a "just-filled" line representing 100% saturation of the inter-lump void space. This line can easily be calculated if the internal shale lump void ratio is known, which can be measured using standard test methods on intact lumps. In general, mixes would not be expected to plot above this line irrespective of the degree of compaction. In this case, it can be seen that the optimum mix ratio (by dry mass) is around 6:1 shale solids–tailings solids, corresponding to a final solids content of approximately 75% by mass.

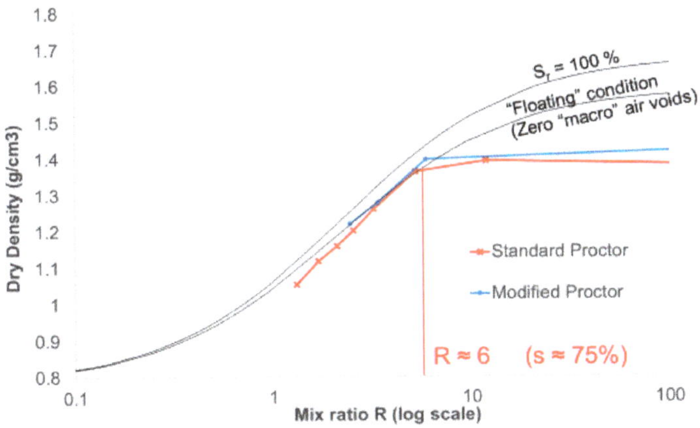

Figure 11. Dry density with respect to mix ratio for oil sands tailings–clay shale blends compacted using the standard and modified Proctor method.

An emerging technology with potential for wide-ranging application to the study of commingled mine wastes is X-ray computed tomography (CT) scanning. CT scanning is widely used in medicine, and its effectiveness has also been demonstrated in geotechnical applications, including the study of shear deformation and fracture features, as well as void ratio evolution during tri-axial testing [33–35]. A CT scanner is capable of measuring X-ray attenuation values for volumetric elements (known as "voxels") within its field of view, effectively creating a high-resolution 3D model of attenuation for the specimen. Attenuation can be linearly correlated with bulk density for the range of X-ray energies typically used in geotechnical applications [36].

Consequently, with calibration, a CT scanner and image processing software can be used to produce a 3D model of a sample of commingled mine wastes, which can be used to measure the volumes of constituent parts and visually evaluate particle structure, blend configuration and degree of mixing. Figure 12 shows an example of a CT scan carried out on a blend of commingled oil sands tailings and clay shale overburden. Figure 12a shows tailings in the voids of the commingled blend and Figure 12b shows shale overburden.

(a) (b)

Figure 12. X-ray computed tomography scan of a sample of blended oil sands tailings and clay shale overburden, showing (**a**) tailings and (**b**) clay shale overburden.

4. Discussion and Conclusions

Commingling of waste rock and filtered tailings is an emerging technology that has advantages over traditional methods of mine waste disposal. Filtered tailings "dry stacking" is becoming an increasingly widely used alternative to traditional slurry tailings. The addition of waste rock to a filtered tailings stack has the potential to improve the stability of the deposit, possibly allowing for rapid stacking without compaction and for the dry stacking to be economical at the biggest mines.

Work by previous researchers, focusing primarily on blends of waste rock and paste tailings, has shown that the geotechnical behavior of a blend is governed by particle packing configuration and that the configuration can be predicted from the mix ratio. However, the established theory is not applicable in the case of filtered tailings and waste rock blends that are placed loose and compacted under self-weight. This paper presents an extended theoretical model to predict the structure and behavior of blends of waste rock and tailings, based upon mix ratio and density. It is based upon the co-disposal mix design theory developed by Wickland et al. [7] but is extended to consider the case of loosely placed materials and the effect of volume change on structural configuration. The extended model may be used to describe the geotechnical behavior of commingled tailings and waste rock, based upon the particle packing arrangement of the tailings. It is postulated that the geotechnical behavior of blends, and the primary mechanism of volume change, are governed by particle configuration. The model may be used to illustrate the structure and behavior of the blend, based on the mix ratio and density. The model may also be used as a framework for further investigations and the development of constitutive models to characterize these materials.

A methodology for the application of ternary diagrams to distinguish blends of waste rock and tailings blends is presented. These diagrams provide a useful method for quickly evaluating the properties of a blend of a given mix ratio and tailings solids content, which may assist in high-level tailings planning and design.

Specific conclusions of this paper include:

- Blends of commingled waste rock and filtered tailings can be characterized using five distinct structural configurations.
- The structural configuration of a specific blend is primarily dependent on the mixture ratio and the density of the blend. The configuration can be predicted based on these and other basic properties.
- It is postulated that the structural configuration of the blend defines the geotechnical behavior of the blend in general terms, specifically the mechanism of volume change in response to an increase in vertical stress. Four distinct mechanisms are proposed.
- Experimental methods such as CT scanning have the potential to allow the direct observation and measurement of blend structural configuration.

Recommendations for further study include the development of a rigorous mathematical model to characterize the behavior of filtered tailings and waste rock blends. The model would be required to model compaction, consolidation and waste rock skeleton deformation processes, and would have to consider both saturated and unsaturated conditions. The theoretical framework developed in this paper could be used as the basis for the model. Field-scale trials should be undertaken to study the stacking behavior of filtered tailings and waste rock blends to verify the experimental findings.

Author Contributions: Development of original theory, R.B.; experimental work, R.B.; writing—original draft preparation, R.B.; writing—review and editing, G.W.W.; project administration, G.W.W.; supervision, G.W.W. All authors have read and agreed to the published version of the manuscript.

Funding: This research was funded through the NSERC/COSIA Industrial Research Chair in Oil Sands Tailings Geotechnique, grant number IRCPJ 460863-18.

Data Availability Statement: Not applicable.

Acknowledgments: The authors would like to acknowledge the support of Mike London and Alberta Innovates for performing the CT scan and producing the image (Figure 12) presented in this paper.

Conflicts of Interest: The authors declare no conflict of interest. The funders had no role in the design of the study; in the collection, analysis, or interpretation of data; in the writing of the manuscript; or in the decision to publish the results.

References

1. Vick, S.G. *Planning, Design, and Analysis of Tailings Dams*; BiTech Publishers Ltd.: Vancouver, BC, Canada, 1990.
2. Morgenstern, N.R.; Vick, S.G.; Van Zyl, D. Report on Mount Polley Tailings Storage Facility Breach. Available online: ReportonMountPolleyTailingsStorageFacilityBreach.pdf (accessed on 10 December 2022).
3. Morgerstern, N.R.; Vick, S.G.; Watts, B.D. Fundão Tailings Dam Review Panel. In *Report on the Immediate Causes of the Failure of the Fundão Dam*. Available online: https://www.google.com.hk/url?sa=t&rct=j&q=&esrc=s&source=web&cd=&ved=2ahUKEwjzxuPikpz9AhVFa94KHdF8COIQFnoECAoQAQ&url=https%3A%2F%2Fpedlowski.files.wordpress.com%2F2016%2F08%2Ffundao-finalreport.pdf&usg=AOvVaw0YzlPAGMkDXdno__cKoDCq (accessed on 10 December 2022).
4. Robertson, P.; de Melo, L.; Williams, D.; Wilson, G. *Report of The Expert Panel on the Technical Causes of the Failure of Feijão Dam 1*. Available online: https://www.google.com.hk/url?sa=t&rct=j&q=&esrc=s&source=web&cd=&ved=2ahUKEwiWkNCak5z9AhXDhMYKHWMtDuQQFnoECAwQAQ&url=https%3A%2F%2Fbdrb1investigationstacc.z15.web.core.windows.net%2Fassets%2FFeijao-Dam-I-Expert-Panel-Report-ENG.pdf&usg=AOvVaw3mw_L3eEEiaAk1q4ru97v8 (accessed on 10 December 2022).
5. Crystal, C.; Hore, C. Filter-Pressed Dry Stacking: Design Considerstions Based on Practial Experience. In Proceedings of the Tailings and Mine Waste 2018, Keystone, CO, USA, 30 September–2 October 2018.
6. Amoah, N. Large-Scale Tailings Filtration and Dry Stacking at Karara Magnetite Iron Ore Operation. In Proceedings of the Tailings and Mine Waste, Vancouver, BC, Canada, 17–20 November 2019.
7. Wickland, B.E.; Wilson, G.W.; Wijewickreme, D.; Klein, B. Design and Evaluation of Mixtures of Mine Waste Rock and Tailings. *Can. Geotech. J.* **2006**, *43*, 928–945. [CrossRef]
8. Khalili, A.; Wijewickreme, D.; Wilson, G.W. Mechanical Response of Highly Gap-Graded Mixtures of Waste Rock and Tailings. Part I: Monotonic Shear Response. *Can. Geotech. J.* **2010**, *47*, 552–565. [CrossRef]
9. Wijewickreme, D.; Khalili, A.; Wilson, G.W. Mechanical Response of Highly Gap-Graded Mixtures of Waste Rock and Tailings. Part II: Undrained Cyclic and Post-Cyclic Shear Response. *Can. Geotech. J.* **2010**, *47*, 566–582. [CrossRef]
10. Jehring, M.M.; Bareither, C.A. Tailings Composition Effects on Shear Strength Behavior of Co-Mixed Mine Waste Rock and Tailings. *Acta Geotech.* **2016**, *11*, 1147–1166. [CrossRef]
11. Hamade, M.; Bareither, C. Consolidated Undrained Shear Behavior of Synthetic Waste Rock and Synthetic Tailings Mixtures. *Geotech. Test. J.* **2019**, *42*, 20180007. [CrossRef]
12. Bareither, C.A.; Gorakhki, J.; Scalia, J.; Jacobs, M. Compression Behaviour of Filtered Tailings and Waste Rock Mixtures: Geowaste. In Proceedings of the Tailings and Mine Waste 2018, Keystone, CO, USA, 30 September–2 October 2018.
13. Burden, R.; Williams, D.J.; Wilson, G.W.; Jacobs, M. The Shear Strength of Filtered Tailings and Waste Rock Blends. In Proceedings of the Tailings and Mine Waste 2018, Keystone, CO, USA, 30 September–2 October 2018; pp. 347–355.
14. Wickland, B.E.; Wilson, G.W. Self-Weight Consolidation of Mixtures of Mine Waste Rock and Tailings. *Can. Geotech. J.* **2005**, *42*, 327–339. [CrossRef]
15. Davies, M.P.; Rice, S. *An Alternative to Conventional Tailings Management–"Dry Stack" Filtered Tailings*; CRC Press: Boca Raton, FL, USA, 2001; pp. 411–420.
16. Jewell, R.J.; Fourie, A.B. *Paste and Thickened Tailings: A Guide*; Australian Centre for Geomechanics, The University of Western Australia: Crawley, Australia, 2006; ISBN 0-9756756-4-8.
17. Williams, M.P.A.; Seddon, K.D.; Fitton, T.G. Surface disposal of paste and thickened tailings—A brief history and current confronting issues. In Proceedings of the 11th International Seminar on Paste and Thickened Tailings (Paste 2008), Kasane, Botswana, 5–9 May 2008; Fourie, A., Jewell, R., Slatter, P., Paterson, A., Eds.; Australian Centre for Geomechanics: Nedlands, Australia; pp. 143–164.
18. Green, P. De-Watering Coal Refuse. *Coal Age* **1981**, *86*, 145–157.
19. AMEC Rosemont Copper Company Filtered Tailings Dry Stacks Current State of Practice Final Report. Available online: https://www.rosemonteis.us/files/technical-reports/012312.pdf (accessed on 10 December 2022).
20. Wisdom, T. Recent Developments in Tailings Dewatering Technology. In *Mine Tailings: Perspectives for a Changing World*; Society for Mining Metallurgy & Exploration: Englewood, CO, USA, 2020; p. 147.
21. Lupo, J.; Hall, J. Dry Stack Tailings—Design Considerations. In *Tailings and Mine Waste 2010*; CRC Press: Boca Raton, FL, USA, 2010; pp. 327–334, ISBN 978-0-415-61455-9.
22. Newman, L.; Arnold, K.; Wittwer, D. Dry Stack Tailings Design for the Rosemont Copper Project. In *Tailings and Mine Waste 2010*; CRC Press: Boca Raton, FL, USA, 2010; pp. 315–326, ISBN 978-0-415-61455-9.
23. Butikofer, D.; Erickson, B.; Marsh, A.; Friedel, R.; Murray, L.; Piggot, M.J. Filtered Tailings Disposal Case History: Operation and Design Considerations Part II. In Proceedings of the Tailings and Mine Waste 2017, Banff, AB, USA, 5–8 November 2017.

24. Robertson, P.K.; da Fonseca, A.V.; Ulrich, B.; Coffin, J. Characterization of Unsaturated Mine Waste: A Case History. *Can. Geotech. J.* **2017**, *54*, 1752–1761. [CrossRef]
25. Brawner, C.O.; Argall, J. Concepts and Experience for Subsurface Storage of Tailings. In Proceedings of the 2nd International Tailings Symposium, Tailings Disposal Today, Denver, CO, USA, May 1978; Miller Freeman Publications Inc.: San Francisco, CA, USA, 1978; pp. 153–177.
26. Williams, D.J. *Effectiveness of Co-Disposing Coal Washery Wastes*; U.S. Department of Energy: Washington, DC, USA, 1997; Volume 97, pp. 375–383.
27. Bussiere, B. Colloquium 2004: Hydrogeotechnical Properties of Hard Rock Tailings from Metal Mines and Emerging Geoenvironmental Disposal Approaches. *Can. Geotech. J.* **2007**, *44*, 1019–1052. [CrossRef]
28. Wisdom, T.; Jacobs, M.; Chaponnel, J. *GeoWasteTM–Continuous Comingled Tailings for Large-Scale Mines*; Jewell, R., Fourie, A., Jewell, R., Fourie, A., Eds.; Australian Centre for Geomechanics: Perth, Australia, 11 April 2018; pp. 465–472.
29. Mikula, R.; Wang, N.; Cleminson, R. Overburden/Tailings Mixtures for Engineered Tailings Deposit Control 2016. Available online: https://patents.google.com/patent/US9352328/en (accessed on 20 December 2022).
30. Wang, N.; Cleminson, R.; Lorentz, J. Co-Mixing of Fluid Fine Tailings and Clearwater Overburden. In Proceedings of the COSIA Oil Sands Innovation Summit, Calgary, AB, USA, 21–22 March 2017.
31. Charles, M.E.; Charles, R.A. The Use of Heavy Media in the Pipeline Transport of Particulate Solids. In *Advances in Solid–Liquid Flow in Pipes and Its Application*; Elsevier: Amsterdam, The Netherlands, 1971; pp. 187–197.
32. Oil Sands Tailings Consortium (OSTC); Canada's Oil Sands Innovation Alliance (COSIA). *Technical Guide for Fluid Fine Tailings Management*; OSTC: Calgary, AB, Canada; COSIA: Calgary, AB, Canada, 2012.
33. Alshibli, K.A.; Alramahi, B.A. Microscopic Evaluation of Strain Distribution in Granular Materials during Shear. *J. Geotech. Geoenviron. Eng.* **2006**, *132*, 80–91. [CrossRef]
34. Wong, R.C. Shear Deformation of Locked Sand in Triaxial Compression. *Geotech. Test. J.* **2000**, *23*, 158–170.
35. Thomson, P.; Wong, R. Specimen Nonuniformities in Water-Pluviated and Moist-Tamped Sands under Undrained Triaxial Compression and Extension. *Can. Geotech. J.* **2008**, *45*, 939–956. [CrossRef]
36. Wellington, S.L.; Vinegar, H.J. X-Ray Computerized Tomography. *J. Pet. Technol.* **1987**, *39*, 885–898. [CrossRef]

MDPI AG
Grossepeteranlage 5
4052 Basel
Switzerland
Tel.: +41 61 683 77 34

Minerals Editorial Office
E-mail: minerals@mdpi.com
www.mdpi.com/journal/minerals

www.ingramcontent.com/pod-product-compliance
Lightning Source LLC
Chambersburg PA
CBHW061805210926
51850CB00006B/618